Human Biology: Current Concepts and Perspectives

Human Biology: Current Concepts and Perspectives

Edited by Alejandra Donaldson

SYRAWOOD
PUBLISHING HOUSE

New York

Published by Syrawood Publishing House,
750 Third Avenue, 9th Floor,
New York, NY 10017, USA
www.syrawoodpublishinghouse.com

Human Biology: Current Concepts and Perspectives
Edited by Alejandra Donaldson

International Standard Book Number: 978-1-68286-721-1 (Hardback)

Cataloging-in-Publication Data

Human biology : current concepts and perspectives / edited by Alejandra Donaldson.
    p. cm.
Includes bibliographical references and index.
ISBN 978-1-68286-721-1
1. Human biology. 2. Biology. I. Donaldson, Alejandra.
QP34.5 .H86 2019
612--dc23

TABLE OF CONTENTS

PREFACE

I am honored to present to you this unique book which encompasses the most up-to-date data in the field. I was extremely pleased to get this opportunity of editing the work of experts from across the globe. I have also written papers in this field and researched the various aspects revolving around the progress of the discipline. I have tried to unify my knowledge along with that of stalwarts from every corner of the world, to produce a text which not only benefits the readers but also facilitates the growth of the field.

Human biology is the scientific study of physiology, anatomy, epidemiology, anthropology and nutrition of humans. Studies of genetics, evolution, population genetics and influence of society and culture on human behavior are also encompassed in this science. It is also concerned with the physical development as well as development of biological systems of humans. This book attempts to understand the multiple branches that fall under the discipline of human biology and how such concepts have practical applications. It presents researches and studies performed by experts across the globe. For all readers who are interested in this field, the case studies included in this book will serve as an excellent guide to develop a comprehensive understanding.

Finally, I would like to thank all the contributing authors for their valuable time and contributions. This book would not have been possible without their efforts. I would also like to thank my friends and family for their constant support.

Editor

Detecting differentially expressed genes of heterogeneous and positively skewed data using half Johnson's modified t-test

I-Shiang Tzeng[1]*, Li-Shya Chen[2], Shy-Shin Chang[3] and Yung-ling Leo Lee[4]

*Corresponding author: I-Shiang Tzeng, Institute of Epidemiology and Preventive Medicine, College of Public Health, National Taiwan University, Taipei, Taiwan
E-mail: g1354502@nccu.edu.tw
Reviewing editor: Jurg Bahler, University College London, UK
Additional information is available at the end of the article

Abstract: *Background*: Microarray technology allows simultaneously detecting thousands of genes within one single experiment. The Student's t-test (for a two-sample situation) can be used to compare the mean expression of a gene, taken from replicate arrays, to detect differential expression under the conditions being studied, such as a disease. However, a general statistical test may have insufficient power to correctly detect differentially expressed genes of heterogeneous and positively skewed data. *Methods*: Here we define a differentially expressed gene as with significantly different expression in means, variances, or both between the two groups of microarray. Monte Carlo simulation shows that the "half Johnson's modified t-test" maintains quite accurate type I error rates in normal and non-normal distributions. And the half Johnson's modified t-test was more powerful than the half Student's t-test overall when the ratio of standard deviations between case and control groups is greater than 1. *Results*: Analysis of a colon cancer data shows that when the false discovery rate (FDR) is controlled at 0.05, the half Johnson's modified t-test can detect 429 differentially expressed genes, which is larger than the number of differentially expressed genes (i.e. 344) detected by the half Student's t. To

ABOUT THE AUTHOR

I-Shiang Tzeng is a doctoral researcher in the National Translational Medicine and Clinical trial Resource Center (NTCRC) composed by Academia Sinica, National Taiwan University and National Yang-Ming University, Taiwan. He currently serves as a bioinformatics and biostatistics consultant in NTCRC. He also is an adjunct assistant professor in the department of statistics, National Taipei University, Taiwan. His area of research includes biostatistics and epidemiologic method and further studies proposing the potential powerful method to detect differential expressed genes. His research interests include the field of age-period-cohort (APC) modeling from social issues to biological issues as well as the analysis of the APC models that arise in all these applications.

PUBLIC INTEREST STATEMENT

Gene expression has been a popular research topic in recent years. Student's t-test is commonly adopted to screen disease-related genes. However, when the researches are focused on heterogeneous and positively skewed expression data, the means of gene expression levels between case and control groups may be similar, and thus, the difference would be insignificant using conventional Student's t-test. This study proposed half Johnson's modified t-test to correctly detect differentially expressed genes of heterogeneous and positively skewed data. Test statistics of half Johnson's modified t-test only considers sample standard deviation of control group, while that of case group is not included. After controlling false discovery rate (cut-off point set at 0.05) of colon cancer gene expression data, half Johnson's modified t-test could detect 364 more significant genes than conventional Student's t-test. Half Student t-test is worth recommending as a method for detecting differentially expressed genes in heterogeneous and positively skewed data.

target 100 priority genes, the half Johnson's modified t only set FDR to 4.28×10^{-8}, but for the half Student's t, it is set to 5.39×10^{-4}. *Conclusions*: The half Johnson's modified t-test is recommended for the detection of differentially expressed genes in heterogeneous and ONLY positively skewed data.

Subjects: Computer Science; Mathematics & Statistics; Medicine, Dentistry, Nursing & Allied Health

Keywords: gene expression; positively skewed data; Johnson's modified t-test

1. Introduction

Microarray technology allows simultaneously detecting thousands of genes within one single experiment (Templin et al., 2002). One of the main goals of microarray data analysis is to detect the differentially expressed genes, which is a two-step process. The first step involves selecting a statistic to rank the genes by expression data. The second step is to set a criterion (critical value) to consider which of the ranked genes is differentially expressed. The overall aim of this process is to identify a number of candidate genes for further studies, such as using molecular biological techniques. Statistical knowledge is often necessary for the analysis of microarray data, as researchers deal with massive amounts of data with various sources of variability in order to identify important genes. For example, fold change is often used in determining change in the expression level of individual gene for detecting differentially expressed genes in a microarray (Chen, Dougherty, & Bittner, 1997). For simplicity, researchers often use Student's t-test to compare the mean expression of a gene, taken from replicate arrays, to detect differential expression under the conditions being studied, such as a disease (Dudoit, Yang, Callow, & Speed, 2002; Pan, 2002).

However, a general statistical test may have insufficient power to correctly detect differentially expressed genes in heterogeneous disease. A heterogeneous disease may encompass a multitude of etiological entities that have different morphological features and clinical behavior. Examples of heterogeneous diseases are otosclerosis (Van Den Bogaert et al., 2002), rheumatoid arthritis (van der Pouw Kraan et al., 2003), primary thyroid lymphoma (Thieblemont et al., 2002), and acute lymphoblastic leukemia (Yeoh et al., 2002). A gene may be overexpressed in some cases, but may be expressed normally or even underexpressed in other cases of heterogeneous diseases. This phenomenon (multimodality) will present itself in a higher variance of case group. That is, the variance (or standard deviation) of gene expression values in diseased individuals (cases) is more than that of non-diseased individuals (controls). This particular gene provides useful information and belongs to the differentially expressed class because of heterogeneity in disease. Further, mean expression values may have a small apparent difference in case and control groups, and the gene expression values may follow a positively skewed distribution (Newton, Kendziorski, Richmond, Blattner, & Tsui, 2001). In such instances, the conventional t-test or the "half Student's t-test" (Hsu & Lee, 2010) would not be applicable to detect the gene. The original t-test may have less power under conditions of heterogeneity, while the "half Student's t-test" may be powerful; however, neither test is suitable for non-normal data. (Note that here we assume there are some patient subgroups, at least more than one entity, but we don't know how many subgroups exist and how to define and characterize each of them. Otherwise, we can reconstruct the diseased subjects according to different "disease entities" rather than simply different "diseases." Then, we can perform a stratified analysis if we have known the patient subgroup structure).

In the statistical genomic field, for the last fifteen years, many researchers have developed innovative alternatives relying upon either parametric or nonparametric approaches (Tusher, Tibshirani, & Chu, 2001) which are based on frequentism or Bayesianism (Smyth, 2004). Moreover, the question of data transformation has been extensively discussed by statisticians (Johnson, 1978; Tukey, 1977) and has been widely considered with highly relevant implications for microarray. In order to determine differentially expressed genes in heterogeneous and positively skewed data, we propose the "half Johnson's modified t-test." The half Johnson's modified t-test is used to correct the t variables for heterogeneity and non-normality of the population distribution, without abandoning the Student's t distribution as a criterion. Here, the null *compliance* hypothesis would be that two groups (i.e. case group and control group) have the same distribution of gene expression data. The alternative hypothesis would be that means, variances, or both for the gene expression data are different between the two groups. (Note that we assume that a case effect on mean response is expected to be accompanied by an increase in variability).

Finally, a Monte Carlo simulation was performed to exhibit the statistical characteristics of the half Johnson's modified *t*-test in this study, and a colon cancer gene expression data-set (Alon et al., 1999) was analyzed for demonstration.

2. Methods

Let the sample size, the sample mean, and the sample standard deviation of gene expression for case group separately be n_1, $\overline{X}_1$, and s_1. The corresponding notations of control group are n_0, $\overline{X}_0$, and s_0, respectively. The ordinary test statistic t_s of the Student's *t*-test is as follows:

$$t_s = \frac{\overline{X}_1 - \overline{X}_0}{s_p \sqrt{\frac{1}{n_1} + \frac{1}{n_0}}}$$

where $s_p = \sqrt{\frac{(n_1-1)s_1^2 + (n_0-1)s_0^2}{n_1+n_0-2}}$ represented the pooled standard deviation. Under normality assumption, t_s follows a Student's *t* distribution with $n_1 + n_0 - 2$ degrees of freedom (d.f.).

Welch's *t*-test does not require the variances to be equal (Welch, 1947). Therefore, it is a more robust test than the original Student's *t*-test. Welch's *t*-test uses case and control groups' standard deviations separately. The test statistic t_w of Welch's *t*-test is as follows:

$$t_w = \frac{\overline{X}_1 - \overline{X}_0}{\sqrt{\frac{s_1^2}{n_1} + \frac{s_0^2}{n_0}}}$$

Under normality assumption, t_w follows a Student's *t* distribution with d.f. being

$$v \approx \frac{\left(\frac{s_1^2}{n_1} + \frac{s_0^2}{n_0}\right)^2}{\frac{\left(\frac{s_1^2}{n_1}\right)^2}{n_1-1} + \frac{\left(\frac{s_0^2}{n_0}\right)^2}{n_0-1}}$$

The two-sample Student's *t*-test can be used in such occasions (i.e. heterogeneous diseases) to gauge statistical significances. However, when the diseases under study are heterogeneous (Thieblemont et al., 2002; Van Den Bogaert et al., 2002; van der Pouw Kraan et al., 2003; Yeoh et al., 2002), t_s or t_w may be underpowered to detect differentially expressed genes.

To tackle the heterogeneity problem, the half Student's *t*-test, t_h, proposed in (Hsu & Lee, 2010) is presented as follows:

$$t_h = \frac{\overline{X}_1 - \overline{X}_0}{s_0 \sqrt{\frac{1}{n_1} + \frac{1}{n_0}}}$$

which only uses the standard deviation of the control group. Hence, the test statistic t_h is named as the half Student's *t*-test. Note that t_h has the same numerator but a different denominator as t_s. Under normality assumptions, t_h follows a Student's *t* distribution with d.f. $= n_0 - 1$.

In case of one sample, the Johnson's modified t-test (Cressie & Whitford, 1986) was proposed to correct t variables (for one sample) if population distribution is not normal, but not abandon the Student's t distribution as a criterion. The form of the Johnson's modified t-test is derived by using Cornish-Fisher expansion and the first few terms of inverse Cornish-Fisher expansion. To correct nonzero skewness of t_w, Johnson's one-sample modified t-test was extended to deal with two-sample test (Johnson, 1978), and the modified test for t_w is:

$$t_{wJ} = \left[(\overline{X}_1 - \overline{X}_0) + \frac{\hat{\mu}_3}{6\left(\sqrt{\frac{s_1^2}{n_1} + \frac{s_0^2}{n_0}}\right)^2} + \frac{\hat{\mu}_3(\overline{X}_1 - \overline{X}_0)^2}{3\left(\sqrt{\frac{s_1^2}{n_1} + \frac{s_0^2}{n_0}}\right)^4} \right] \cdot \left(\frac{s_1^2}{n_1} + \frac{s_0^2}{n_0} \right)^{-1/2}$$

where $\hat{\mu}_3 = \frac{1}{n_1^2}\sum_{i=1}^{n_1}\frac{(X_{1i}-\bar{X}_1)^3}{n_1} - \frac{1}{n_0^2}\sum_{i=1}^{n_0}\frac{(X_{0i}-\bar{X}_0)^3}{n_0}$ is the sample third central moment for $\overline{X}_1 - \overline{X}_0$, while $\sum_{i=1}^{n_1}\frac{(X_{1i}-\bar{X}_1)^3}{n_1}$ and $\sum_{i=1}^{n_0}\frac{(X_{0i}-\bar{X}_0)^3}{n_0}$ are the sample third central moments for the case and control groups, respectively. The d.f. of t_{wJ} is the same as that for t_w.

For the two-sample situation (one case group vs. one control group) (Johnson, 1978), in order to integrate the features of the aforementioned two modified tests, t_{wJ} and t_h, we propose to only consider the standard deviation of control group in t_{wJ}. Then, the half Johnson's modified t-test would be as follows:

$$t_{hJ} = \left[(\overline{X}_1 - \overline{X}_0) + \frac{\hat{\mu}_3}{6\left(s_0\sqrt{\frac{1}{n_1} + \frac{1}{n_0}}\right)^2} + \frac{\hat{\mu}_3(\overline{X}_1 - \overline{X}_0)^2}{3\left(s_0\sqrt{\frac{1}{n_1} + \frac{1}{n_0}}\right)^4} \right] \cdot \left[s_0^2\left(\frac{1}{n_1} + \frac{1}{n_0} \right) \right]^{-1/2}$$

The rejection region is $t_{hJ} > t_{n_0-1,\alpha/2}$ or $t_{hJ} < -t_{n_0-1,\alpha/2}$. The significance level is denoted as α in this study.

2.1. Monte Carlo simulation

We used free R software (R Development Core Team, 2008) for testing and analysis. The two test procedures studied were the half Student's t-test and the half Johnson's modified t-test. The analyses were performed on two sample sizes: 40 ($n_0 = n_1 = 20$) and 120 ($n_0 = n_1 = 60$). The difference in means of gene expression data between case and control groups was denoted as d, being set to 0, 15, and 25. The standard deviation ratio of case group to control group was denoted as r, being set to 1, 1.5, 2, and 2.5. The standard deviation for the control group was set to 30. Let $\gamma_3 = E(X - \mu)^3/\sigma^3$ be the skewness coefficient. In addition, $\gamma_3 > 1$, $0.6 < \gamma_3 \leq 1$, and $0 < \gamma_3 \leq 0.6$ correspond to high, moderate, and minor positive skewness, respectively. For the completeness of study, a normality scenario and three non-normality scenarios were incorporated: (1) normal distribution; (2) uniform distribution (non-normal but symmetric distribution); (3) Gamma distribution (positively skewed and $\gamma_3 = 0.6$); and (4) negatively skewed distribution ($\gamma_3 = -0.6$). Note that to generate from (4), a random number from (3) was first simulated, multiplied by −1, and added by twice of the mean of Gamma distribution.

For each scenario, the half Student's t-test and the half Johnson's modified t-test were performed under 1,000,000 simulations. It is essential to understand that the null hypothesis corresponds to the ratio and difference of $r = 1$ and $d = 0$. As for other settings, any exception of the null hypothesis would be the alternative one.

2.2. A colon cancer example

A colon cancer data-set consists of 40 tumor tissue samples (case group) and 22 normal colon tissue samples (control group). Oligonucleotide arrays provide a broad picture of the state of the cell through monitoring the expression level of thousands of genes simultaneously. Tissue and hybridization were analyzed by using an Affymetrix oligonucleotide Hum6000 array complementary to more than 6,500 human genes. Probes being complementary to the sequence of interest are perfect match (PM), while mismatch (MM) happens for homomeric base change at a specific position. A probe pair is a combination of a PM and an MM. Each probe pair in a probe set plays potential role in determining the signal value. The real signal value is estimated by taking LOG transformation of the PM intensity after subtracting the slide estimates. Affymetrix arrays give absolute expression values for a given gene. 2,000 genes were further analyzed as they crossed the minimal intensity across samples. The average sample skewness of the case group is 1.39, while it is 0.74 for the control group. We also use R software to demonstrate the application of the proposed test.

3. Results

3.1. Main findings

Table 1 shows type I error rates that were calculated under significance level (α-level) of 0.05, 0.01, 0.005, and 0.001. The half Johnson's modified t-test and the half Student's t-test maintained fairly precise type I error rates in all four distributions and at each significance level when sample size of case and control groups was larger ($n_1 = n_0 = 60$). When sample size of case and control groups was small ($n_1 = n_0 = 20$), the half Student's t-test maintained fairly precise type I error rates under normal and uniform distribution. Under uniform distribution, type I error rates of half Johnson's modified t-test are much smaller than significance levels for small samples, whereas they are close to but still smaller than significance levels for larger sample sizes. Although the type I error rate of both tests

Table 1. Type I error rates for the half Student's t-test and half Johnson's modified t-test

Significance level	$n_0 = n_1 = 20$		$n_0 = n_1 = 60$	
	Half Student's t-test	Half Johnson's modified t-test	Half Student's t-test	Half Johnson's modified t-test
Normal distribution				
0.05	0.0492	0.0494	0.0513	0.0510
0.01	0.0102	0.0106	0.0114	0.0113
0.005	0.0051	0.0056	0.0056	0.0057
0.001	0.0010	0.0013	0.0012	0.0012
Non-normal but symmetric distribution ($\gamma_3 = 0$)				
0.05	0.0470	0.0382	0.0496	0.0464
0.01	0.0094	0.0053	0.0096	0.0079
0.005	0.0047	0.0020	0.0047	0.0037
0.001	0.0010	0.0002	0.0010	0.0007
Positively skewed distribution ($\gamma_3 = 0.6$)				
0.05	0.0534	0.0557	0.0514	0.0528
0.01	0.0126	0.0166	0.0104	0.0115
0.005	0.0068	0.0107	0.0052	0.0062
0.001	0.0017	0.0043	0.0010	0.0016
Negatively skewed distribution ($\gamma_3 = -0.6$)				
0.05	0.0529	0.0570	0.0508	0.0523
0.01	0.0122	0.0166	0.0113	0.0122
0.005	0.0071	0.0106	0.0059	0.0068
0.001	0.0021	0.0048	0.0015	0.0020

was mildly inflated at small significance levels such as 0.005 or 0.001 under skewed distributions, such outcome was in line with our expectations due to departure from normality and homogeneity of variance (Adusah & Brooks, 2011).

Figure 1 presents the statistical powers of the half Johnson's modified t-test (solid lines) and the half Student's t-test (dashed lines) with normal distribution. For $r > 1$ and $d > 0$, the half Student's t-test was more powerful than the half Johnson's modified t-test overall. Note that the maximal difference in power between these two tests was 9%. Also note that under $d = 0$, both tests had some power for detecting difference between variances, with power increasing in r. For $d > 0$, powers of both tests decreased marginally as r increased, except for a condition ($d = 15$, $n_0 = n_1 = 20$), power increased as r increased.

Under non-normal but symmetric such as uniform distribution, Figure 2 shows the statistical powers of the two tests. When $n_0 = n_1 = 20$, the half Student's t-test was more powerful than the half Johnson's modified t-test for $r > 1$, and the largest difference in power was 19%. When $n_0 = n_1 = 60$, both tests had almost the same power when $d > 0$.

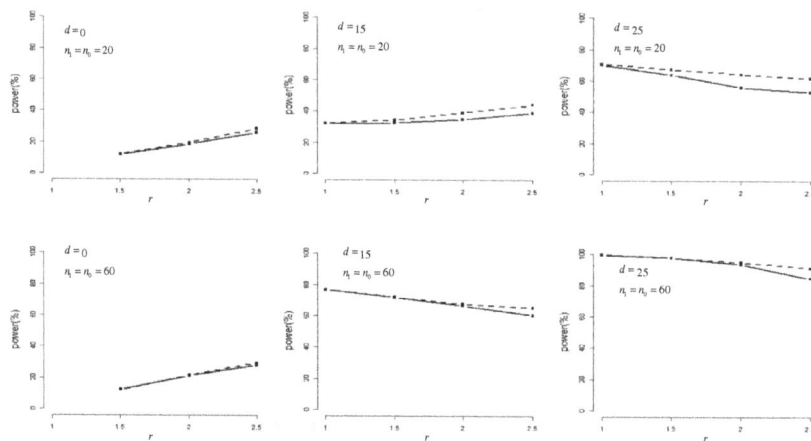

Figure 1. Statistical power in normal distribution.

Notes: Solid line: half Johnson's modified t; dash line: half Student's t. The difference in means between case and control groups (denot d) was set to 0, 15, and 25. The ratio of standard deviations for case group to control group (denot r) was set to 1, 1.5, 2, and 2.5.

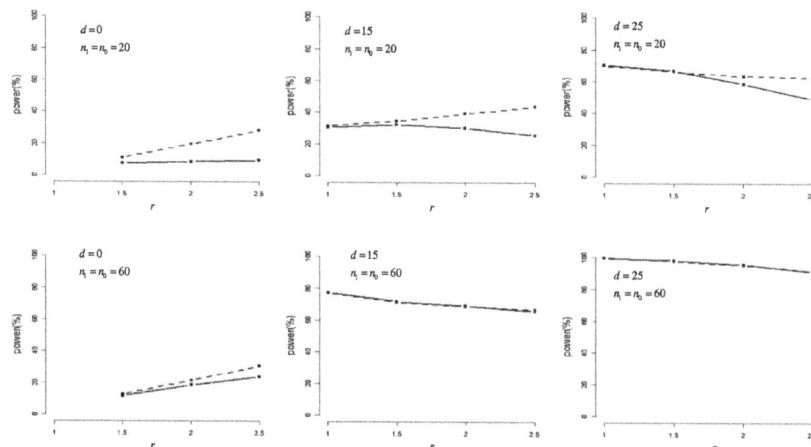

Figure 2. Statistical power in non-normal but symmetric distribution.

Notes: Solid line: half Johnson's modified t; dash line: half Student's t. The difference in means between case and control groups (denoted as d) was set to 0, 15, and 25. The ratio of standard deviations for case group to control group (denoted as r) was set to 1, 1.5, 2, and 2.5.

Figure 3 summarizes the statistical powers under positively skewed distributions. What was note-worthy was that the half Johnson's modified t-test was more powerful than the half Student's t-test when $r > 1$ under each of other settings, with the largest difference in power being 12%. The power performances of the two tests were similar when $r = 1$. For $d = 0$, both tests had some power to de-tect difference between variances, with power increasing in r for both tests. For $d > 0$, powers of both tests decreased with r increased, except for a case ($d = 15$, $n_0 = n_1 = 20$), power increased in r.

Figure 4 shows the statistical powers under negatively skewed distributions. For $r > 1$ and $d > 0$, the half Student's t-test was more powerful than the half Johnson's modified t-test overall. For $d = 0$, both tests also had some power for detecting the difference in variances between case and control groups with power increasing in r, and half Johnson's modified t-test had a little more power than half Student's t-test under $n_0 = n_1 = 20$. However, the half Johnson's modified t-test could not do so when $d > 0$ and $r > 1.5$.

3.2. Extensive study results
We also conducted extensive simulations to evaluate the performance of different tests. The results are summarized below. (For more details, refer to Supplementary Methods).

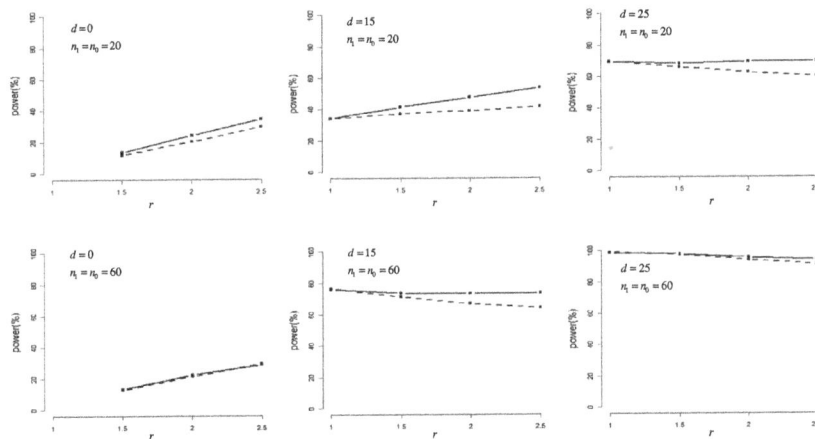

Figure 3. Statistical power in positively skewed distribution.

Notes: Solid line: half Johnson's modified t; dash line: half Student's t. The difference in means between case and control groups (denoted as d) was set to 0, 15, and 25. The ratio of standard deviations for case group to control group (denoted as r) was set to 1, 1.5, 2, and 2.5.

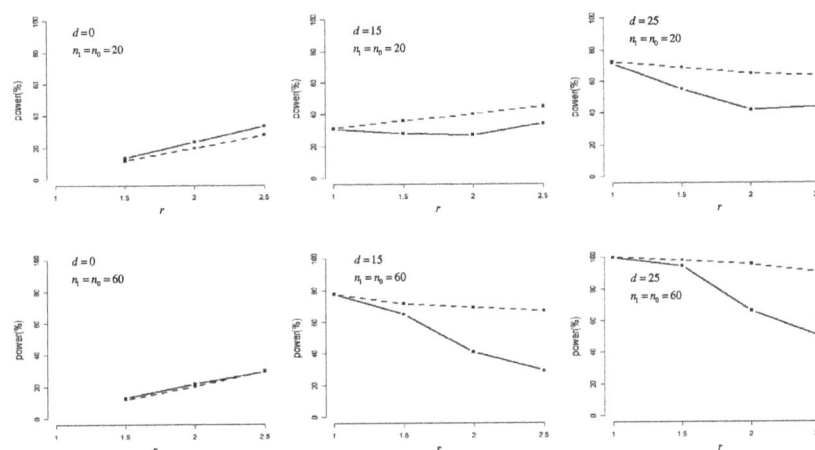

Figure 4. Statistical power in negatively skewed distribution.

Notes: Solid line: half Johnson's modified t; dash line: half Student's t. The difference in means between case and control groups (denoted as d) was set to 0, 15, and 25. The ratio of standard deviations for case group to control group (denoted as r) was set to 1, 1.5, 2, and 2.5.

3.2.1. Unequal sample sizes

We examined the situations of unequal sample sizes. We found that the half Johnson's modified t-test also maintained fairly precise type I error rates under four situations of equal and unequal sample sizes. The half Johnson's modified t-test was also more powerful than the half Student's t-test in a positively skewed scenario for both equal and unequal sample sizes. Notice that type I error rates of both tests was marginally inflated at a small significance level of 0.005 or 0.001 under skewed distribution with small control sample and large case sample. It has been discussed that the type I error would be inflated at the nominal significance levels for unequal sample sizes (Adusah & Brooks, 2011).

3.2.2. Unequal skewness

We examined the situation of increasing difference in skewness between the control and case groups (skewness of the case group is greater than skewness of the control group in most situations). We found that difference in power (between half Johnson's t-test and half Student's modified t-test) increased as difference in skewness increased. However, it should be cautioned that type I error rates were inflated in scenarios (of r and d) departing from normality and homogeneity of variance. We also observed that half Johnson's modified t-test was more powerful than the other tests. The results suggested that half Johnson's modified t-test can overcome heterogeneity and non-normality simultaneously when $0.3 \leq \gamma_3 \leq 0.6$ for the control group.

3.2.3. Powers of tests

We examined the power performances of Student's t-test, half Student's t-test, Johnson's modified t-test, and half Johnson's modified t-test. We found that half Johnson's modified t performs best among these tests. It's no surprise to understand that half Johnson's modified t-test can overcome heterogeneity (a higher variance for case group) and non-normality simultaneously under minor positive skewness ($0.3 < \gamma_3 < 0.6$ for the control group).

3.2.4. Combined test

We examined the power performances of a combined test which simultaneous testing means (using Student's t-test) and variances (using F-test) for the control and case group with a Bonferroni correction. For comparison of this combined test and two half tests, we found that the Johnson's modified t doesn't outperform the half Student's t under $r \leq 1.5$ for small sample size normal data. The half Student's t outperforms this combined test when there is a difference in means between the case group and the control group under $r \leq 1.4$.

3.3. Main findings for colon cancer data

Table 2 presents the numbers (percentages) of differentially expressed genes detected using the half Johnson's modified t, half Student's t, Welch's t, and Wilcoxon rank-sum tests (Wilcoxon, 1945), respectively. We set significance levels at 0.05, 0.01, 0.005, and 0.001 in this study. The half Johnson's

Table 2. Number (percentage) of differentially expressed genes of studied test methods in colon cancer data

	Test methods			
	Half Student's t-test	Half Johnson's modified t-test	Welch's t-test	Wilcoxon test
Significance level				
0.05	577 (28.9%)	632 (31.6%)	355 (17.8%)	275 (13.8%)
0.01	360 (18.0%)	422 (21.1%)	192 (9.60%)	141 (7.05%)
0.005	299 (15.0%)	368 (18.4%)	147 (7.35%)	99 (4.95%)
0.001	214 (10.7%)	292 (14.6%)	74 (3.70%)	45 (2.25%)
False discovery rate				
0.05	344 (17.2%)	429 (21.5%)	117 (5.85%)	48 (2.40%)
0.005	185 (9.25%)	278 (13.9%)	3 (0.15%)	5 (0.25%)

modified t-test detected more differentially expressed genes than the half Student's t-test at all significance levels.

Since a total of 2,000 genes were selected, we consider controlling the false discovery rate (FDR) (Benjamini & Hochberg, 1995; Storey & Tibshirani, 2003) to reduce the problem of multiple testing. The FDR (Benjamini & Hochberg, 1995; Storey & Tibshirani, 2003) was set to 0.05 and 0.005, respectively. From Table 2, half Johnson's modified t-test still detected more differentially expressed genes than the other tests. For instance, setting FDR to be 0.05, there are 429 differentially expressed genes determined by half Johnson's modified t-test, while 344 differentially expressed genes are determined by half Student's t-test.

4. Discussion

We found that half Johnson's modified t-test maintains the nominal α level and is fairly precise for normal and skewed distributions when standard deviation of case group is larger than that of control group. Further, half Johnson's modified t-test is more powerful than half Student's t-test for a positively skewed distribution. This means that half Johnson's modified t-test is suitable for studying positively skewed microarray gene expression data of heterogeneous diseases. In a heterogeneous disease, there is more than one entity that causes various clinical pictures and etiologies. Thus, the ratio of standard deviation between case group and control group is greater than 1 (that is the case group's standard deviation is larger). However, if the expression data for a heterogeneous disease is not positively skewed distributed, half Johnson's modified t will not achieve good power; instead, there may be power loss. Theoretically, half Johnson's modified t can test for the difference in the means and the difference in the variances simultaneously. From simulation results (refer to Supplementary Methods for details), half Johnson's modified t has very low power for testing the difference in two variances when means are equal or about the same (even with less power than half Student's t). But this shortcoming can be overcome since one can combine Student's t and F-test to achieve better power when the case group and the control group differ mainly in their variances. Therefore, half Johnson's modified t is mainly a test of equality of two means in heterogeneous diseases.

An overlap analysis was designed to match the baseline (selected as the Student's t-test) detection outcome (Supplement Table 4). These methods detected at least 92% overlap in differentially expressed genes. Under significance level of 0.05, the half Johnson's modified t-test had a 95.6% overlap and detected the most novel differentially expressed genes (i.e. 260). Researchers may be concerned about FDR settings when studying a large number of genes. Half Johnson's modified t-test provided a more rigorous FDR setting than the half Student's t-test for targeting the same number of priority genes. For example, to target 100 priority genes, the FDR for half Johnson's modified t-test was set to 4.28×10^{-8}, but for the half Student's t, it was set to 5.39×10^{-4}.

In practice, one may calculate the standard deviation of case and control to determine the status prior to applying the proposed test, the half Johnson's modified t-test. We suggest researchers carry out both half Student's t-test and the half Johnson's modified t-test to compare their results for heterogeneous and minor skewed gene expression data. However, if researchers have no prior idea about heterogeneity and skewness of gene expression data, then we suggest not using both tests simultaneously in the beginning. When detecting differentially expressed genes, type I error rate of both tests is mildly inflated at a strict significance level under skewed distribution. Slight inflation of type I error rate had no bearing on the findings of the present study. If researchers have enough resources to investigate more genes, we suggest they initially choose a moderate significance level (i.e. 0.05) for detecting differentially expressed genes.

In conclusion, half Johnson's modified t-test maintains fairly precise type I error rates in simulation scenarios, when the ratio of standard deviation between case group and control group is large ($r > 1$), and the distribution of gene expression in each group has positive skewness. In summary, half Johnson's modified t-test is recommended for detecting differentially expressed genes in heterogeneous and ONLY positively skewed data.

Funding

This study was partly supported by a grant from the Ministry of Science and Technology, Taiwan (MOST 105-2325-B-002-027).

Competing Interest

The authors declare no competing interest.

Author details

I-Shiang Tzeng[1]
E-mail: g1354502@nccu.edu.tw
ORCID ID: http://orcid.org/0000-0002-9047-8141
Li-Shya Chen[2]
E-mail: lschen@nccu.edu.tw
Shy-Shin Chang[3]
E-mail: sschang0529@gmail.com
Yung-ling Leo Lee[4]
E-mail: leolee@ntu.edu.tw

[1] Institute of Epidemiology and Preventive Medicine, College of Public Health, National Taiwan University, Taipei, Taiwan.
[2] Department of Statistics, National Chengchi University, Taipei, Taiwan.
[3] Department of Family Medicine, Chang Gung Memorial Hospital, Taipei, Taiwan.
[4] Graduate Institute of Epidemiology and Preventive Health, College of Public Health, National Taiwan University, Taipei, Taiwan.

References

Adusah, A. K., & Brooks, G. P. (2011). Type I error inflation of the separate-variances Welch *t*-test with very small sample sizes when assumptions are met. *Journal of Modern Applied Statistical Methods, 10*, 362–372.

Alon, U., Barkai, N., Notterman, D. A., Gish, K., Ybarra, S., Mack, D., & Levine, A. J. (1999). Broad patterns of gene-expression revealed by clustering analysis of tumor and normal colon tissues probed by oligonucleotide arrays. *Proceedings of the National Academy of Sciences, 96*, 6745–6750.
http://dx.doi.org/10.1073/pnas.96.12.6745

Benjamini, Y., & Hochberg, Y. (1995). Controlling the false discovery rate: A practical and powerful approach to multiple testing. *Journal of the Royal Statistical Society: Series B, 57*, 289–300.

Chen, Y., Dougherty, E. R., & Bittner, M. L. (1997). Ratio-based decisions and the quantitative analysis of cDNA microarray images. *Journal of Biomedical Optics, 2*, 364–374.
http://dx.doi.org/10.1117/12.281504

Cressie, N. A. C., & Whitford, H. J. (1986). How to use the two samplet-test *Biometrical Journal, 28*, 131–148.
http://dx.doi.org/10.1002/(ISSN)1521-4036

Dudoit, S., Yang, Y. H., Callow, M. J., & Speed, T. P. (2002). Statistical methods for identifying differentially expressed genes in replicated cDNA microarray experiments. *Statistica Sinica, 12*, 111–139.

Hsu, C. L., & Lee, W. C. (2010). Detecting differentially expressed genes in heterogeneous diseases using half

Student's t-test. *International Journal of Epidemiology, 39*, 1597–1604.
http://dx.doi.org/10.1093/ije/dyq093

Johnson, N. J. (1978). Modified t tests and confidence intervals for asymmetrical populations. *Journal of the American Statistical Association, 73*, 536–544.

Newton, M. A., Kendziorski, C. M., Richmond, C. S., Blattner, F. R., & Tsui, K. W. (2001). On differential variability of expression ratios: Improving statistical inference about gene expression changes from microarray data. *Journal of Computational Biology, 8*, 37–52.
http://dx.doi.org/10.1089/106652701300099074

Pan, W. (2002). A comparative review of statistical methods for discovering differentially expressed genes in replicated microarray experiments. *Bioinformatics, 18*, 546–554.
http://dx.doi.org/10.1093/bioinformatics/18.4.546

R Development Core Team. (2008). *R: A language and environment for statistical computing*. Vienna: R Foundation for Statistical Computing. ISBN 3-900051-07-0. Retrieved from http://www.R-project.org.

Smyth, G. K. (2004). Linear models and empirical bayes methods for assessing differential expression in microarray experiments. *Statistical Applications in Genetics and Molecular Biology, 3*, 1–25.

Storey, J. D., & Tibshirani, R. (2003). Statistical significance for genomewide studies. *Proceedings of the National Academy of Sciences, 100*, 9440–9445.
http://dx.doi.org/10.1073/pnas.1530509100

Templin, M. F., Stoll, D., Schrenk, M., Traub, P. C., Vöhringer, C. F., & Joos, T. O. (2002). Protein microarray technology. *Drug Discovery Today, 7*, 815–822.
http://dx.doi.org/10.1016/S1359-6446(00)01910-2

Thieblemont, C., Mayer, A., Dumontet, C., Barbier, Y., Callet-Bauchu, E., Felman, P., ... Coiffier, B. (2002). Primary thyroid lymphoma is a heterogeneous disease. *The Journal of Clinical Endocrinology & Metabolism, 87*, 105–111.
http://dx.doi.org/10.1210/jcem.87.1.8156

Tukey, J. W. (1977). *Exploratory data analysis*. Reading, PA: Addison-Wesley.

Tusher, V. G., Tibshirani, R., & Chu, G. (2001). Significance analysis of microarrays applied to the ionizing radiation response. *Proceedings of the National Academy of Sciences, 98*, 5116–5121.
http://dx.doi.org/10.1073/pnas.091062498

Van Den Bogaert, K., Govaerts, P. J., De Leenheer, E. M., Schatteman, I., Verstreken, M., Chen, W., ... Van Camp, G. (2002). Otosclerosis: A genetically heterogeneous disease involving at least three different genes. *Bone, 30*, 624–630.
http://dx.doi.org/10.1016/S8756-3282(02)00679-8

van der Pouw Kraan, T. C., van Gaalen, F. A., Kasperkovitz, P. V., Verbeet, N. L., Smeets, T. J., Kraan, M. C., ... Verweij, C. L. (2003). Rheumatoid arthritis is a heterogeneous disease: Evidence for differences in the activation of the STAT-1 pathway between rheumatoid tissues. *Arthritis & Rheumatism, 48*, 2132–2145.
http://dx.doi.org/10.1002/(ISSN)1529-0131

Welch, B. L. (1947). The generalization of "Student's" problem when several different population variances are involved. *Biometrika, 34*, 28–35.

Wilcoxon, F. (1945). Individual comparisons by ranking methods. *Biometrics Bulletin, 1*, 80–83.
http://dx.doi.org/10.2307/3001968

Yeoh, E. J., Ross, M. E., Shurtleff, S. A., Williams, W. K., Patel, D., Mahfouz, R., ... Downing, J. R. (2002). Classification, subtype discovery, and prediction of outcome in pediatric acute lymphoblastic leukemia by gene-expression profiling. *Cancer Cell, 1*, 133–143.
http://dx.doi.org/10.1016/S1535-6108(02)00032-6

2

Protective role of green tea on diabetic nephropathy

Md. Mohabbulla Mohib[1], S.M. Fazla Rabby[1], Tasfiq Zaman Paran[1], Md. Mehedee Hasan[2], Iqbal Ahmed[1], Nahid Hasan[1], Md. Abu Taher Sagor[1]* and Sarif Mohiuddin[3]

*Corresponding author: Md. Abu Taher Sagor, Department of Pharmaceutical Sciences, North South University, Dhaka 1229, Bangladesh
E-mail: sagor2008nsu@gmail.com
Reviewing editor: Tsai-Ching Hsu, Chung Shan Medical University, Taiwan
Additional information is available at the end of the article

Abstract: Nowadays, diabetes and diabetes-mediated dysfunctions are overwhelming at every nook and corner of the world which has been a sober concern to the current health care professionals. However, chronic hyperglycemic subjects often suffer from hypertension, atherosclerosis, insulin resistance, brain injury, and other dysfunctions due to high glucose level which lead to kidney failure. Altogether, diabetic nephritis, fibrosis, stenosis, iron overload, and hypertrophy may often escort towards diabetic nephropathy. Furthermore, hyperglycemia-generated free radical-mediated oxidative stress plays the central role to aggravate the condition. Oxidative stress also inhibits production of several natural antioxidant genes like Nrf2, Sirt1, PGC-1α, superoxide desmutase, and catalase. Similarly, production of pro-inflammatory cytokines such as IL, TNF-α, MIP, α-SMA, and NF-$\kappa\beta$ has been observed high in the diabetic subjects. High glucose inside the body also activates mitogen-activated protein kinase family and facilitates diabetic nephropathy. On the other hand, green tea (GT) is a widely used drink which has several protective functions. This plant possesses multiple cathecin, theoflavins, flavonoids, flavinol, caffeine, and other biological active components. Thus, this review will try to explain how GT-derived molecules prevent diabetic nephropathy, both by reducing free radical generation and improving insulin secretion. The molecular interactions between antioxidant genes and free radical-mediated oxidative stress will be explained

ABOUT THE AUTHORS

Our group endeavors to explore the role of oxidative stress in different diseases and the potential role of antioxidant derived from natural sources. In our current study, we tried to identify the role of green tea as a potential source of antioxidant against diabetic nephropathy. We focused on possible mechanism by means of which the molecules from green tea can suppress the oxidative damage as well as how it can restore the normal antioxidant level in the system. Our study was led by Sarif Mohiuddin, lecturer at Pioneer Dental College and Hospital. His area of specialization is Diabetology, Natural Antioxidant, and Angiotensin-II. He has completed his MBBS from Dhaka University and has been awarded FRSPH from The Royal Society for Public Health (RSPH) UK. He has also completed certified course on Diabetology, and extension of diabetic Care Course from BIRDEM Hospital, Dhaka, Bangladesh.

PUBLIC INTEREST STATEMENT

Diabetes is one of the most growing concerns of the world, and everyday more and more people are getting affected by diabetes. Thus, the risks of diabetes associated diseases are also on the rise. diabetic nephropathy is one of the most common diabetes-associated diseases, and just by developing a small habit people who are suffering from diabetes can avoid diabetic nephropathy. In this study, we tried to emphasize on how green tea is conducive to prevent diabetic nephropathy.

through activation of AMPK and mTOR. Finally, this study will try to correlate the possible therapy strategy and molecular mechanism of GT to reduce the pathogenesis of diabetic nephropathy.

Subjects: Bioscience; Health and Social Care; Medicine, Dentistry, Nursing & Allied Health

Keywords: diabetic nephropathy; free radicals; AMPK; Nrf2; Sirt1; green tea

1. Introduction

In this current era, it is often heeded that consumption of high fat diet and fructose-containing beverages, cigarette smoking, alcoholism, less or no physical exercise are rising in urban areas at an alarming rate which lead to several disorders like insulin resistance, obesity, hyperlipidemia, metabolic syndrome, and diabetes (Guh, Zhang, Bansback, Amarsi, Birmingham, & Anis, 2009). Diabetes, a heterogeneous disorder which is primarily characterized by impaired hormone secretion, in addition it is also caused by several impairments like protein, fat, and carbohydrate metabolism by either insufficient amount of insulin production or reduced sensitivity of tissue to insulin (Pistrosch et al., 2015). According to WHO report 2011, 9% of the total population above 18 years are suffering from Diabetes Mellitus (DM) (Alwan, 2011). If this scenario continues, the projected number of diabetic patients would be approximately 552 million in 2030 (Reno et al., 2015; Whiting, Guariguata, Weil, & Shaw, 2011). Evidences also documented that around one-third of diabetic subjects suffer from diabetic nephropathy (DN) resulting in the overall cost of the treatment beyond reach (Atkins & Zimmet, 2010).

DN is being considered as one of the major microvascular complications of DM and it has been claimed as a primary cause of end-stage renal diseases (Jin et al., 2012). Hyperglycemia-induced DN creates long-term complications which lead to high mortality and morbidity rate (Kim, Davis, Zhang, He, & Mathews, 2009). Several studies suggested that diabetes is also affiliated with other complications like retinopathy, cardio-myopathy, neuropathy, atherosclerosis, systemic hypertension, stroke, coronary ischemia, and most importantly diabetic kidney failure (Kupelian, Araujo, Wittert, & McKinlay, 2015; Rutter & Nesto, 2011). However, study also reported that renin–angiotensin system (RAS) plays a pivotal role in the pathogenesis of DN (Peti-Peterdi, Kang, & Toma, 2008). In the diabetic subjects, hyperglycemia often stimulates pro-inflammatory cytokines, neutrophil infiltration, and other pathogenic factors (Chow, Ozols, Nikolic-Paterson, Atkins, & Tesch, 2004) which generate reactive oxygen species that further exacerbates the situation (Ha, Yu, Choi, Kitamura, & Lee, 2002). On top of that, recent studies proved that diabetic subjects often lack antioxidant activities which may begin defenseless oxidative stress and progress diabetic complexity (Nourooz-Zadeh et al., 1997; Santini et al., 1997). Sometimes hypertension may develop DN by influencing inflammatory cytokines as well as generating free radicals (Lopes de Faria, Silva, & Lopes de Faria, 2011). In fact, DN kidney mostly lacks of AMPK/Sirt1 expression on experimented animal model (Chuang et al., 2011). Besides, diabetic kidneys also suffer from low level of TIMP3 and FoxO1; conversely, STAT1 level was noticed high (Fiorentino et al., 2013).

In recent years, green tea (GT) has become a very popular drink in several regions like South-East Asia (Wolfram, 2007). GT extracts possess several antioxidant group of molecules like flavonoids, flavonols, polyphenols, theaflavins, tannins and other important components (Lin, Juan, Chen, Liang, & Lin, 1996) which control several biological mechanisms (Polychronopoulos et al., 2008) like increased expression of antioxidant genes (Nomura et al., 2015), protect glomerulas (Peng et al., 2011), promote insulin sensitivity (Nomura et al., 2015), suppress pro-inflammatory cytokines (Kim, Murakami, Miyamoto, Tanaka, & Ohigashi, 2010), prevent RAS (Kurita, Maeda-Yamamoto, Tachibana, & Kamei, 2010), augment insulin production (Ortsäter, Grankvist, Wolfram, Kuehn, & Sjöholm, 2012), decrease α-amylase level (Gao, Xu, Wang, Wang, & Hochstetter, 2013), lower lipids levels (Ramadan, El-Beih, & Abd El-Ghffar, 2009), prevent free radical generation (Yokozawa, Noh, & Park, 2012), cyto-protective (Shin, Chung, Lee, & Kim, 2009), improve and protect podocyte production (Peixoto et al., 2015), enhance mitochondrial biogenesis (Rehman et al., 2013), stabilize cellular signaling (Kim, Quon, & Kim, 2014), protect genetic materials (Glei & Pool-Zobel, 2006), and inhibit cancer (Darvesh & Bishayee, 2013). In addition, experiment revealed that GT extract was able to reduce proteinurea on tacrolimus-induced nephrotoxic mice (Back et al., 2015). Reduced p-ERK1/2, MAPKp38, p-JNK, and p-AKT have been showed when EGCG 50 mg/kg/day was given to rats-induced crescentic glomerulonephritis (Ye et al., 2015). Similarly, another study described that long-term dietary antioxidant treatment lowers kidney inflammatory cytokines and oxidative stress markers on diabetic mice (Park, Park, & Lim, 2011). Restoration of antioxidant genes can be targeted as pharmacological approach for DN which can help in cell survival against diabetes-mediated dysfunctions (He et al., 2010). Therefore, this review will try to make a correlation among hyperglycemia, antioxidant genes, free radicals, and GT.

2. Pathology of diabetic nephropathy

Diabetes is often known as metabolic disorder which explains the inability of endocrine glands or hormonal secretion. Several approaches have been explained to develop diabetes inside a subject. Study described that diabetes is the outcome of either improper hormone secretion or insufficient and defective hormone production. However, it is also explained that improper Ca++ signaling or defective insulin mRNA are responsible for the development of diabetes (Kabir et al., 2015). Not only the clinical features of DN are 3P (Polyurea, Poly phasia, and Polydypsia) but also showed higher albumin elimination, abnormal glomerular filtration rate, and rapid decreasing renal functions which finally lead to end-stage renal failure. Besides, hyperglycemia may also induce oxidative stress by generating free radicals, advanced glycation end-products and activating protein kinase C to further aggravate diabetic kidney (Giacco & Brownlee, 2010). With the help of free radical, advanced glycation end-products (Lacmata et al., 2012) are formed that later interacts with its receptor RAGE and develop DN. It is suggested that blocking of RAGE or deletion of RAGE can be an effective approach in preventing diabetes-mediated complications at initial stage (Tan et al., 2010; Wendt et al., 2003). It has been evaluated that higher glucose in the body often stimulates diacylglycerol to increase the vascular permeability for inviting immune cell infiltration like neutrophil, monocyte, leukocytes, macrophage, and others. Taken together, protein kinase C also participates to activate local myofibroblastic cells which further secret collagen and extra cellular matrix that leads to kidney fibrosis. Furthermore, these pathways also regulate cell growth, cytokine and chemokine release, vasoconstrictions, apoptosis, and finally cell death (Noh & King, 2007). It was also noticed that RAS is highly responsible for DN by changing hemodynamic alteration and activating iNOS as well as endothelin (ET-1) in diabetic mellitus subjects (Har et al., 2013; Mohib et al., 2016; Ruggenenti, Cravedi, & Remuzzi, 2010). Likewise, hyperglycemia also found responsible for nephropathy by controlling blood flow and blocking small vessels (Elmarakby & Sullivan, 2012). Not only glucose or insulin is involved in the patho-physiology of DN, but family history and environmental factors are also responsible (Martini, Eichinger, Nair, & Kretzler, 2008). Patients who are suffering from DN found with excess accumulation of p62/SQSTM1 (Sequestosome 1) protein in proximal tubular cells from their kidney biopsy (Yamahara et al., 2013). Drug-induced kidney dysfunctions are on rise (Sagor, Mohib, Tabassum, Ahmed, & Reza, 2016) and these complications are making difficulties for diabetic patients to achieve an effective therapy.

3. Green tea history, traditional uses and its functional actives

The tea plant was first noticed in China and this plant was then cultivated by the Chinese traditional inhabitant from the ancient times (International Tea Committee, 2009). It has been reported that GT was first exported to Japan from India during seventieth century. Around 2.5 million tons of tea is produced currently and 20% of that leaves are being processed for GT which are generally consumed by USA, Europe, Asia, and some places of North Africa (Chacko, Thambi, Kuttan, & Nishigaki, 2010). Tea is the second most popular drink after water (Haidari, Shahi, Zarei, Rafiei, & Omidian, 2012). In last few decades, several beneficial properties on human health have been noticed from the consumption of GT (Cabrera, Artacho, & Giménez, 2006). Traditionally, tea is classified into GT and black tea. However, tea (*Camellia sinensis*) belongs to Theaceae family; oolong where GT is prepared from the young green leaves. On the other hand, black tea is made by steaming. The tree normally looks green and can grow around 30 feet high in wild environment, but for commercial cultivation it is pruned to 2–5 feet. In adult age, the leaves become dark green, seen oval shape, and appear singly or cluster. After collecting the tea leaves, black tea undergoes fermentation, in contrast GT is not

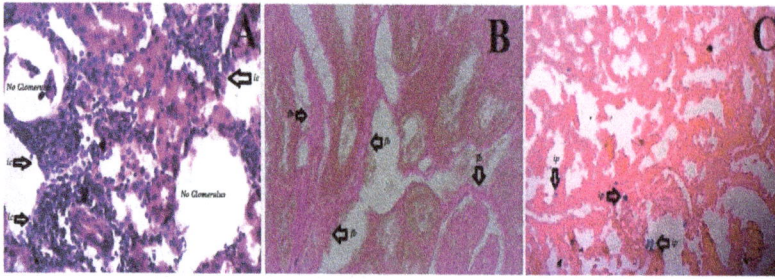

Figure 1. Histology represents kidney dysfunctions induced by Alloxan. (A) Which was stained by Hematoxylene and Eosin shows various inflammatory components infiltration, most of the glomerulas have been destroyed by diabetes, (B) which was stained using sirius red and picric acid shows collagen deposition (red parts), and (C) which was stained Prusian blue to determine the iron overload. Inside figures *ic*—inflammatory cells, *fb*—fibrosis, and *ip*—iron pigments. The histology was prepared in our lab.

(Hamilton-Miller, 1995). Collected young leaves undergo steaming process at higher temperatures which inactivates the oxidizing enzymes thereby polyphenols remain intact (Alschuler, 1998). Traditionally, GT has been used on several purposes to treat viral diseases (Weber, Ruzindana-Umunyana, Imbeault, & Sircar, 2003), antibacterial infections (Sudano Roccaro, Blanco, Giuliano, Rusciano, & Enea, 2004), inflammation (Dona et al., 2003), cardiovascular diseases (Sueoka et al., 2001), obesity, and lipid lowering (Raederstorff, Schlachter, Elste, & Weber, 2003), angiogenesis (Sartippour et al., 2002), cancer (Kavanagh et al., 2001), neuro-protective (Weinreb, Mandel, Amit, & Youdim, 2004), anti-arthritic (Haqqi et al., 1999), antioxidant (Osada et al., 2001), and other health-related disorders. GT possesses diverse types of bioactive molecules which have several protective mechanisms. So far, various types of flavonoids, polyphenols, and tannins have been isolated from GT leaves (Graham, 1992). Classically, four types of catechins have been studied from GT extract and those potent catechins are epigallocatechin-3-gallate (EECG), (+)-catechin (CE), epicatechin-3-gallate (ECG), and epigallocatechin (EGC) (Zaveri, 2006). All the catechins and other isolated molecules (Figure 2) from GT were found to be effective against diabetic-mediated kidney dysfunctions (Al-Attar & Zari, 2010).

4. Diabetes and free radical biology

Free radicals are parts of the body, sometimes they are called byproducts. They are normally generated either during ATP production or cellular degeneration or by any harmful stimuli (Sagor, Tabassum, Potol, & Alam, 2015). They often produce oxidative stress which further leads to organ dysfunctions by inducing pro-inflammatory cytokines, transcription factors for fibrosis (Figure 1(B)), hemeoxygenase (HO) for iron overload (Figure 1(C)), mast cell accumulation, and other detrimental factors (Reza, Sagor, & Alam, 2015; Reza et al., 2016). Interestingly, they are responsible for cell membrane damage by oxidizing membrane lipids through malondehyde (MDA) activity. Later, free radicals react with nucleus, DNA, alter genetic codes, and oxidized regulatory protein which results in the production of advanced oxidative protein products (AOPP) that further damage various components of cell (Abu Taher et al., 2016; Sagor, Chowdhury, et al., 2015). However, nitric oxide is potent vasodilator produced from nitric oxide synthetase (NOS) can be very harmful to a system once it is produced from induced nitric oxide synthetase (iNOS) (Gupta et al., 2005). iNOS, a potent oxidant, is generally processed by free radical which damages cytoplasm and has a powerful role to activate NF-$\kappa\beta$, which further stimulate several other pro-inflammatory cytokine (Alam, Chowdhury, Jain, Sagor, & Reza, 2015). Unfortunately, chronic hyperglycemia is often blamed for generation of free radical-mediated oxidative stress too (Tse, Anderson, Ganini, & Mason, 2015). Moreover, study also explored that oxidative stress has a significant role on production of advanced glycation end products (AGE) (Pazdro & Burgess, 2012). It is highly suggested that NAD(P)H-mediated oxidative stress helps in disease progression of DN (Jha et al., 2014). NAD(P)H oxidase has been also claimed for damaging cell membrane, cytoplasm, and mitochondria through oxidative-mediated stress as the expression of NOX-4 is always found high in DN (Eid et al., 2010; Sedeek et al., 2010). Another study recommended that suppression of NOX-4 reduced glomerulas damage and protected overall kidney functions on db/db BLKS mice by decreasing albuminuria production (Figure 3) (Sedeek et al., 2013).

5. Role of inflammation on diabetic nephropathy

Immunity is the ultimate hero of a biological system which fights against any foreign invaders. Most importantly, inflammation is considered as primary defensive mechanism and is generally induced by any noxious or harmful stimuli. Unfortunately, it is sometimes activated against its own host (Aldhahi & Hamdy, 2003). The relation between DN and inflammation is common and often inflammation plays a pivotal role to develop DN, although it is very difficult to show a single molecular mechanism to understand the patho-physiology of DN through inflammatory molecules and pro-inflammatory cytokines (Figure 1(A)) (Ruggenenti et al., 2010). It is extensively studied that diabetic status of a subject often attracts several inflammatory and pro-inflammatory cytokines like nuclear factor for (NF)-$\kappa\beta$, tumor necrosis factor (TNF)-α, interleukin (IL)-6, interleukin (IL)-1β, α-smooth muscle actin (SMA), macrophage inflammatory protein (MIP), T-lymphocyte, matrix metalloproteinase (MMP), cyclooxygenase-2 (Cox, Abu-Ghannam, & Gupta, 2010), mast cell, monocytes, macrophages,

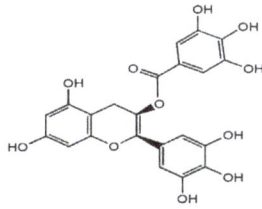

Epigallocatechin gallate (EGCG)

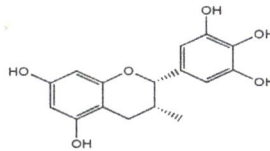

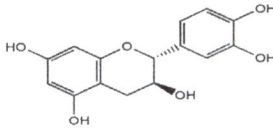

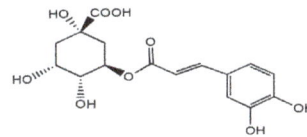

Epigallocatechin (EGC)

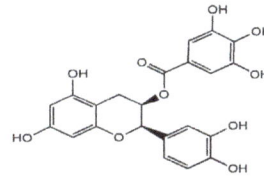

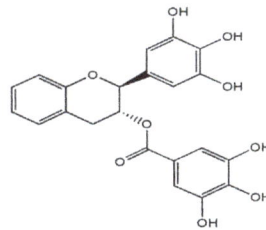

Epicatechin gallate (ECG)

Epicatechin (EC)

Gallocatechin gallate (GCG)

Chlorogenic acid

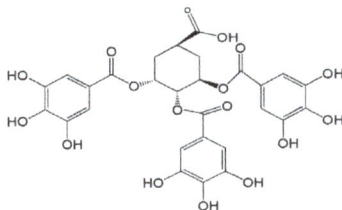

3-galloylquinic acid

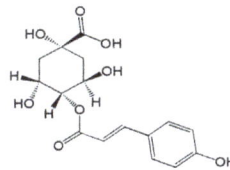

3-p-coumaroylquinic acid

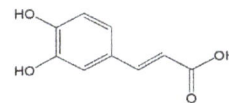

Caffeic acid

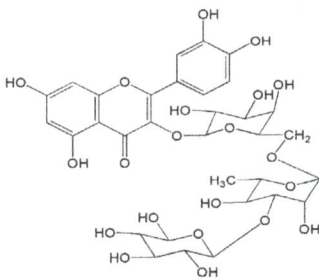

Quercetin-rhamnosylgalactoside

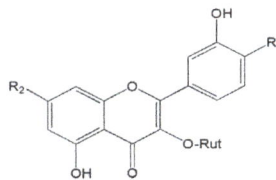

Quercetin-3-rutinoside

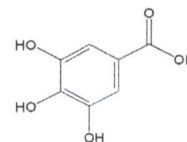

Gallic Acid

(Continued)

Figure 2. *(Continued).*

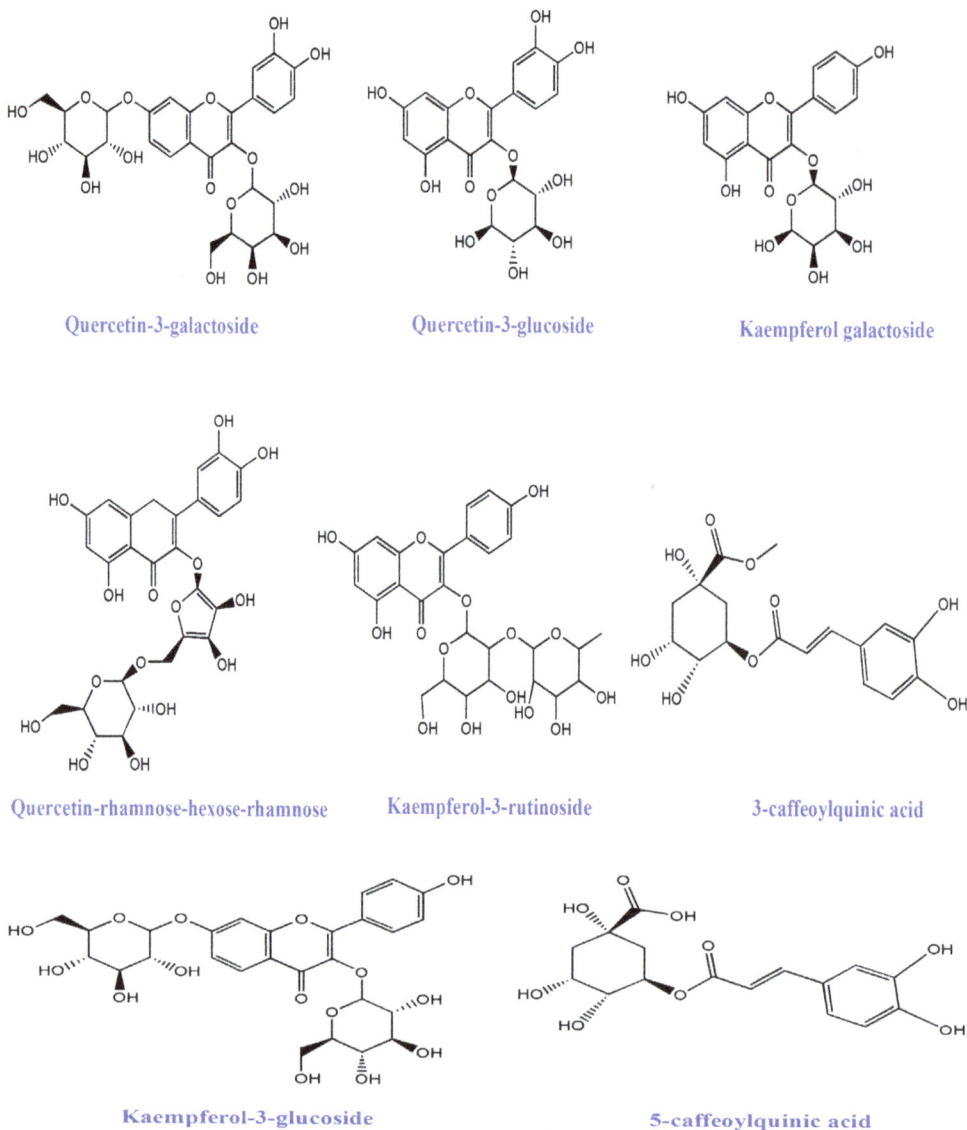

Quercetin-3-galactoside   Quercetin-3-glucoside   Kaempferol galactoside

Quercetin-rhamnose-hexose-rhamnose   Kaempferol-3-rutinoside   3-caffeoylquinic acid

Kaempferol-3-glucoside   5-caffeoylquinic acid

Figure 2. Molecules, which have been isolated from green tea and have biological activity.

myloperoxidase (Rojas, Ochoa, Ocampo, & Muñoz, 2006), PGE_2, interferon (INF)-ϒ, and others, some of them are associated with TLR4-MyDD88 interactions (Donate-Correa, Martín-Núñez, Muros-de-Fuentes, Mora-Fernández, & Navarro-González, 2015). Taken together, a study found out that hyperglycemia induces PKC which further activates pro-inflammatory cytokines. Likewise, high glucose also induces oxidative stress which plays a critical role inside kidney generating various harmful free radicals (Mora & Navarro, 2004). Most probably, inflammatory markers like IL-1 and TNF-α were first claimed for nephropathy in diabetic condition (Ienaga & Kondo, 1991). Another study explained that, mRNA expression of IL-6 found high in human renal sample on a diabetic study (Suzuki et al., 1995). A clinical study showed that, expression of IL-18 was found remarkably high in diabetic patients (Fantuzzi, Reed, & Dinarello, 1999) which further linked to mitogen-activated protein kinase (MAPK) (Miyauchi, Takiyama, Honjyo, Tateno, & Haneda, 2009). In addition, TNF-α was also claimed to generate free radicals in rat glomeruli via MAPK and protein kinase C (PKC) pathway (Koike, Takamura, & Kaneko, 2007). Reduction of antioxidant gene like Nrf-2 leads to chronic kidney diseases (CKD) through oxidative stress-mediated inflammation (Ruiz, Pergola, Zager, & Vaziri, 2013). It was also noticed that disruption in Nrf-2 gene signaling may also create lupus-like autoimmune nephritis and induce diabetic inflammation along with nephropathy (Yoh et al., 2001, 2008).

6. Role of green tea on diabetic nephropathy

In DN, cell membrane gets damaged due to diabetic inflammation which is a result of excess glucose concentration inside the organ (Giacco et al., 2014). High glucose concentration always hampers multiple cellular signaling like GLUT, MAPK, and PKC (Dronavalli, Duka, & Bakris, 2008). On top of that, it also increases AGEs, inflammatory cytokines, tumor necrosis factors, nuclear factor-$\kappa\beta$, interleukin-1β, and pro-inflammatory cytokines are often induced by methylglyoxal signaling (Wang, Meng, Gordon, Khandwala, & Wu, 2007). (+)−Catechins have been highly effective against DN and related other complications. One of the studies found that administration of (+)−catechin for 16 weeks was quite effective against DN by diminishing renal damage and methylglyoxal signaling on db/db mice. Similarly, catechin was found to be very effective against DN on human endothelium-derived cells (Zhu et al., 2014). Altogether, diabetic subjects possess free radicals generation, AGEs production, HbA1C level and free glucose concentration which always hamper kidney functions by preventing antioxidant genes like Nrf2, Sirt1, PGC-α, SOD, FOXO (Ding & Choi, 2015), and others which are mostly mediated through mTOR (Zoncu, Efeyan, & Sabatini, 2011) and AMPK pathway (Lee, Park, Takahashi, & Wang, 2010; Price et al., 2012). It is significantly observed that GT protects DN by either increasing antioxidant genes (Wang et al., 2015), or reducing free radical production (Khan et al., 2009) or suppressing pro-inflammatory mediators (Figure 4) (Sachdeva, Kuhad, Tiwari, Arora, & Chopra, 2010).

GT possesses several types of cathechins and flavonoids which generally exert multiple mechanisms to protect from DN-related complexities (Funamoto et al., 2016). Nrf2, the most protective gene, known as the master regulator of a cell that fights against inflammation; fibrosis and free radicals mediated oxidative stress in chronic kidney diseases (Aminzadeh et al., 2014). Nuclear factor-erythroid-2-related factor 2 (Nrf2) plays the central role to regulate and coordinate in the induction of more than 250 gene which protect cell through several signaling. The genes which are expressed by the help of Nrf2 are generally known as antioxidant or protective gene and some of the potent components are SOD, heme oxygenase-1, glutamate cysteine ligase, catalase, NAD(P)H: quinone oxidoreductase-1 (NQO1), thioredoxin, glutathione S-transferase, and glutathione peroxidase which mainly work by either restoring antioxidant property or help in harmful substances metabolism like phase-II drug enzymes (Li et al., 2008; Wakabayashi, Slocum, Skoko, Shin, & Kensler, 2010). Activation and production of antioxidant genes like Nrf2, ARE, and SOD can protect a cell from any kind of stress which make them a target for drug molecules (Sriram, Kalayarasan, & Sudhandiran, 2009). It is often observed that nephropathy subjects always lack Nrf2 production in kidney (Aminzadeh, Nicholas, Norris, & Vaziri, 2013). Experiment showed that expression of Nrf2 also ameliorates oxidative stress, reduces inflammation, and kidney fibrosis in animal model via Nrf2-keap1 signaling (Soetikno et al., 2013). It is observed that Nrf2-mediated pathway reduces NF-$\kappa\beta$-inflammatory signaling and thus initiates apoptosis (Li et al., 2008). GT flovonoids have been proposed for production of Nrf2 mRNA on DN subjects (Na & Surh, 2008; Yoon et al., 2014). One of the most potent cathecins, epigallocatechin-3-gallate protects from Cisplatin-induced nephrotoxic rats through Nrf2/HO-1signaling pathway (Sahin et al., 2010). In addition, another study revealed that epigallocatechin-3-gallate stops lupus nephritis development by enhancing Nrf2 and inhibiting NLRP3 (Tsai et al., 2011). Several protective effects of GT molecules have been summarized in Tables 1 and 2.

Another important cell saving component is superoxide dismutase which protects cytoplasm, mitochondria, and nucleus of a cell (Wassmann, Wassmann, & Nickenig, 2004). Hyperglycemia generates several free radicals like superoxide anion which contributes in the development of DN progression (Ha & Kim, 1999) and damage kidney podocyte as well as glomerulas (United States Renal Data System, 2011). It has been also noticed that interaction of AT_1R with Ang-II may generate superoxide anion which is a positive signal for pathogenesis of chronic kidney diseases through NAD(P)H oxidase (Kim, Sato, Rodriguez-Iturbe, & Vaziri, 2011; Vaziri, Dicus, Ho, Boroujerdi-Rad, & Sindhu, 2003). Diabetic kidney often suffers from various disturbances due to over production of

Table 1. Recent protective findings of Green tea on diabetic animal studies

Models	Outcomes of the study	References
Model: Albino mice of MF1 strain *Wt of model:* 26.4–32.2 gram *Disease induced by:* Streptozotocin *Dose of GT:* 0.5 mL/day *Route:* Oral *Duration:* 30 days	• Slight decreased serum triglycerides, cholesterol, total protein, creatinine, urea, and uric acid levels	Al-Attar and Zari (2010)
Model: db/db mice *Age of model:* 7 weeks *Disease induced by:* N/A *Dose of GT:* 1% (w/w) of EECG *Route:* Oral *Duration:* 10 weeks	• Reduced expression of DNA-damage inducible transcript 3 (Ddit3), DNA damage inducible protein 34 (Ppp1r15a) and cyclin dependent kinase inhibitor 1a (Cdkn1a) • The treatment also increased insulin sensitivity	Faria et al. (2012)
Model: Spontaneous Hypertensive Rats *Age of model:* 12 weeks Disease induced by: Streptozotocin *Dose of GT:* 5 g GT/kg body weight/day *Route:* Oral *Duration:* 12 weeks	• Oral GT up regulated eNOS expression by reducing phospho-Thr495 eNOS and phospho-Ser1177 eNOS expression • It also enhanced eNOS uncoupling	Faria, Papadimitriou, Silva, Lopes de Faria, and Lopes de Faria (2012)
Model: Spontaneous Hypertensive Rats *Wt of model:* 250 gram *Disease induced by:* N/A *Dose of GT:* 13.3 g /L with water *Route:* Oral *Duration:* 12 weeks as	• Reduced oxidative stress by decreasing 8-hydroxy-2′-deoxyguanosine (8-OHdG) and NAD(P)H oxidase-dependent superoxide generation • It also down regulated NAD(P)H oxidase (Nox)-4 and nitrotyrosine gene expression	Ribaldo et al. (2009)
Model: Male C57BL/Ks db/db mice *Wt of model:* N/A *Disease induced by:* N/A *Dose of GT:* 10% non-fermented GT extract *Route:* Oral *Duration:* 24 weeks	• The numbers of Madin-Darby canine kidney epithelial Cells (MDCK Line) were increased	Kang et al. (2012)
Model: Male Sprague–Dawley rats *Wt of model:* 200–230 gram *Disease induced by:* Streptozotocin *Dose of GT:* 16% (w/w) GT *Route:* Oral *Duration:* 12 weeks	• Green tea caused reduction in serum glucose, glycosylated protein, as well as blood urea nitrogen excretion and glycogen-filled tubules found decreased	Renno, Abdeen, Alkhalaf, and Asfar (2008)
Model: Male Wistar albino rats *Wt of model:* 120 gram *Disease induced by:* Alloxen *Dose of GT:* 50 or 100 mg/kg body weight GT extract daily *Route:* Oral *Duration:* 4 weeks	• Oral GT extract decreased clotting time, serum glucose, total direct bilirubin ratio, urea and creatinine levels	Ramadan, El-Beih, and Abd El-Ghffar (2009)

(Continued)

Table 1. *(Continued)*

Models	Outcomes of the study	References
Model: Female Wister albino rats *Wt of model:* 150–200 gram *Disease induced by:* Gentamicin *Dose of GT:*300 mg/kg/d GT extract *Route:* Oral *Duration:* 7 days	• The level of glutathione, superoxide dismutase, and catalase(CAT) were increased and decreased thiobarbituric acid reactive substance (TBARS) levels in renal tissue	Abdel-Raheem, El-Sherbiny, and Taye (2010)
Model: Male Wistar rats *Wt of model:* 150–180 gram *Disease induced by:* Cisplatin *Dose of GT:* GT extract (3%, w/v) with drinking water *Route:* Oral *Duration:* 25 days	• Increased phosphate (Pi) uptake in the presence of a Na-gradient in the uphill phase (30s), • It also enhanced activity of hexokinase (HK), malate dehydrogenase (MDH) in the renal cortex and medulla	Khan et al. (2009)
Model: Male Wistar rats *Wt of model:* 200–215 gram *Disease induced by:* Cisplatin *Dose of GT:* 100 mg/kg/day EGCG *Route:* Orally *Duration:* 12 days	• The expression of Nrf2 and HO-1 were induced • Decreased NF-$\kappa\beta$ p65 and 4-Hydroxynonenal (HNE) expression • The contents of SOD, catalase (CAT), Gpx, and GSH in the kidney were also restored	Sahin et al. (2010)
Model: Musmusculusvar albino mice *Wt of model:* 25–30 gram *Disease induced by:* Cisplatin *Dose of GT:* 100 mg/kg *Route:* Oral *Duration:* 7 days	• Inhibited the thickening of the glomerular basement membrane, degeneration and necrosis of tubular epithelial cells slightly • It also increased • GSH level and reduced MDA level in the kidney	Yapar, Çavuşoğlu, Oruç, and Yalçin (2009)
Model: Male sprague-dawley rats *Wt of model:* 200–250 gram *Disease induced by:* Cyclosporin A *Dose of GT:* 80 mg/kg/day extract *Route:* Oral *Duration:* 21 day	• Increased mitochondrial DNA (mtDNA) numbers, ATP synthase-β (AS-β) and NADH dehydrogenase-3 (ND3), • The treatment activated peroxisome proliferator-activated receptor-Iα co-activator (PGC)-1α and renal mitochondrial transcription factor-A (TFAM) mRNA • It also attenuated renal injury and improved renal function by enhancing brush border activity	Rehman et al. (2013)
Model: Male 129/svJ mice *Age of model:* 8 weeks *Disease induced by:* Immune-mediated glomerulonephritis *Dose of GT:* 50 mg/kg/day EGCG) *Route:* Oral *Duration:* 14 days	• Reduced proteinuria, lowered H_2O_2 level and • In addition of EECG further reduced the p65/ nuclear factor (NF)-$\kappa\beta$ expression • Also increased Peroxisome proliferator-activated receptor gamma (PPARγ) activity	Peng et al. (2011)

(Continued)

Table 1. *(Continued)*

Models	Outcomes of the study	References
Model: Sprague–Dawley rats *Wt of model*: 200–250 gram *Disease induced by*: Cyclosporine *Dose of GT*: 100 mg/kg GT extract *Route*: S.C. injection *Duration*: 21 days	• Normalized the proximal tubular necrosis and mild interstitial inflammation • It also suppressed plasma renin activity (PRA) and serum aldosterone levels	Ryu et al. (2011)
Model: Male C57 BLKS/J (db/db) mice *Age of model*: 6 weeks *Disease induced by*: *Dose of GT*: 15, 30, and 60 mg Catechin /kg *Route*: Oral *Duration*: 16 weeks	• Normalized glomerular cell loss (in a dose dependent way), plasma and renal methylglyoxal levels • It also reduced glomerular hypertrophy • Furthermore, it inhibited advanced glycation end products-receptor for advanced glycation end products (AGE–RAGE) mediated inflammatory pathway by attenuating nuclear factor kappa beta (NF-$\kappa\beta$) activation	Zhu et al. (2014)
Model: Wistar rats *Wt of model*: 120–130 gram *Disease induced by*: Nephrectomy along with streptozotocin *Dose of GT*: 25, 50, and 100 mg/kg/day EECG *Route*: Oral *Duration*: 50 days	• Decreased albumin and AGE accumulation (in a dose-dependent way) • It also down regulated cyclooxygenase (COX-2) • Furthermore, the treatment decreased phosphorylated nuclear factor of kappa light polypeptide gene enhancer in B-cells inhibitor alpha (IkB-α), transforming growth factor beta (TGF-β_1) and fibronectin expressions	Yamabe, Yokozawa, Oya, and Kim (2006)
Model: ICR mice *Wt of model*: 23.3 ± 1.2 gram *Disease induced by*: Streptozotocin *Dose of GT*: 50, 100, and 200 mg/kg EGCG *Route*: S.C. injection *Duration*: 1 week	• Decreased proteinuria level and expression of osteopontin (OPN) • Renal histo-pathological observations of glomerular mesangial matrix found to be reduced	Yoon et al. (2014)
Model: Male C57BL/6 mice *Wt of model*: 18–22 gram *Disease induced by*: Unilateral Ureteral Obstruction-Induced Renal Interstitial fibrosis *Dose of GT*: 5 mg/kg EECG *Route*: IP *Duration*: 14 days	• Decreased extracellular matrix, mRNA expression of MCP-1 and 3, TNF- α, IL-1β and phosphorylations of Smad 2 and 3 by reducing TGF-β_1	Wang et al. (2015)
Model: Male db/db mice *Age of model*: 5 weeks *Disease induced by*: *Dose of GT*: 2.5, 5, or 10 g/kg of diet EGCG supplement *Route*: Oral *Duration*: 5 weeks	• Improved oral glucose tolerance (in a dose-dependent manner) • It also increased plasma concentrations of insulin and reduced expression of phosphoenol pyruvate carboxykinase (PEPCK) level	Wolfram et al. (2006)

(Continued)

Table 1. *(Continued)*

Models	Outcomes of the study	References
Model: Male albino Wistar rats *Wt of model*: 160–180 gram *Disease induced by*: Fluoride intoxication *Dose of GT*: 40 mg/kg EECG *Route*: Oral *Duration*: 4 weeks	• Decreased the levels of tumor necrosis factor (TNF)-α, nitric Oxide (NO), interleukin (IL)-6 and nuclear factor (NF)-κβ p65, lipid hydroperoxide (LOOH) and protein carbonyl content (PCC) • It also increased vitamins C and E contents, β-cell lymphoma (Bcl-2) levels in kidneys • Restored expression of Nrf2, HO-1, and GST	Thangapandiyan and Miltonprabu (2014)
Model- Female CD:1 mice *Age of model*: 5 weeks *Disease induced by*: Streptozotocin *Dose of GT*: 0.01% of GT extract with drinking water *Route*: Oral *Duration*: 12 weeks	• Suppressed the insulin resistance • Improved the function of pancreatic β cell	Tang, Li, Liu, Huang, and Ho (2013)
Model: Female db/db-mice*Wt of model*: 30.1±0.5 gram*Disease induced by*: N/A *Dose of GT*: 1 g/kg GT extract *Route*: Oral *Duration*: 28 days	• Reduced plasma soluble cell adhesion molecules (sICAM) concentration	Wein et al. (2013)
Model: Female obese, diabetic KK-ay and C57BL/6 J mice*Wt of model*: N/A *Disease induced by*: *Dose of GT*: 300 mg Catechins /kg/day *Route*: Oral *Duration*: 4 weeks	• Increased GLUT-4 content in plasma and insulin sensitivity by suppressing the c-Jun N-terminal kinase (JNK) pathway • It also suppressed the effect of Dexamethasone	Yan, Zhao, Suo, Liu, and Zhao (2012)
Model: Male ICR mice *Wt of model*: 17–19 gram *Disease induced by* Dextran sulfate sodium-induced colitis *Dose of GT*: 1% (w/w) Polyphenols *Route*: Oral *Duration*: 6 days	• Normalized the heat shock protein (HSP) production in the kidney	Inoue et al. (2011)
Model: KK-Ay Diabetic Mice *Age of model*: 5 weeks *Disease induced by*: N/A *Dose of GT*: Fermented tea (0.4%) *Route*: Oral *Duration*: 90 days	• The level of glycated hemoglobin (HbA1c) and insulin resistance were reduced	Lee, Park, Nam, Yi, and Lim (2013)

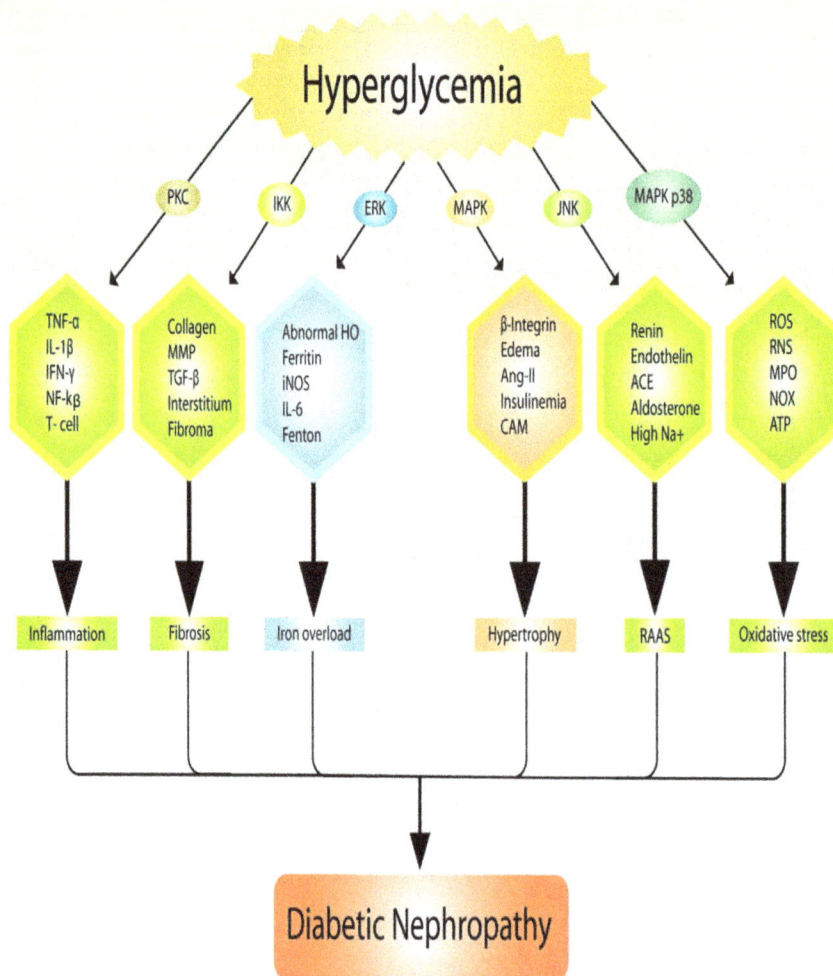

Figure 3. How hyperglycemia induces diabetic nephropathy and related other complications.

superoxide anion and hydrohen per-oxide (Koya et al., 2003). SOD family exerts its protective function by residing inside membrane (EC-SO, mitochondria (Mn-SOD), and cytoplasm (Cu-Zn-SOD) (Bartz & Piantadosi, 2010; Rosenthal & Nocera, 2007). Production of mitochondrial SOD can also decrease hyperglycemia and related complications (Nishikawa et al., 2000).

Recently Sirt1 has achieved several protective functions like DNA repairing; neutralizing oxidants, maintaining glucose homeostasis, hormone secretion and others. Hence, it is now called silent information regulator (Weber et al., 2003) gene. It also regulates in the transcription of non-histone protein including FOXO1/3, p53, PPAR, and PGC-1α (Morris, 2013). Study also showed that expression of Sirt1 in the proximal tubules enhances glomerular function and defends against diabetes-induced renal damage (Hasegawa et al., 2013). A recent experiment explored that GT polyphenol (−)-epigallocatechin-3-gallate (EGCG, 50 mg/kg BW/day × 3 weeks) restored Sirt1 expression, resulting in the reduction of serum creatinine, proteinurea which ultimately protects kidney cell. The study also linked that supplementation of GT polyphenol reduced p-ERK1/2, p-Akt, p-JNK, and p-P38 signaling along with activating PPARγ activity (Ye et al., 2015). GT isoflavones have been investigated to induce Sirt1 and other protein content like TFAM for mitochondrial biogenesis when rat's kidney were treated with cyclosporine A (Rehman et al., 2013).

Role of Green Tea on Diabetic Kidney Cell's

Cell Membrane

-Inhibit MDA production
-Reduce LPO level
-Repair the cell membrane
-Maintain proper isotonicity
-Monitor Glucose
-Maintain GLUT channel
-Reduce TLR activity

Mitochondria

-Increase AMPK production
-Induce Nrf2 via ARE-keap1 signaling
- Increase PGC-1α concentration
-Induce TFAM production
-Increase electron transport system activity
-Stabilize the energy production
-Help in mitochondrial biogenesis

Cytoplasm

-Reduce ROS production
-Block the cleaving of Golgi apparatus
-Protect NADPH system
-Initiate productions of MnSOD
-Reduce albuminuria production
-Abate uric acid production
-Control creatinin level

Nucleus

-Protect Nuclear membrane
-Block faulty gene expression
-Prevent abnormal replications
-Stabilize the DNA strand
-Retain normal cell signaling
-Bases become stabilize
-Reduce NF-κβ production

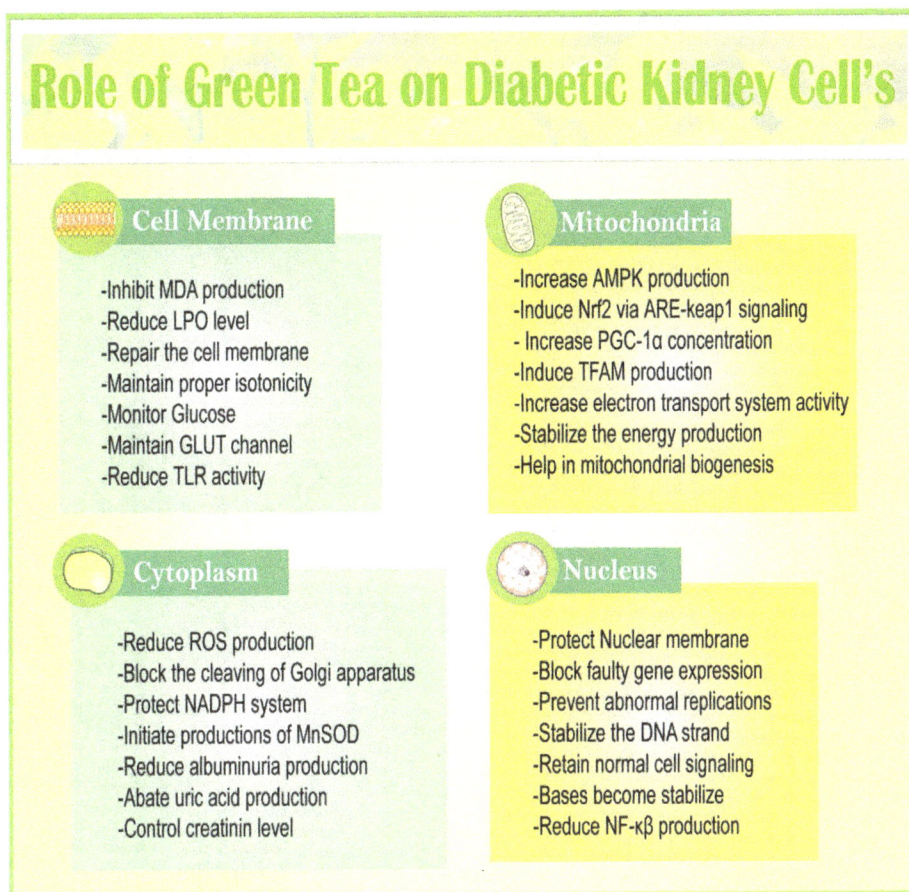

Figure 4. An overview of green tea on diabetic kidney.

AMPK, another master regulator, known as a nutrient sensor which is generally activated under energy-depleted conditions by phosphorylation of a preserved threonine residue (T172) (Alers, Wesselborg, & Stork, 2012). In type-1 and type-2 diabetic patients, AMPK phosphorylation and other activities were noticed lower inside both glomeruli and tubules (Kitada, Kume, Imaizumi, & Koya, 2011). GT treatment found to be protective against cyclosporine A-induced renal injury. Increased AMPK and PGC-1α help in the mitochondrial biogenesis and thus prevent drug-induced kidney dysfunctions (Rehman et al., 2013).

Beside these, epigallocatechin-3-Gallate treatment also protected the kidney function by reducing MCP-1, MCP-3, and TGF-β content on fibrotic rat model (Wang et al., 2015). Another study showed that GT normalizes renal kidney injury by down regulating NOX-4 production on diabetic rats (Ribaldo et al., 2009). It was also found out that water extracts of GT lowered DN by controlling glomerular hypertrophy and interstitial inflammation on alloxan induced diabetic rats (Shokouhi et al., 2015). Recent study also revealed that GT polyphenol epigallocatechin-3-gallate up-regulated heme oxygenase-1through phosphatidyl inositol 3-kinase/Akt and ERK and thereby prevent kidney damage (Wu et al., 2006). A proposed molecular mechanism has been explained in the (Figure 5) to show the possible interaction between GT and DN.

7. Problem with green tea treatment

A potent molecule must exert its action on biological system being stable, at the same time it should be safe and not interfere with other signaling pathways. Sometimes it is showed that high dose of a component often produces unwanted side effects, life-threatening ADR which may lead to necrosis-mediated cell death (Elsherbiny et al., 2012). Similarly, a study has also found that 1% GT polyphenols with food produced nephrotoxicity by down-regulating antioxidant protein expression and heat shock protein expression (Inoue et al., 2011). Likewise, another study explored that high dose of GT

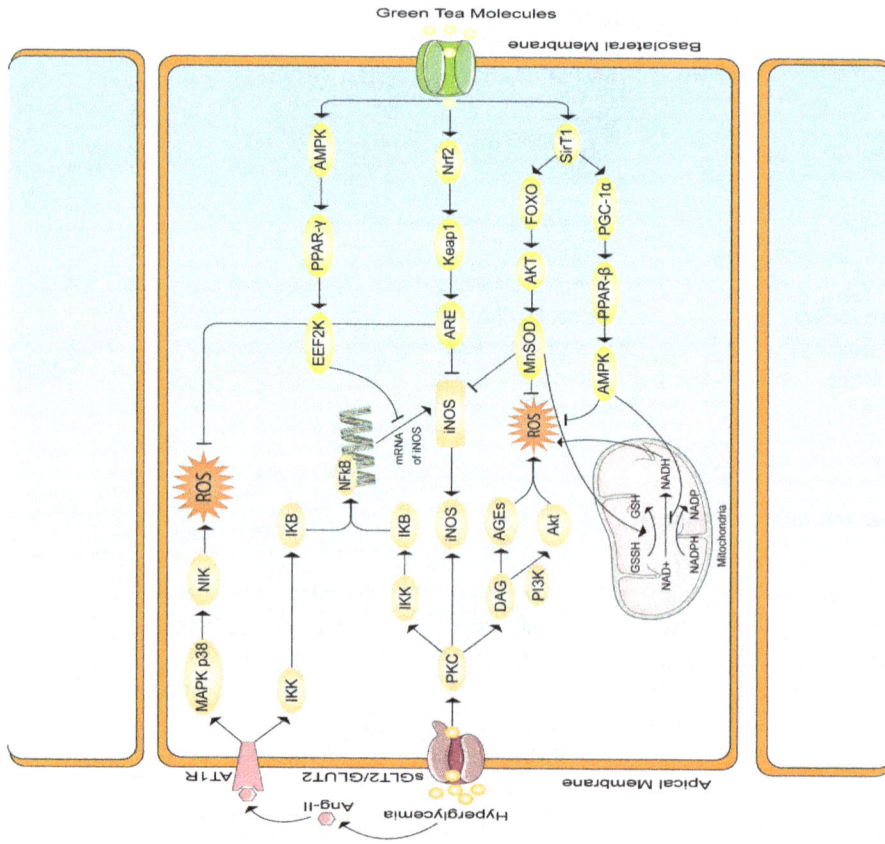

Figure 5. Proposed mechanism explained that, Hyperglycemia induces ROS in numerous ways, such as by activating Mitochondria, or by increasing the accumulation of Transforming Growth Factor Beta (TGF–β), or by triggering protein kinase C (PKC), it also enhances the accumulation of tumor necrosis factor alpha (TNF–α) and up regulates angiotensin-2 production. (1) Inside the mitochondria, hyperglycemia causes electron imbalance which converts oxygen molecule to a reactive one. (2) TGF–β generates ROS via PI3K-Smad2/3 pathway. (3) Activation of PKC by high level of glucose causes formation of ROS via different ways, PKC also possess interleukin 6, which regulates iNOS that is directly involved in the generation of ROS, PKC itself can help in the formation of iNOS directly; PKC causes activation of nuclear factor κβ via IKK-IKB pathway. (4) In addition, hyperglycemia can cause accumulation of TNF–α adjacent the cell membrane, which binds TNF–α receptor that further activates NF–κβ inside the DNA to produce iNOS gene that ultimately generates ROS. (5) Higher glucose level may facilitate the production of Angiotensin-II by Renin-Angiotensin-Aldosterone system, Ang–II binds with Ang–II Type 1 receptor(AT₁R), which causes accumulation of ROS via MAPK p38-NIK (NF–κβ inducing kinase) pathway which further activates NF–κβ via IKK-IKB pathway. Green tea flavonoids and polyphenols block the activation/accumulation/generation of ROS by enhancing 5′ AMP-activated protein kinase (AMPK), nuclear factor-erythroid-2-related factor 2 (Nrf2) and Sirtuin-1 (Sirt1). (1) AMPK generally activates both Nrf2 and Sirt1, it further inhibits the expression of iNOS through Peroxisome proliferator-activated receptor gamma (PPAR–γ) in Eukaryotic elongation factor-2 kinase (EEF-2K) pathway, AMPK may also nullify the effect of TNF–α and PKC. (2) Green tea polyphenols and flavonoids induce Nrf2 which blocks both iNOS production and ROS accumulation by activating Keap1-ARE (antioxidant-responsive element) signaling. (3) Sirt1 up regulates Mn superoxide dismutase (MnSOD) production via Forkhead box O3 FOXO3-AKT (Protein kinase) pathway which prevents ROS generation, furthermore Sirt1 inhibits TNF–α and TGF–β by producing PGC-1α (Peroxisome proliferator-activated receptor gamma co-activator 1-alpha)—PPARβ (peroxisome proliferator-activated receptors-β).

Table 2. Recent protective findings of green tea on diabetic patients through clinical trails

Subjects	Outcomes of the study	References
68 subjects, aging 20–65 years with type 2 diabetes	• Reduced in HOMA-IR index, and insulin level • Increased in the level of ghrelin	Hua et al. (2011)
155 pre-diabetic individuals aging from 35 to 65 years	• Reduced ALT level • Restored the antioxidant capacity and improved eGFR	Toolsee et al. (2013)
78 obese women aged between 16 and 60 years	• Reduced LDL-cholesterol and triglyceride • Increased in the level of HDL-cholesterol, adiponectin and ghrelin	(Hsu et al., 2008)
54 subjects diagnosed of type 2 diabetes mellitus aging 65±6 years	• Increased in HbA1c and it had no hypoglycemic effect	MacKenzie, Leary, and Brooks (2007)
237 overweight and obese participants	• Decreased fasting serum insulin concentration	Dostal et al. (2016)
56 obese, hypertensive subjects aging 30 to 60 years	• Reduced fasting serum glucose, insulin levels, and insulin resistance, • Serum tumor necrosis factor α and C-reactive protein observed low	Bogdanski et al. (2012)
100 type 2 diabetes patients	• Increased HDL-c and improve β-cell function, • Decreased fasting blood insulin, homeostasis model assessment of insulin resistance	Mozaffari-Khosravi, Ahadi, and Tafti (2014)
42 diabetic subjects aging between 60–65 years	• Reduced albuminuria, serum DKK-1, and tumor necrosis factor -α • Limited podocyte apoptosis by activating Wnt pathway	Borges et al. (2016)
300 men and women aging 65–100 years	• Reduced fasting blood glucose levels	Polychronopoulos et al. (2008)
46 obese patients aging 30–60 years	• Lowered Glucose, iron, total cholesterol, LDL, and triglyceride • Increases in total antioxidant level and in zinc concentration; induce HDL cholesterol level	Suliburska et al. (2012)

(EECG) induced hepatic dysfunction by producing liver4-hydroxynonenal (4-HNE) and malonyldial-dehyde (MDA) (Lambert et al., 2010). In addition, similar kind of hepatic dysfunctions have been reported when people take daily GT supplements (Mazzanti et al., 2009). Furthermore, a cell culture study also investigated that EECG induced oxidative stress and also enhanced the production of pro-matrix metalloproteinase-7 (Kim, Murakami, & Ohigashi, 2007). Another report noticed that 1% GT polyphenol treatment induced carcinoma by increasing pro-inflammatory cytokines along with reducing antioxidants like SOD and catalase (Kim et al., 2010). A recent meta-analysis showed that intake of GT also increased the risk of endometrial cancer (Zhou et al., 2016). It has been also reported that GT component EGCG, has poor absorption via oral route due to metabolization by gut microbiata (Chan, Zhang, & Zuo, 2007) and also found very shorter half-life (maximum 30 min in cell culture and 30–60 min on animals) in both animal and human study (Lee et al., 2002). It is normally conjugated with glucuronidation and easily metabolized. Study showed that one-third of the ECGE gets eliminated from the body within first 24 h of its administration (Manach, Williamson, Morand, Scalbert, & Rémésy, 2005). Oxidation of ECGE may produce several free radicals like superoxide anion which later hampers multiple cellular signaling by binding on epidermal growth factor receptor (Hou et al., 2005; Smith, 2011; Wein, Schrader, Rimbach, & Wolffram, 2013). EGCE also blocks insulin secretion from β-cell by inhibiting glutamate dehydrogenase (GDH)-mediated signaling (Li et al., 2006). However, a cross talk has been shown from a clinical trial where 35 obese subjects were given GT (4 cups/a day), and after 8 weeks no significant alteration in serum biomarkers of inflammation including adiponectin, C-reactive protein (CRP), interleukin-(IL-6), interleukin-(IL-1β), soluble vascular cell adhesion molecule-(sVCAM-1), leptin, or leptin: adiponectin ratio were observed, although the authors noticed reduced plasma serum amyloid alpha (Basu et al., 2011; Eid et al., 2010). Most of the studies were conducted with small number of subjects. Therefore, it is very difficult to conclude anything concrete against the obtained results. Clinical trial on a large population must be carried out to get a noble drug molecule from GT.

8. Conclusion and future directions

This study describes possible patho-physiology for kidney dysfunctions, especially in diabetic condition. As the complications of diabetes are increasing tremendously, a safe and alternative treatment must be approached. Beneficial roles of GT cathecins on diabetic-induced kidney dysfunctions have been well studied. Regular administration of GT supplements can be an effective treatment approach for the diabetic subjects. This study also provides strong evidences in favor of GT treatment in DN. The authors highly recommend for large clinical trial to establish the exact molecular mechanism of action of GT against DN.

Funding
The authors received no direct funding for this research.

Competing Interests
The authors declare no competing interest.

Author details
Md. Mohabbulla Mohib[1]
E-mail: mohib_nsu007@yahoo.com
S.M. Fazla Rabby[1]
E-mail: smfazla.rabby@northsouth.edu
Tasfiq Zaman Paran[1]
E-mail: tasfiq175@gmail.com
Md. Mehedee Hasan[2]
E-mail: mehedeehasan13@yahoo.com
Iqbal Ahmed[1]
E-mail: iqa@bpl.net
Nahid Hasan[1]
E-mail: nahidstar09@yahoo.com
ORCID ID: http://orcid.org/0000-0003-3378-4537
Md. Abu Taher Sagor[1]
E-mail: sagor2008nsu@gmail.com
Sarif Mohiuddin[3]
E-mail: sharif.smch@gmail.com

[1] Department of Pharmaceutical Sciences, North South University, Dhaka 1229, Bangladesh.
[2] Department of Pharmacy, State University of Bangladesh, Dhaka 1205, Bangladesh.
[3] Department of Anatomy, Pioneer Dental College and Hospital, Dhaka 1229, Bangladesh.

References
Abdel-Raheem, I. T., El-Sherbiny, G. A., & Taye, A. (2010). Green tea ameliorates renal oxidative damage induced by gentamicin in rats. *Pakistan Journal of Pharmaceutical Sciences, 23*, 21–28.
Abu Taher, S., Hasan Mahmud, R., Nabila, T., Biswajit, S., Anayt, U., Nusrat, S., … Ashraful, A. (2016). Supplementation of rosemary leaves (*Rosmarinus officinalis*) powder attenuates oxidative stress, inflammation and fibrosis in carbon tetrachloride (CCl4) treated rats. *Current Nutrition & Food Science, 12*, 1–8. doi:10.2174/1573401312666160816154610

Alam, M. A., Chowdhury, M. R. H., Jain, P., Sagor, M. A. T., Reza, H. M. (2015). DPP-4 inhibitor sitagliptin prevents inflammation and oxidative stress of heart and kidney in two kidney and one clip (2K1C) rats. *Diabetology & Metabolic Syndrome, 7*(1), 1–10. doi:10.1186/s13098-015-0095-3

Al-Attar, A. M., & Zari, T. A. (2010). Influences of crude extract of tea leaves, Camellia sinensis, on streptozotocin diabetic male albino mice. *Saudi Journal of Biological Sciences, 17,* 295–301. http://dx.doi.org/10.1016/j.sjbs.2010.05.007

Aldhahi, W., & Hamdy, O. (2003). Adipokines, inflammation, and the endothelium in diabetes. *Current Diabetes Reports, 3,* 293–298. http://dx.doi.org/10.1007/s11892-003-0020-2

Alers, S., Wesselborg, S., & Stork, B. (2012). Role of AMPK-mTOR-Ulk1/2 in the regulation of autophagy: cross talk, shortcuts, and feedbacks. *Molecular and Cellular Biology, 32,* 2–11. http://dx.doi.org/10.1128/MCB.06159-11

Alschuler, L. (1998). Green tea: Healing tonic. *American Journal of Natural Medicine, 5,* 28–31.

Alwan, A. (2011). *Global status report on noncommunicable diseases 2010.* Geneva: World Health Organization.

Aminzadeh, M. A., Nicholas, S. B., Norris, K. C., & Vaziri, N. D. (2013). Role of impaired Nrf2 activation in the pathogenesis of oxidative stress and inflammation in chronic tubulo-interstitial nephropathy. *Nephrology Dialysis Transplantation, 28,* 2038–2045. http://dx.doi.org/10.1093/ndt/gft022

Aminzadeh, M. A., Reisman, S. A., Vaziri, N. D., Khazaeli, M., Yuan, J., & Meyer, C. J. (2014). The synthetic triterpenoid RTA dh404 (CDDO-dhTFEA) restores Nrf2 activity and attenuates oxidative stress, inflammation, and fibrosis in rats with chronic kidney disease. *Xenobiotica, 44,* 570–578. http://dx.doi.org/10.3109/00498254.2013.852705

Atkins, R. C., & Zimmet, P. (2010). Diabetic kidney disease: act now or pay later—World Kidney Day, 11 March 2010. *Therapeutic Apheresis and Dialysis, 14*(1), 1–4.

Back, J., Ryu, H. H., Hong, R., Han, S. A., Yoon, Y. M., Kim, D. H., ... Kwon, Y. E. (2015). Antiproteinuric effects of green tea extract on tacrolimus-induced nephrotoxicity in mice. In *Transplantation Proceedings* (Vol. 47, pp. 2032–2034). Elsevier. http://dx.doi.org/10.1016/j.transproceed.2015.06.008

Bartz, R. R., & Piantadosi, C. A. (2010). Clinical review: Oxygen as a signaling molecule. *Critical Care, 14,* 234. http://dx.doi.org/10.1186/cc9185

Basu, A., Du, M., Sanchez, K., Leyva, M. J., Betts, N. M., Blevins, S., ... Lyons, T. J. (2011). Green tea minimally affects biomarkers of inflammation in obese subjects with metabolic syndrome. *Nutrition, 27,* 206–213. http://dx.doi.org/10.1016/j.nut.2010.01.015

Bogdanski, P., Suliburska, J., Szulinska, M., Stepien, M., Pupek-Musialik, D., & Jablecka, A. (2012). Green tea extract reduces blood pressure, inflammatory biomarkers, and oxidative stress and improves parameters associated with insulin resistance in obese, hypertensive patients. *Nutrition Research, 32,* 421–427. http://dx.doi.org/10.1016/j.nutres.2012.05.007

Borges, C. M., et al. (2016). The use of green tea polyphenols for treating residual albuminuria in diabetic nephropathy: A double-blind randomised clinical trial. *Scientific Reports, 6.*

Cabrera, C., Artacho, R., & Giménez, R. (2006). Beneficial effects of green tea—A review. *Journal of the American College of Nutrition, 25,* 79–99. http://dx.doi.org/10.1080/07315724.2006.10719518

Chacko, S. M., Thambi, P. T., Kuttan, R., & Nishigaki, I. (2010). Beneficial effects of green tea: A literature review. *Chinese Medicine, 5*(1), 1–9. doi:10.1186/1749-8546-5-13

Chan, K., Zhang, L., & Zuo, Z. (2007). Intestinal efflux transport kinetics of green tea catechins in Caco-2 monolayer model. *Journal of Pharmacy and Pharmacology, 59,* 395–400. http://dx.doi.org/10.1211/jpp.59.3.0009

Chow, F., Ozols, E., Nikolic-Paterson, D. J., Atkins, R. C., & Tesch, G. H. (2004). Macrophages in mouse type 2 diabetic nephropathy: Correlation with diabetic state and progressive renal injury. *Kidney International, 65,* 116–128. http://dx.doi.org/10.1111/j.1523-1755.2004.00367.x

Chuang, P. Y., Dai, Y., Liu, R., He, H., Kretzler, M., Jim, B., Cohen, C. D., ... Dryer, S. E. (2011). Alteration of forkhead box o (foxo4) acetylation mediates apoptosis of podocytes in diabetes mellitus. *PLoS ONE, 6,* e23566. http://dx.doi.org/10.1371/journal.pone.0023566

Cox, S., Abu-Ghannam, N., & Gupta, S. (2010). An assessment of the antioxidant and antimicrobial activity of six species of edible Irish seaweeds. *International Food Research Journal, 17,* 205–220.

Darvesh, A. S., & Bishayee, A. (2013). Chemopreventive and therapeutic potential of tea polyphenols in hepatocellular cancer. *Nutrition and Cancer, 65,* 329–344. http://dx.doi.org/10.1080/01635581.2013.767367

Ding, Y., & Choi, M. E. (2015). Autophagy in diabetic nephropathy. *Journal of Endocrinology, 224,* R15–R30.

Dona, M., Dell'Aica, I., Calabrese, F., Benelli, R., Morini, M., Albini, A., & Garbisa, S. (2003). Neutrophil restraint by green tea: Inhibition of inflammation, associated angiogenesis, and pulmonary fibrosis. *The Journal of Immunology, 170,* 4335–4341. http://dx.doi.org/10.4049/jimmunol.170.8.4335

Donate-Correa, J., Martín-Núñez, E., Muros-de-Fuentes, M., Mora-Fernández, C., & Navarro-González, J. F. (2015). Inflammatory cytokines in diabetic nephropathy. *Journal of Diabetes Research, 2015,* Article ID: 948417. doi:10.1155/2015/948417

Dostal, A. M., Samavat, H., Espejo, L., Arikawa, A. Y., Stendell-Hollis, N. R., & Kurzer, M. S. (2016). Green tea extract and catechol-O-methyltransferase genotype modify fasting serum insulin and plasma adiponectin concentrations in a randomized controlled trial of overweight and obese postmenopausal women. *Journal of Nutrition, 146,* 38–45. http://dx.doi.org/10.3945/jn.115.222414

Dronavalli, S., Duka, I., & Bakris, G. L. (2008). The pathogenesis of diabetic nephropathy. *Nature Clinical Practice Endocrinology & Metabolism, 4,* 444–452. http://dx.doi.org/10.1038/ncpendmet0894

Eid, A. A., Ford, B. M., Block, K., Kasinath, B. S., Gorin, Y., Ghosh-Choudhury, G., & Abboud, H. E. (2010). AMP-activated protein kinase (AMPK) negatively regulates Nox4-dependent activation of p53 and epithelial cell apoptosis in diabetes. *Journal of Biological Chemistry, 285,* 37503–37512. http://dx.doi.org/10.1074/jbc.M110.136796

Elmarakby, A. A., & Sullivan, J. C. (2012). Relationship between oxidative stress and inflammatory cytokines in diabetic nephropathy. *Cardiovascular Therapeutics, 30,* 49–59. http://dx.doi.org/10.1111/cdr.2012.30.issue-1

Elsherbiny, N. M., El Galil, K. H. A., Gabr, M. M., Al-Gayyar, M. M.,Eissa, L. A., & El-Shishtawy, M. M. (2012). Reno-protective effect of NECA in diabetic nephropathy: Implication of IL-18 and ICAM-1. *European Cytokine Network, 23,* 78–86. doi:10.1684/ecn.2012.0309

Fantuzzi, G., Reed, D. A., & Dinarello, C. A. (1999). IL-12–induced IFN-γ is dependent on caspase-1 processing of the IL-18 precursor. *Journal of Clinical Investigation, 104,* 761–767. http://dx.doi.org/10.1172/JCI7501

Faria, A. M., Papadimitriou, A., Silva, K. C., Lopes de Faria, J. M., & Lopes de Faria, J. B. (2012). Uncoupling endothelial nitric oxide synthase is ameliorated by green tea in experimental diabetes by re-establishing tetrahydrobiopterin levels. *Diabetes, 61,* 1838–1847. http://dx.doi.org/10.2337/db11-1241

Fiorentino, L., Cavalera, M., Menini, S., Marchetti, V., Mavilio, M., Fabrizi, M., ... Federici, M. (2013). Loss of TIMP3 underlies diabetic nephropathy via FoxO1/STAT1 interplay. *EMBO Molecular Medicine, 5*, 441–455. http://dx.doi.org/10.1002/emmm.201201475

Funamoto, M., Masumoto, H., Takaori, K., Taki, T., Setozaki, S., Yamazaki, K., ... Sakata, R. (2016). Green tea polyphenol prevents diabetic rats from acute kidney injury after cardiopulmonary bypass. *The Annals of Thoracic Surgery, 101*, 1507–1513. doi:10.1016/j.athoracsur.2015.09.080

Gao, J., Xu, P., Wang, Y., Wang, Y., & Hochstetter, D. (2013). Combined effects of green tea extracts, green tea polyphenols or epigallocatechin gallate with acarbose on inhibition against α-amylase and α-glucosidase in vitro. *Molecules, 18*, 11614–11623. http://dx.doi.org/10.3390/molecules180911614

Giacco, F., & Brownlee, M. (2010). Oxidative stress and diabetic complications. *Circulation Research, 107*, 1058–1070. http://dx.doi.org/10.1161/CIRCRESAHA.110.223545

Giacco, F., Du, X., D'Agati, V. D., Milne, R., Sui, G., Geoffrion, M., & Brownlee, M. (2014). Knockdown of glyoxalase 1 mimics diabetic nephropathy in nondiabetic mice. *Diabetes, 63*, 291–299. http://dx.doi.org/10.2337/db13-0316

Glei, M., & Pool-Zobel, B. (2006). The main catechin of green tea,(−)-epigallocatechin-3-gallate (EGCG), reduces bleomycin-induced DNA damage in human leucocytes. *Toxicology in Vitro, 20*, 295–300. http://dx.doi.org/10.1016/j.tiv.2005.08.002

Graham, H. N. (1992). Green tea composition, consumption, and polyphenol chemistry. *Preventive Medicine, 21*, 334–350. http://dx.doi.org/10.1016/0091-7435(92)90041-F

Guh, D. P., Zhang, W., Bansback, N., Amarsi, Z., Birmingham, C. L., & Anis, A. H. (2009). The incidence of co-morbidities related to obesity and overweight: A systematic review and meta-analysis. *BMC Public Health, 9*(1), 1–20. doi:10.1186/1471-2458-9-88

Gupta, M., Solís, P. N., Calderón, A. I., Guinneau-Sinclair, F., Correa, M., Galdames, C., ... Ocampo, R. (2005). Medical ethnobotany of the teribes of bocas del toro, panama. *Journal of Ethnopharmacology, 96*, 389–401. http://dx.doi.org/10.1016/j.jep.2004.08.032

Ha, H., & Kim, K. H. (1999). Pathogenesis of diabetic nephropathy: The role of oxidative stress and protein kinase C. *Diabetes Research and Clinical Practice, 45*, 147–151. http://dx.doi.org/10.1016/S0168-8227(99)00044-3

Ha, H., Yu, M. R., Choi, Y. J.,, Kitamura, M., & Lee, H. B. (2002). Role of high glucose-induced nuclear factor-κB activation in monocyte chemoattractant protein-1 expression by mesangial cells. *Journal of the American Society of Nephrology, 13*, 894–902.

Haidari, F., Shahi, M. M., Zarei, M., Rafiei, H., & Omidian, K. (2012). Effect of green tea extract on body weight, serum glucose and lipid profile in streptozotocin-induced diabetic rats. A dose response study. *Saudi Medical Journal, 33*, 128–133.

Hamilton-Miller, J. (1995). Antimicrobial properties of tea (Camellia sinensis L.). *Antimicrobial Agents and Chemotherapy, 39*, 2375–2377. http://dx.doi.org/10.1128/AAC.39.11.2375

Haqqi, T. M., Anthony, D. D., Gupta, S., Ahmad, N., Lee, M.-S., Kumar, G. K., & Mukhtar, H. (1999). Prevention of collagen-induced arthritis in mice by a polyphenolic fraction from green tea. *Proceedings of the National Academy of Sciences, 96*, 4524–4529. http://dx.doi.org/10.1073/pnas.96.8.4524

Har, R., Scholey, J. W., Daneman, D., Mahmud, F. H., Dekker, R., Lai, V., ... Cherney, D. Z. I. (2013). The effect of renal hyperfiltration on urinary inflammatory cytokines/chemokines in patients with uncomplicated type 1 diabetes mellitus. *Diabetologia, 56*, 1166–1173. http://dx.doi.org/10.1007/s00125-013-2857-5

Hasegawa, K., Wakino, S., Simic, P., Sakamaki, Y., Minakuchi, H., Fujimura, K., ... Itoh, H. (2013). Renal tubular Sirt1 attenuates diabetic albuminuria by epigenetically suppressing Claudin-1 overexpression in podocytes. *Nature Medicine, 19*, 1496–1504. http://dx.doi.org/10.1038/nm.3363

He, W., Wang, Y., Zhang, M.-Z., You, L., Davis, L. S., Fan, H., ... Hao, C.-M. (2010). Sirt1 activation protects the mouse renal medulla from oxidative injury. *Journal of Clinical Investigation, 120*, 1056–1068. http://dx.doi.org/10.1172/JCI41563

Hou, Z., Sang, S., You, H., Lee, M. J., Hong, J., Chin, K. V., & Yang, C. S. (2005). Mechanism of action of (−)-epigallocatechin-3-gallate: Auto-oxidation-dependent inactivation of epidermal growth factor receptor and direct effects on growth inhibition in human esophageal cancer KYSE 150 cells. *Cancer Research, 65*, 8049–8056. doi:10.1158/0008-5472.CAN-05-0480

Hsu, C.-H., Tsai, T.-H., Kao, Y.-H., Hwang, K.-C., Tseng, T.-Y., & Chou, P. (2008). Effect of green tea extract on obese women: A randomized, double-blind, placebo-controlled clinical trial. *Clinical Nutrition, 27*, 363–370. http://dx.doi.org/10.1016/j.clnu.2008.03.007

Hua, C. H., Liao, Y. L., Lin, S. C., Tsai, T. H., Huang, C. J., & Chou, P. (2011). Does supplementation with green tea extract improve insulin resistance in obese type 2 diabetics? A randomized, double-blind, and placebocontrolled clinical trial. *Alternative Medicine Review, 16*, 157–163.

Ienaga, K., & Kondo, M. (1991). Possible role of tumor necrosis factor and interleukin-1 in the development of diabetic nephropathy. *Kidney international, 40*, 1007–1012.

Inoue, H., Akiyama, S., Maeda-Yamamoto, M., Nesumi, A., Tanaka, T., & Murakami, A. (2011). High-dose green tea polyphenols induce nephrotoxicity in dextran sulfate sodium-induced colitis mice by down-regulation of antioxidant enzymes and heat-shock protein expressions. *Cell Stress and Chaperones, 16*, 653–662. http://dx.doi.org/10.1007/s12192-011-0280-8

International Tea Committee. (2009). *International tea committee: Annual bulletin of statistics*. London: Author.

Jha, J. C., Gray, S. P., Barit, D., Okabe, J., Namikoshi, T., Thallas-Bonke, V., ... Jandeleit-Dahm, K. A. (2014). Genetic targeting or pharmacologic inhibition of NADPH oxidase Nox4 provides renoprotection in long-term diabetic nephropathy. *Journal of the American Society of Nephrology, 25*, 1237–1254. http://dx.doi.org/10.1681/ASN.2013070810

Jin, D. C., Ha, I. S., Kim, N. H., Lee, S. W., & Yoon, S. R.Kim, B. S. (2012). Brief report: Renal replacement therapy in Korea, 2010. *Kidney Research and Clinical Practice, 31*, 62–71. http://dx.doi.org/10.1016/j.krcp.2012.01.005

Kabir, A. U., Samad, M. B., Ahmed, A., Akhter, F., Jahan, M. R., Akhter, F., ... Hannan, J. M. A. (2015). aqueous fraction of beta vulgaris ameliorates hyperglycemia in diabetic mice due to enhanced glucose stimulated insulin secretion, mediated by acetylcholine and GLP-1, and elevated glucose uptake via increased membrane bound GLUT4 transporters. *PLOS ONE, 10*, e0116546. http://dx.doi.org/10.1371/journal.pone.0116546

Kang, M.-Y., Park, Y. H., Kim, B. S., Seo, S. Y., Jeong, B. C., Kim, J.-I., & Kim, H. H. (2012). Preventive effects of green tea (Camellia sinensis var. Assamica) on diabetic nephropathy. *Yonsei Medical Journal, 53*, 138–144. http://dx.doi.org/10.3349/ymj.2012.53.1.138

Kavanagh, K. T., Hafer, L. J., Kim, D. W., Mann, K. K., Sherr, D. H., Rogers, A. E., & Sonenshein, G. E. (2001). Green tea extracts decrease carcinogen-induced mammary tumor burden in rats and rate of breast cancer cell proliferation in culture. *Journal of Cellular Biochemistry, 82*, 387–398. http://dx.doi.org/10.1002/(ISSN)1097-4644

Khan, S. A., Priyamvada, S., Khan, W., Khan, S., Farooq, N., & Yusufi, A. N. K. (2009). Studies on the protective effect of green tea against cisplatin induced nephrotoxicity. *Pharmacological Research, 60*, 382–391. http://dx.doi.org/10.1016/j.phrs.2009.07.007

Kim, H. J., Sato, T., Rodriguez-Iturbe, B., & Vaziri, N. D. (2011). Role of intrarenal angiotensin system activation, oxidative stress, inflammation, and impaired nuclear factor-erythroid-2-related factor 2 activity in the progression of focal glomerulosclerosis. *Journal of Pharmacology and Experimental Therapeutics, 337*, 583–590. http://dx.doi.org/10.1124/jpet.110.175828

Kim, H.-S., Quon, M. J., & Kim, J.-A. (2014). New insights into the mechanisms of polyphenols beyond antioxidant properties; lessons from the green tea polyphenol, epigallocatechin 3-gallate. *Redox Biology, 2*, 187–195. http://dx.doi.org/10.1016/j.redox.2013.12.022

Kim, M., Murakami, A., & Ohigashi, H. (2007). Modifying effects of dietary factors on (–)-epigallocatechin-3-gallate-induced pro-matrix metalloproteinase-7 production in HT-29 human colorectal cancer cells. *Bioscience, Biotechnology, and Biochemistry, 71*, 2442–2450. http://dx.doi.org/10.1271/bbb.70213

Kim, M., Murakami, A., Miyamoto, S., Tanaka, T., & Ohigashi, H. (2010). The modifying effects of green tea polyphenols on acute colitis and inflammation-associated colon carcinogenesis in male ICR mice. *Biofactors, 36*, 43–51. doi:10.1002/biof.69

Kim, T., Davis, J., Zhang, A. J., He, X., & Mathews, S. T. (2009). Curcumin activates AMPK and suppresses gluconeogenic gene expression in hepatoma cells. *Biochemical and Biophysical Research Communications, 388*, 377–382. http://dx.doi.org/10.1016/j.bbrc.2009.08.018

Kitada, M., Kume, S., Imaizumi, N., & Koya, D. (2011). Resveratrol improves oxidative stress and protects against diabetic nephropathy through normalization of Mn-SOD dysfunction in AMPK/SIRT1-independent pathway. *Diabetes, 60*, 634–643. http://dx.doi.org/10.2337/db10-0386

Koike, N., Takamura, T., & Kaneko, S. (2007). Induction of reactive oxygen species from isolated rat glomeruli by protein kinase C activation and TNF-α stimulation, and effects of a phosphodiesterase inhibitor. *Life Sciences, 80*, 1721–1728. http://dx.doi.org/10.1016/j.lfs.2007.02.001

Koya, D., Hayashi, K., Kitada, M., Kashiwagi, A., Kikkawa, R., & Haneda, M. (2003). Effects of antioxidants in diabetes-induced oxidative stress in the glomeruli of diabetic rats. *Journal of the American Society of Nephrology, 14*(suppl 3), 250S–253. http://dx.doi.org/10.1097/01.ASN.0000077412.07578.44

Kupelian, V., Araujo, A. B., Wittert, G. A., & McKinlay, J. B. (2015). Association of moderate to severe lower urinary tract symptoms with incident type 2 diabetes and heart disease. *The Journal of Urology, 193*, 581–586. http://dx.doi.org/10.1016/j.juro.2014.08.097

Kurita, I., Maeda-Yamamoto, M., Tachibana, H., & Kamei, M. (2010). Antihypertensive effect of benifuuki tea containing O -methylated EGCG. *Journal of Agricultural and Food Chemistry, 58*, 1903–1908. http://dx.doi.org/10.1021/jf904335g

Lacmata, S. T., Kuete, V., Dzoyem, J. P., Tankeo, S. B., Teke, G. N., Kuiate, J. R., & Pages, J. M. (2012). Antibacterial activities of selected Cameroonian plants and their synergistic effects with antibiotics against bacteria expressing MDR phenotypes. *Evidence-Based Complementary and Alternative Medicine, 2012*, Article ID: 623723. doi:10.1155/2012/623723

Lambert, J. D., Kennett, M. J., Sang, S., Reuh, K. R., Ju, J., & Yang, C. S. (2010). Hepatotoxicity of high oral dose (–)-epigallocatechin-3-gallate in mice. *Food and Chemical Toxicology, 48*, 409–416. http://dx.doi.org/10.1016/j.fct.2009.10.030

Lee, M. J., Maliakal, P., Chen, L.,Meng, X., Bondoc, F. Y., Prabhu, S., ... Yang, C. S. (2002). Pharmacokinetics of tea catechins after ingestion of green tea and (–)-epigallocatechin-3-gallate by humans formation of different metabolites and individual variability. *Cancer Epidemiology Biomarkers & Prevention, 11*, 1025–1032.

Lee, J. W., Park, S., Takahashi, Y., & Wang, H.-G. (2010). The association of AMPK with ULK1 regulates autophagy. *PLoS ONE, 5*, e15394. http://dx.doi.org/10.1371/journal.pone.0015394

Lee, S.-Y., Park, S.-Y., Nam, Y.-D., Yi, S.-H., & Lim, S.-I. (2013). Anti-diabetic effects of fermented green tea in KK-A y diabetic mice. *Korean Journal of Food Science and Technology, 45*, 488–494. http://dx.doi.org/10.9721/KJFST.2013.45.4.488

Li, C., Allen, A., Kwagh, J., Doliba, N. M., Qin, W., Najafi, H., ... Smith, T. J. (2006). Green tea polyphenols modulate insulin secretion by inhibiting glutamate dehydrogenase. *Journal of Biological Chemistry, 281*, 10214–10221. http://dx.doi.org/10.1074/jbc.M512792200

Li, W., Khor, T. O., Xu, C., Shen, G., Jeong, W.-S., Yu, S., & Kong, A. N. (2008). Activation of Nrf2-antioxidant signaling attenuates NFκB-inflammatory response and elicits apoptosis. *Biochemical Pharmacology, 76*, 1485–1489. http://dx.doi.org/10.1016/j.bcp.2008.07.017

Lin, Y.-L., Juan, I.-M., Chen, Y.-L., Liang, Y.-C., & Lin, J.-K. (1996). Composition of polyphenols in fresh tea leaves and associations of their oxygen-radical-absorbing capacity with antiproliferative actions in fibroblast cells. *Journal of Agricultural and Food Chemistry, 44*, 1387–1394. http://dx.doi.org/10.1021/jf950652k

Lopes de Faria, J. B. L., Silva, K. C., & Lopes de Faria, J. M. L. (2011). The contribution of hypertension to diabetic nephropathy and retinopathy: The role of inflammation and oxidative stress. *Hypertension Research, 34*, 413–422. http://dx.doi.org/10.1038/hr.2010.263

MacKenzie, T., Leary, L., & Brooks, W. B. (2007). The effect of an extract of green and black tea on glucose control in adults with type 2 diabetes mellitus: double-blind randomized study. *Metabolism, 56*, 1340–1344. http://dx.doi.org/10.1016/j.metabol.2007.05.018

Manach, C., Williamson, G., Morand, C., Scalbert, A., & Rémésy, C. (2005). Bioavailability and bioefficacy of polyphenols in humans. I. Review of 97 bioavailability studies. *The American journal of clinical nutrition, 81*(1), 230S–242S.

Martini, S., Eichinger, F., Nair, V., & Kretzler, M. (2008). Defining human diabetic nephropathy on the molecular level: Integration of transcriptomic profiles with biological knowledge. *Reviews in Endocrine and Metabolic Disorders, 9*, 267–274. http://dx.doi.org/10.1007/s11154-008-9103-3

Mazzanti, G., Menniti-Ippolito, F., Moro, P. A., Cassetti, F., Raschetti, R., Santuccio, C., & Mastrangelo, S. (2009). Hepatotoxicity from green tea: A review of the literature and two unpublished cases. *European Journal of Clinical Pharmacology, 65*, 331–341. http://dx.doi.org/10.1007/s00228-008-0610-7

Miyauchi, K., Takiyama, Y., Honjyo, J., Tateno, M., & Haneda, M. (2009). Upregulated IL-18 expression in type 2 diabetic subjects with nephropathy: TGF-β1 enhanced IL-18 expression in human renal proximal tubular epithelial cells. *Diabetes Research and Clinical Practice, 83*, 190–199. http://dx.doi.org/10.1016/j.diabres.2008.11.018

Mohib, M. M., Hasan, I., Chowdhury, W. K., Chowdhury, N. U., Mohiuddin, S., Sagor, M. A. T., ... Alam, M. A. (2016). Role of angiotensin ii in hepatic inflammation through MAPK pathway: A review. *Journal of Hepatitis, 2*, 13–20.

Mora, C., & Navarro, J. F. (2004). Inflammation and pathogenesis of diabetic nephropathy. *Metabolism, 53*, 265–266. http://dx.doi.org/10.1016/j.metabol.2003.11.005

Morris, B. J. (2013). Seven sirtuins for seven deadly diseases ofaging. *Free Radical Biology and Medicine, 56*, 133–171. http://dx.doi.org/10.1016/j.freeradbiomed.2012.10.525

Mozaffari-Khosravi, H., Ahadi, Z., & Tafti, M. F. (2014). The effect of green tea versus sour tea on insulin resistance, lipids profiles and oxidative stress in patients with type 2 diabetes mellitus: A randomized clinical trial. *Iranian Journal of Medical Sciences, 39*, 424–432.

Na, H.-K., & Surh, Y.-J. (2008). Modulation of Nrf2-mediated antioxidant and detoxifying enzyme induction by the green tea polyphenol EGCG. *Food and Chemical Toxicology, 46*, 1271–1278. http://dx.doi.org/10.1016/j.fct.2007.10.006

Nishikawa, T., Edelstein, D., Du, X. L., Yamagishi, S. I., Matsumura, T., Kaneda, Y., ... Giardino, I. (2000). Normalizing mitochondrial superoxide production blocks three pathways of hyperglycaemic damage. *Nature, 404*, 787–790. doi:10.1038/35008121

Noh, H., & King, G. (2007). The role of protein kinase C activation in diabetic nephropathy. *Kidney International, 72*, S49–S53. http://dx.doi.org/10.1038/sj.ki.5002386

Nomura, S., Monobe, M., Ema, K., Matsunaga, A., Maeda-Yamamoto, M., & Horie, H. (2015). Effects of flavonol-rich green tea (*Camellia sinensis* L. cv. Sofu) on blood glucose and insulin levels in diabetic mice. *Integrative Obesity and Diabetes, 1*, 109–111.

Nourooz-Zadeh, J., Rahimi, A., Tajaddini-Sarmadi, J., Tritschler, H., Rosen, P., Halliwell, B., & Betteridge, D. J. (1997). Relationships between plasma measures of oxidative stress and metabolic control in NIDDM. *Diabetologia, 40*, 647–653. http://dx.doi.org/10.1007/s001250050729

Ortsäter, H., Grankvist, N., Wolfram, S., Kuehn, N., & Sjöholm, Å. (2012). Diet supplementation with green tea extract epigallocatechin gallate prevents progression to glucose intolerance in db/db mice. *Nutrition & Metabolism, 9*(1), 1. doi:10.1186/1743-7075-9-11

Osada, K., et al. (2001). Tea catechins inhibit cholesterol oxidation accompanying oxidation of low density lipoprotein in vitro. *Comparative Biochemistry and Physiology Part C: Toxicology & Pharmacology, 128*, 153–164.

Park, N.-Y., Park, S.-K., & Lim, Y. (2011). Long-term dietary antioxidant cocktail supplementation effectively reduces renal inflammation in diabetic mice. *British Journal of Nutrition, 106*, 1514–1521. http://dx.doi.org/10.1017/S0007114511001929

Pazdro, R., & Burgess, J. R. (2012). The antioxidant 3H-1, 2-dithiole-3-thione potentiates advanced glycation end-product-induced oxidative stress in SH-SY5Y cells. *Experimental Diabetes Research, 2012*, Article ID: 137607. doi:10.1155/2012/137607

Peixoto, E., Papadimitriou, A., Teixeira, D. A. T., Montemurro, C., Duarte, D. A., Silva, K. C., ... Lopes de Faria, J. B. (2015). Reduced LRP6 expression and increase in the interaction of GSK3β with p53 contribute to podocyte apoptosis in diabetes mellitus and are prevented by green tea. *The Journal of Nutritional Biochemistry, 26*, 416–430. http://dx.doi.org/10.1016/j.jnutbio.2014.11.012

Peng, A., Ye, T., Rakheja, D., Tu, Y., Wang, T., Du, Y., ... Zhou, X. J. (2011). The green tea polyphenol (−)-epigallocatechin-3-gallate ameliorates experimental immune-mediated glomerulonephritis. *Kidney International, 80*, 601–611. http://dx.doi.org/10.1038/ki.2011.121

Peti-Peterdi, J., Kang, J. J., & Toma, I. (2008). Activation of the renal renin–angiotensin system in diabetes—new concepts. *Nephrology Dialysis Transplantation, 23*, 3047–3049. http://dx.doi.org/10.1093/ndt/gfn377

Pistrosch, F., Ganz, X., Bornstein, S. R., Birkenfeld, A. L., Henkel, E., & Hanefeld, M. (2015). Risk of and risk factors for hypoglycemia and associated arrhythmias in patients with type 2 diabetes and cardiovascular disease: A cohort study under real-world conditions. *Acta Diabetologica, 52*, 889–895. doi:10.1007/s00592-015-0727-y

Polychronopoulos, E., Zeimbekis, A., Kastorini, C.-M., Papairakleous, N., Vlachou, I., Bountziouka, V., & Panagiotakos, D. B. (2008). Effects of black and green tea consumption on blood glucose levels in non-obese elderly men and women from Mediterranean Islands (MEDIS epidemiological study). *European Journal of Nutrition, 47*, 10–16. http://dx.doi.org/10.1007/s00394-007-0690-7

Price, N. L., Gomes, A. P., Ling, A. J. Y., Duarte, F. V., Martin-Montalvo, A., Agarwal, B., ... Sinclair, D. A. (2012). SIRT1 is required for AMPK activation and the beneficial effects of resveratrol on mitochondrial function. *Cell Metabolism, 15*, 675–690. http://dx.doi.org/10.1016/j.cmet.2012.04.003

Raederstorff, D. G., Schlachter, M. F., Elste, V., & Weber, P. (2003). Effect of EGCG on lipid absorption and plasma lipid levels in rats. *The Journal of Nutritional Biochemistry, 14*, 326–332. http://dx.doi.org/10.1016/S0955-2863(03)00054-8

Ramadan, G., Nadia, .M., & El-Ghffar, E. A. A. (2009). Modulatory effects of black v. green tea aqueous extract on hyperglycaemia, hyperlipidaemia and liver dysfunction in diabetic and obese rat models. *British Journal of Nutrition, 102*, 1611–1619.

Ramadan, G., El-Beih, N. M., & Abd, E. A. (2009). El-Ghffar, *Modulatory effects of black v. green tea aqueous extract on hyperglycaemia, hyperlipidaemia and liver dysfunction in diabetic and obese rat models. British Journal of Nutrition, 102*, 1611–1619.

Rehman, H., Krishnasamy, Y., Haque, K., Thurman, R. G., Lemasters, J. J., Schnellmann, R. G., & Zhong, Z. (2013). Green tea polyphenols stimulate mitochondrial biogenesis and improve renal function after chronic cyclosporin a treatment in rats. *PLoS One, 8*(6), e65029. doi:10.1371/journal.pone.0065029

Rehman, H., Krishnasamy, Y., Haque, K., Thurman, R. G., Lemasters, J. J., Schnellmann, R. G., & Zhong, Z. (2013). Green tea polyphenols stimulate mitochondrial biogenesis and improve renal function after chronic cyclosporin A treatment in rats. *PLoS ONE, 8*, e65029. http://dx.doi.org/10.1371/journal.pone.0065029

Renno, W. M., Abdeen, S., Alkhalaf, M., & Asfar, S. (2008). Effect of green tea on kidney tubules of diabetic rats. *British Journal of Nutrition, 100*, 652–659. http://dx.doi.org/10.1017/S0007114508911533

Reno, F. E., et al. (2015). A novel nasal powder formulation of glucagon: Toxicology studies in animal models. *BMC Pharmacology and Toxicology, 16*(1), 937. http://dx.doi.org/10.1186/s40360-015-0026-9

Reza, H. M., Sagor, M. A. T., & Alam, M. A. (2015). Iron deposition causes oxidative stress, inflammation and fibrosis in carbon tetrachloride-induced liver dysfunction in rats. *Bangladesh Journal of Pharmacology, 10*, 152–159.

Reza, H. M., Tabassum, N., Sagor, M. A. T., Chowdhury, M. R. H., Rahman, M., Jain, P., & Alam, M. A. (2016). Angiotensin-converting enzyme inhibitor prevents oxidative stress, inflammation, and fibrosis in carbon tetrachloride-treated rat liver. *Toxicology Mechanisms and Methods, 26*, 46–53. doi:10.3109/15376516.2015.1124956

Ribaldo, P. D., Souza, D. S., Biswas, S. K., Block, K., de Faria, J. M. L., & de Faria, J. B. L. (2009). Green tea (*Camellia sinensis*) attenuates nephropathy by downregulating Nox4 NADPH oxidase in diabetic spontaneously hypertensive rats. *The Journal of Nutrition, 139*, 96–100.

Rojas, J. J., Ochoa, V. J., Ocampo, S. A., & Muñoz, J. F (2006). Screening for antimicrobial activity of ten medicinal plants used in Colombian folkloric medicine: A possible alternative in the treatment of non-nosocomial infections. *BMC Complementary and Alternative Medicine, 6*(1), 43. http://dx.doi.org/10.1186/1472-6882-6-2

Rosenthal, J., & Nocera, D. G. (2007). Role of proton-coupled electron transfer in O–O bond activation. *Accounts of Chemical Research, 40*, 543–553. http://dx.doi.org/10.1021/ar7000638

Ruggenenti, P., Cravedi, P., & Remuzzi, G. (2010). The RAAS in the pathogenesis and treatment of diabetic nephropathy. *Nature Reviews Nephrology, 6*, 319–330. http://dx.doi.org/10.1038/nrneph.2010.58

Ruiz, S., Pergola, P. E., Zager, R. A., & Vaziri, N. D. (2013). Targeting the transcription factor Nrf2 to ameliorate oxidative stress and inflammation in chronic kidney disease. *Kidney International, 83*, 1029–1041. http://dx.doi.org/10.1038/ki.2012.439

Rutter, M. K., & Nesto, R. W. (2011). Blood pressure, lipids and glucose in type 2 diabetes: How low should we go? Rediscovering personalized care *European Heart Journal, 32*, 2247–2255. http://dx.doi.org/10.1093/eurheartj/ehr154

Ryu, H. H., Kim, H. L., Chung, J. H., Lee, B. R., Kim, T. H., & Shin, B. C. (2011). Renoprotective effects of green tea extract on renin-angiotensin-aldosterone system in chronic cyclosporine-treated rats. *Nephrology Dialysis Transplantation, 26*, 1188–1193. http://dx.doi.org/10.1093/ndt/gfq616

Sachdeva, A. K., Kuhad, A., Tiwari, V., Arora, V., & Chopra, K. (2010). Protective effect of epigallocatechin gallate in murine water-immersion stress model of chronic fatigue syndrome. *Basic & Clinical Pharmacology & Toxicology, 106*, 490–496. http://dx.doi.org/10.1111/pto.2010.106.issue-6

Sagor, A. T., Chowdhury, M. R. H., Tabassum, N., Hossain, H., Rahman, M. M., & Alam, M. A (2015). Supplementation of fresh ucche (Momordica charantia L. var. muricata Willd) prevented oxidative stress, fibrosis and hepatic damage in CCl4 treated rats. *BMC Complementary and Alternative Medicine, 15*, 474. http://dx.doi.org/10.1186/s12906-015-0636-1

Sagor, M. A. T., Tabassum, N., Potol, M. A., & Alam, M. A. (2015). Xanthine oxidase inhibitor, allopurinol, prevented oxidative stress, fibrosis, and myocardial damage in isoproterenol induced aged rats. *Oxidative Medicine and Cellular Longevity, 2015*, 9 p. Article ID 478039. doi:10.1155/2015/47803

Sagor, M. A. T., Mohib, M. M., Tabassum, N., Ahmed, I., & Reza, H. M. (2016). Fresh seed supplementation of *Syzygium cumini* attenuated oxidative stress, inflammation, fibrosis, iron overload, hepatic dysfunction and renal injury in acetaminophen induced rats. *Journal of Drug Metabolism & Toxicology, 7*, 2. doi:10.4172/2157-7609.1000208

Sahin, K., Tuzcu, M., Gencoglu, H., Dogukan, A., Timurkan, M., Sahin, N., … Kucuk, O. (2010). Epigallocatechin-3-gallate activates Nrf2/HO-1 signaling pathway in cisplatin-induced nephrotoxicity in rats. *Life Sciences, 87*, 240–245. http://dx.doi.org/10.1016/j.lfs.2010.06.014

Santini, S. A., Marra, G., Giardina, B., Cotroneo, P., Mordente, A., MartoranaG. E., … Ghirlanda, G. (1997). Defective plasma antioxidant defenses and enhanced susceptibility to lipid peroxidation in uncomplicated IDDM. *Diabetes, 46*, 1853–1858. http://dx.doi.org/10.2337/diab.46.11.1853

Sartippour, M. R., Shao, Z. M., Heber, D., Beatty, P., Zhang, L., Liu, C., … Brooks, M. N. (2002). Green tea inhibits vascular endothelial growth factor (VEGF) induction in human breast cancer cells. *The Journal of Nutrition, 132*, 2307–2311.

Sedeek, M., Callera, G., Montezano, A., Gutsol, A., Heitz, F., Szyndralewiez, C., … Hebert, R. L. (2010). Critical role of Nox4-based NADPH oxidase in glucose-induced oxidative stress in the kidney: implications in type 2 diabetic nephropathy. *AJP: Renal Physiology, 299*, F1348–F1358. http://dx.doi.org/10.1152/ajprenal.00028.2010

Sedeek, M., Gutsol, A., Montezano, A. C., Burger, D., Nguyen Dinh Cat, A., Kennedy, C. R. J., … Touyz, R. M. (2013). Renoprotective effects of a novel Nox1/4 inhibitor in a mouse model of Type 2 diabetes. *Clinical Science, 124*, 191–202. http://dx.doi.org/10.1042/CS20120330

Shin, B. C., Chung, J. H., Lee, B. R., & Kim, H. L. (2009). The protective effects of green tea extract against L-arginine toxicity to cultured human mesangial cells. *Journal of Korean Medical Science, 24*(Suppl 1), S204–S209. http://dx.doi.org/10.3346/jkms.2009.24.S1.S204

Shokouhi, B., AbediGaballu, F., Khyavy, O. M., Mamandy, H., Rasoulian, H., Hajizadeh, N., & Mardomi, A. (2015). Effects of *Nigella sativa* and *Camellia sinensis* water extracts on alloxan induced diabetic nephropathy in rats. *Advances in Bioresearch, 6*, 128–132.

Smith, T. J. (2011). Green tea polyphenols in drug discovery: A success or failure? *Expert Opinion on Drug Discovery, 6*, 589–595. http://dx.doi.org/10.1517/17460441.2011.570750

Soetikno, V., Sari, F. R., Lakshmanan, A. P., Arumugam, S., Harima, M., Suzuki, K., & Watanabe, K. (2013). Curcumin alleviates oxidative stress, inflammation, and renal fibrosis in remnant kidney through the Nrf2–keap1 pathway. *Molecular Nutrition & Food Research, 57*, 1649–1659. http://dx.doi.org/10.1002/mnfr.v57.9

Sriram, N., Kalayarasan, S., & Sudhandiran, G. (2009). Epigallocatechin-3-gallate augments antioxidant activities and inhibits inflammation during bleomycin-induced experimental pulmonary fibrosis through Nrf2–Keap1 signaling. *Pulmonary Pharmacology & Therapeutics, 22*, 221–236. http://dx.doi.org/10.1016/j.pupt.2008.12.010

Sudano Roccaro, A. S., Blanco, A. R., Giuliano, F., Rusciano, D., & Enea, V. (2004). Epigallocatechin-gallate enhances the activity of tetracycline in staphylococci by inhibiting its efflux from bacterial cells. *Antimicrobial Agents and Chemotherapy, 48*, 1968–1973. http://dx.doi.org/10.1128/AAC.48.6.1968-1973.2004

Sueoka, N., Suganuma, M., Sueoka, E., Okabe, S., Matsuyama, S., Imai, K., … Fujiki, H. (2001). A new function of green tea: Prevention of lifestyle-related diseases. *Annals of the New York Academy of Sciences, 928*, 274–280.

Suliburska, J., Bogdanski, P., Szulinska, M., Stepien, M., Pupek-Musialik, D., & Jablecka, A. (2012). Effects of green tea supplementation on elements, total antioxidants, lipids, and glucose values in the serum of obese patients. *Biological Trace Element Research, 149*, 315–322. http://dx.doi.org/10.1007/s12011-012-9448-z

Suzuki, D., Miyazaki, M., Naka, R., Koji, T., Yagame, M., Jinde, K., Sakai, H., … Nomoto, Y. (1995). *In Situ* hybridization of interleukin 6 in diabetic nephropathy. *Diabetes, 44*, 1233–1238. http://dx.doi.org/10.2337/diab.44.10.1233

Tan, A. L., Sourris, K. C., Harcourt, B. E, Thallas-Bonke, V., Penfold, S., Andrikopoulos, S., … Forbes, J. M. (2010). Disparate effects on renal and oxidative parameters following RAGE deletion, AGE accumulation inhibition, or dietary AGE control in experimental diabetic nephropathy. *AJP: Renal Physiology, 298*, F763–F770. http://dx.doi.org/10.1152/ajprenal.00591.2009

Tang, W., Li, S., Liu, Y., Huang, M.-T., & Ho, C.-T. (2013). Anti-diabetic activity of chemically profiled green tea and black tea extracts in a type 2 diabetes mice model via different mechanisms. *Journal of Functional Foods, 5*, 1784–1793. http://dx.doi.org/10.1016/j.jff.2013.08.007

Thangapandiyan, S., & Miltonprabu, S. (2014). Epigallocatechin gallate supplementation protects against renal injury induced by fluoride intoxication in rats: Role of Nrf2/HO-1 signaling. *Toxicology Reports, 1*, 12–30. http://dx.doi.org/10.1016/j.toxrep.2014.01.002

Toolsee, N. A., Aruoma, O. I., Gunness, T. K., Kowlessur, S., Dambala, V., Murad, F., & Bahorun, T. (2013). Effectiveness of green tea in a randomized human cohort: Relevance to diabetes and its complications. *BioMed Research International, 2013*, Article ID: 412379. doi:10.1155/2013/412379

Tsai, P.-Y., Ka, S.-M., Chang, J.-M., Chen, H.-C., Shui, H.-A., Li, C.-Y., … Chen, A. (2011). Epigallocatechin-3-gallate prevents lupus nephritis development in mice via enhancing the Nrf2 antioxidant pathway and inhibiting NLRP3 inflammasome activation. *Free Radical Biology and Medicine, 51*, 744–754. http://dx.doi.org/10.1016/j.freeradbiomed.2011.05.016

Tse, H., Anderson, B., Ganini, D., & Mason, R. (2015). Immuno-spin trapping to detect immune-derived free radicals in type 1 diabetes (TECH2P. 903). *The Journal of Immunology, 194*(1 Supplement), 206–213.

United States Renal Data System. (2011). *Annual data report: atlas of chronic kidney disease and end-stage renal disease in the United States.* Bethesda, MD: National Institutes of Health, National Institute of Diabetes and Digestive and Kidney Diseases.

Vaziri, N. D., Dicus, M., Ho, N. D., Boroujerdi-Rad, L., & Sindhu, R. K. (2003). Oxidative stress and dysregulation of superoxide dismutase and NADPH oxidase in renal insufficiency. *Kidney International, 63,* 179–185. http://dx.doi.org/10.1046/j.1523-1755.2003.00702.x

Wakabayashi, N., Slocum, S. L., Skoko, J. J., Shin, S., & Kensler, T. W. (2010). When NRF2 Talks, who's listening? *Antioxidants & Redox Signaling, 13,* 1649–1663. http://dx.doi.org/10.1089/ars.2010.3216

Wang, H., Meng, Q. H., Gordon, J. R., Khandwala, H., & Wu, L. (2007). Proinflammatory and proapoptotic effects of methylglyoxal on neutrophils from patients with type 2 diabetes mellitus. *Clinical Biochemistry, 40,* 1232–1239. http://dx.doi.org/10.1016/j.clinbiochem.2007.07.016

Wang, Y., Wang, B., Du, F., Su, X., Sun, G., Zhou, G., … Liu, N. (2015). Epigallocatechin-3-gallate attenuates unilateral ureteral obstruction-induced renal interstitial fibrosis in mice. *Journal of Histochemistry & Cytochemistry, 63,* 270–279. http://dx.doi.org/10.1369/0022155414568019

Wang, Y., Wang, B., Du, F., Su, X., Sun, G., Zhou, G., … Liu, N. (2015). Epigallocatechin-3-gallate attenuates oxidative stress and inflammation in obstructive nephropathy via NF-κB and Nrf2/HO-1 signalling pathway regulation. *Basic & Clinical Pharmacology & Toxicology, 117,* 164–172. http://dx.doi.org/10.1111/bcpt.2015.117.issue-3

Wassmann, S., Wassmann, K., & Nickenig, G. (2004). Modulation of oxidant and antioxidant enzyme expression and function in vascular cells. *Hypertension, 44,* 381–386. http://dx.doi.org/10.1161/01.HYP.0000142232.29764.a7

Weber, J. M., Ruzindana-Umunyana, A., Imbeault, L., & Sircar, S. (2003). Inhibition of adenovirus infection and adenain by green tea catechins. *Antiviral Research, 58,* 167–173. http://dx.doi.org/10.1016/S0166-3542(02)00212-7

Wein, S., Schrader, E., Rimbach, G., & Wolffram, S. (2013). Oral green tea catechins transiently lower plasma glucose concentrations in female db/db mice. *Journal of Medicinal Food, 16,* 312–317. http://dx.doi.org/10.1089/jmf.2012.0205

Weinreb, O., Mandel, S., Amit, T., & Youdim, M. B. H. (2004). Neurological mechanisms of green tea polyphenols in Alzheimer's and Parkinson's diseases. *The Journal of Nutritional Biochemistry, 15,* 506–516. http://dx.doi.org/10.1016/j.jnutbio.2004.05.002

Wendt, T. M., Tanji, N., Guo, J., Kislinger, T. R., Qu, W., Lu, Y., … Schmidt, A. N. (2003). RAGE drives the development of glomerulosclerosis and implicates podocyte activation in the pathogenesis of diabetic nephropathy. *The American Journal of Pathology, 162,* 1123–1137. http://dx.doi.org/10.1016/S0002-9440(10)63909-0

Whiting, D. R., Guariguata, L., Weil, C., & Shaw, J. (2011). IDF diabetes atlas: Global estimates of the prevalence of diabetes for 2011 and 2030. *Diabetes Research and Clinical Practice, 94,* 311–321. http://dx.doi.org/10.1016/j.diabres.2011.10.029

Wolfram, S. (2007). Effects of Green Tea and EGCG on Cardiovascular and Metabolic Health. *Journal of the American College of Nutrition, 26,* 373S–388S. http://dx.doi.org/10.1080/07315724.2007.10719626

Wolfram, S., Raederstorff, D., Preller, M., Wang, Y., Teixeira, S. R., Riegger, C., & Weber, P. (2006). Epigallocatechin gallate supplementation alleviates diabetes in rodents. *The Journal of Nutrition, 136,* 2512–2518.

Wu, C., Hsu, M. C., Hsieh, C. W., Lin, J. B., Lai, P. H., & Wung, B. S. (2006). Upregulation of heme oxygenase-1 by Epigallocatechin-3-gallate via the phosphatidylinositol 3-kinase/Akt and ERK pathways. *Life Sciences, 78,* 2889–2897. http://dx.doi.org/10.1016/j.lfs.2005.11.013

Yamabe, N., Yokozawa, T., Oya, T., & Kim, M. (2006). Therapeutic potential of (-)-epigallocatechin 3-o-gallate on renal damage in diabetic nephropathy model rats. *Journal of Pharmacology and Experimental Therapeutics, 319,* 228–236. http://dx.doi.org/10.1124/jpet.106.107029

Yamahara, K., Kume, S., Koya, D., Tanaka, Y., Morita, Y., Chin-Kanasaki, M., … Matsusaka, T. (2013). Obesity-mediated autophagy insufficiency exacerbates proteinuria-induced tubulointerstitial lesions. *Journal of the American Society of Nephrology.* p. ASN. 2012111080.

Yan, J., Zhao, Y., Suo, S., Liu, Y., & Zhao, B. (2012). Green tea catechins ameliorate adipose insulin resistance by improving oxidative stress. *Free Radical Biology and Medicine, 52,* 1648–1657. http://dx.doi.org/10.1016/j.freeradbiomed.2012.01.033

Yapar, K., Çavuşoğlu, K., Oruç, E., & Yalçin, E. (2009). Protective effect of royal jelly and green tea extracts effect against cisplatin-induced nephrotoxicity in mice: A comparative study. *Journal of Medicinal Food, 12,* 1136–1142. http://dx.doi.org/10.1089/jmf.2009.0036

Ye, T., Zhen, J., Du, Y., Peng, A., Vaziri, N. D., Mohan, C., & ZhouX. J. (2015). Green tea polyphenol (−)-epigallocatechin-3-gallate restores Nrf2 activity and ameliorates crescentic glomerulonephritis. *PLOS ONE, 10,* e0119543. http://dx.doi.org/10.1371/journal.pone.0119543

Yoh, K., Itoh, K., Enomoto, A., Hirayama, A., Yamaguchi, N., Kobayashi, M., … Takahashi, S. (2001). Nrf2-deficient female mice develop lupus-like autoimmune nephritis11See Editorial by Byrd and Thomas, p. 1606. *Kidney International, 60,* 1343–1353. http://dx.doi.org/10.1046/j.1523-1755.2001.00939.x

Yoh, K., Hirayama, A., Ishizaki, K., Yamada, A., Takeuchi, M., Yamagishi, S., … Yamamoto, M. (2008). Hyperglycemia induces oxidative and nitrosative stress and increases renal functional impairment in Nrf2-deficient mice. *Genes to Cells, 13,* 1159–1170. doi:10.1111/j.1365-2443.2008.01234.x

Yokozawa, T., Noh, J. S., & Park, C. H. (2012). Green tea polyphenols for the protection against renal damage caused by oxidative stress. *Evidence-Based Complementary and Alternative Medicine, 2012,* Article ID: 845917. doi:10.1155/2012/845917

Yoon, S. P., Hong, R., Lee, B. R., Kim, C. G., Kim, H. L., Chung, J. H., & Shin, B. C. (2014). Protective effects of epigallocatechin gallate (EGCG) on streptozotocin-induced diabetic nephropathy in mice. *Acta Histochemica, 116,* 1210–1215. http://dx.doi.org/10.1016/j.acthis.2014.07.003

Zaveri, N. T. (2006). Green tea and its polyphenolic catechins: Medicinal uses in cancer and noncancer applications. *Life Sciences, 78,* 2073–2080. http://dx.doi.org/10.1016/j.lfs.2005.12.006

Zhou, Q., Li, H., Zhou, J. G., Ma, Y., Wu, T., & Ma, H. (2016). Green tea, black tea consumption and risk of endometrial cancer: A systematic review and meta-analysis. *Archives of Gynecology and Obstetrics, 293,* 143–155. http://dx.doi.org/10.1007/s00404-015-3811-1

Zhu, D., Wang, L., Zhou, Q., Yan, S., Li, Z., Sheng, J., & Zhang, W. (2014). (+)-Catechin ameliorates diabetic nephropathy by trapping methylglyoxal in type 2 diabetic mice. *Molecular Nutrition & Food Research, 58,* 2249–2260. http://dx.doi.org/10.1002/mnfr.201400533

Zoncu, R., Efeyan, A., & Sabatini, D. M. (2011). mTOR: From growth signal integration to cancer, diabetes and ageing. *Nature Reviews Molecular Cell Biology, 12,* 21–35. http://dx.doi.org/10.1038/nrm3025

3

Investigation of pre-delirium and delirium in patients with alcohol withdraw

Lyudmila Neykova-Vasileva[1], Tony Donchev[2] and Krasimir Kostadinov[2]*

*Corresponding author: Krasimir Kostadinov, Psychiatry Clinic, Military Medical Academy, Georgi Sofiiski 3, Sofia 1606, Bulgaria
E-mail: kkostadinov@ymail.com

Reviewing editor: Hani El-Nezami, University of Eastern Finland, Finland
Additional information is available at the end of the article

Abstract: *Introduction*: Alcohol withdraw syndrome (AWS) and delirium tremens (DTs) are part of the clinical picture of alcohol dependence (AD). The mechanisms by which these conditions develop are not entirely understood, but there are data suggesting that several factors are largely responsible for the emergence of the AWS and DTs. According to some authors, with the prolongation of the AD and increasing tolerance, the AWS aggravates, somatic and neurovegetative manifestations worsen. Russian authors accept that the degree of clinical manifestations depends on the individual characteristics of the response to ethanol. *Materials and methods*: The present work is a systematic study of 28 alcohol-dependent patients that developed clinical signs and symptoms of pre-delirium and delirium during alcohol detoxification treatment. *Results*: The relationship between sex, age, type and duration for abuse, type of alcoholic beverage, and the clinical picture has been analyzed and discussed as well as the criteria for their diagnosis and follow-up. *Conclusion*: Our analysis of the development and the manifestations of pre-delirium and DTs, during detoxification of AD patients, underlines the protracted course of the illness and the need for multidisciplinary intensive care. Dynamic monitoring of the patients and timely correction of the therapeutic program guarantee the success of the detoxification.

Subjects: Toxicology; Clinical Toxicology; Addiction - Alcohol - Adult; Psychiatry

Keywords: toxicology; alcohol dependence; withdraw; pre-delirium; delirium tremens; psychiatry; addiction

1. Introduction

Alcohol withdraw syndrome (AWS) and delirium tremens (DTs) are part of the clinical picture of alcohol dependence (AD)—condition due to a number of pathological processes in the central nervous system

ABOUT THE AUTHORS

The Military Medical Academy (MMA)—Sofia is a large complex for medical treatment, as well as education, located in Sofia, Bulgaria. It has several branches and smaller clinics in other cities in the country. The current investigation was carried out by the Department of Emergency Toxicology of The MMA—Sofia, led by Lyudmila Neykova-Vasileva and in close collaboration with Tony Donchev and Krasimir Kostadinov of the Department of Psychiatry, MMA-Sofia. Our team, as members of the both clinics, is dedicated to providing quality care for patients with substance abuse disorders both in acute and chronic settings.

PUBLIC INTEREST STATEMENT

Alcohol withdraw is a serious clinical syndrome and also a part of the clinical picture of Alcohol use disorder. As alcohol use is widely spread and socially acceptable in most countries around the world, it is important for health care professionals to recognize its consequences and be able to provide adequate management of such patients based on reliable assessments of its severity. In our article, the relationship between sex, age, type and duration of use, type of alcoholic beverage, and the clinical picture has been analyzed and discussed.

Table 1. Change in the activity of the affected neurotransmitter systems by ethanol			
Neurotransmitters	Receptors	Alcoholic intoxication-healthy persons	Chronic intoxication-AD and AWS
Glutamate	NMDA, secondary importance: AMPA, KA	↓	↑
GABA	$GABA_A$ and $GABA_B$	↑	↓
Dopamine	D_1, D_2	↑	↑ or ↓
Serotonin	$5\text{-}HT_{1A}$, $5\text{-}HT_{1B}$, $5\text{-}HT_{2C}$, $5\text{-}HT_3$	↑	↓
Homocysteine	NMDA	↓	↑
Aspartate			↑
Acetylcholine	nAChR	↓	↑
Endocanabinoids	CB_1	↓	↑
Adenosine	A_1	↑	↓

(CNS) determined by the abuse of ethanol and its specific metabolism. The fast metabolization of ethanol determines the build-up of acetaldehyde in the organism (Gerevich, 2007). In addition to the direct toxic effects on neurons, which is functionally metabolic, active metabolites from its degradation lead to morphologically destructive changes, turning the process into exo- and endo-intoxication. The mechanisms by which these conditions develop are not entirely understood, but there are data suggesting that several factors are largely responsible for the emergence of the AWS and DTs (Gerevich, 2007).

1.1. Impact on the receptors in the central nervous system
The ethanol attacks the lipid component built mainly from phospholipids and covering the main protein ingredient of GABA-receptors. A disharmony occurs in the permeability of the cell membrane, the stimulation of secondary messengers and impairments in central brain neurotransmission and the hormonal secretion as well (Tatebayashi, Motomura, & Narahashi, 1998).

The blocking of vasopresin leads to general dehydratation effect as well as in the cells of the brain and on the cells of the stroma (Mander et al., 1985).

Acetaldehyde (a metabolite that is much more aggressive than ethanol) plays a very important role in the emergence of the AWS and DTs, mainly through its propensity to condensation with neurotransmitters in the CNS (Koob & Le Moal, 2006; Licata & Renshaw, 2010).

In Table 1 are shown the neurotransmitter systems (NM) which are affected by the acute and chronic (systemic) action of ethanol (Koob & Le Moal, 2006; Licata & Renshaw, 2010).

The oxidation of ethanol, apart from the synthesis of acetaldehyde, is a process leading to the intracellular accumulation of NADH 2 and NADPH 2 and is a prerequisite for a number of pathological syntheses which start intra-cellulary (Thurman, Glassman, Handler, & Forman, 1989).

Energy source of these processes (syntheses) comes from the high caloricity of alcoholic beverages—the separation of large quantity of hydrogen as well as from the opportunity to be used in the cycle of the Krebs as products of ketogenesis (Khanna & Israel, 1980).

1.2. Interaction of acetaldehyde with neurotransmitters in the CNS
To begin with, acetaldehyde condenses with dopamine (DA) and a condensate is formed—salsalinol, which has hallucinogenic properties. Apart from that during his amine oxidation, tetra-hydro-isoquinolines (THIQ) are formed. The latter are extremely stable compounds, practically they do not degrade, but accumulate in the CNS and play an important role in the occurrence of withdraw reactions (Naoi, Maruyama, & Nagy, 2004).

The harmaline is degraded extremely slow, it stimulates the release of DA in the CNS; it is a cholinest-erase inhibitor and also blocks MAO-inhibitors. Decreased synthesis of melatonin and is a powerful in-dole stimulator in the CNS. It determines the agitation and hallucinatory symptoms (Collins, 1988).

Of course, other contributing factors for the development of the AWS and DTs cannot be ignored, such as amyl alcohol—the result of the degradation of the carbon skeleton of the branched chain amino acids. It can be found in almost all alcoholic beverages (Westerhoff, Groen, & Wanders, 1984).

There are also many other factors, coming mainly from impairments of the hepatocyte metabo-lism: reduced synthesis of aldehyde-dehydrogenase; high homocysteine levels; change in acid–base balance, leaning toward acidosis; dehydration of body; general enzyme deficiency, as well as hypo-vitaminosis of water soluble vitamins and the insufficiency of zinc (Zn) (Seeto, Fenn, & Rockey, 2000).

There are two phases in the course of the AWS—in the beginning, after an interruption of the al-coholic use, symptoms (not at the same time) amplify and reach a maximum after which they grad-ually (not at the same time) decrease. The tremor, asteno-adynamic syndrome, dyspeptic nad cardio-vascular symptoms develop during the first 6–12 h after the last alcohol consumption, hal-lucinations and seizures—after 24–28 h and DTs—3–5 days after the last drink (Bayard, McIntyre, Hill, & Woodside, 2004).

According to some authors, with the prolongation of the alcohol dependence and the increasing of the tolerance, the AWS aggravates, and the somatic and neurovegetative manifestations worsen (Entin, 1990). Russian authors accept that the degree of clinical manifestations depends on the indi-vidual characteristics of the response to ethanol (Ivanetz & Igonin, 1987).

There are authors who claim that the severity of AWS is not age-dependent (Wetterling, 2001). Delirium tremens is the worst expression of AWS and the most frequent type of alcohol- induced disorder of cognition. It develops in 5–15% of alcoholic dependent patients. It is observed when people abusing mainly with concentrated drinks, more often after 30 years of age (Lofwall, Schuster, & Strain, 2008).

Delirium can be observed during a long intoxication and in the course of withdraw (American Psychiatric Association, 2000; Esel, 2006). People with alcohol dependence rarely stop drinking abruptly. Frequently in the presence of the level of alcohol in the blood, the first symptoms of AWS emerge along with symptoms of alcohol intoxication (Elholm, Larsen, Hornnes, Zierau, & Becker, 2010). Clinical practice indicates that the objective assessment of the existence and the severity of the abstinence are not uncommonly hampered by a number of nonmedical factors. Patient anxiety, low insight, craving for alcohol, social stigma , pressure from relatives for aggressive treatment or no treatment at all are common factors that distort the course of gathering accurate information about the condition and the real alcohol consumption of the patients. For this reason, the standardized assessment with specially developed scales has a number of advantages.

Few studies have validated scales for the evaluation of the AWS (Elholm, Larsen, Hornnes, Zierau, & Becker, 2011). At this stage, there is no consensus on the standardized monitoring of AWS and DTs as well (Seeto et al., 2000). The most widely used and the most well studied in scale for the evalua-tion of the AWS is Clinical Institute Withdrawal Assessment for Alcohol Scale (CIWA-Ar) (Sullivan, Sykora, Schneiderman, Naranjo, & Sellers, 1989). It contains 10 items with a weight of 0–7 by which a score is calculated, that gives the clinician an idea of the severity of the AWS. One of its advantages is that there are different protocols for guidance of the treatment plan (especially the dosage of benzodiazepines) based on the score of the scale even in multiple patient assessments (Ng, Dahri, Chow, & Legal, 2011).

When a disorder of consciousness is suspected there are a number of standardized clinical tools for its evaluation and for the verification of the presence of a delirium (irrespective of the etiology). Such are the

Richmond Agitation-Sedation Scale; Delirium Triage Screen (DTS); brief Confusion Assessment Method; Confusion Assessment Method is (CAM); Confusion Assessment Method is for the ICU (CAM-ICU).

2. Aim

The current study analyses of the various factors—sex, type and duration of alcohol use, type of drink (high concentration and/or low concentration of alcohol), in order to predict the dynamics and severity of the alcoholic delirium in the period of detoxification.

3. Materials and methods

Target group are patients in pre-delirium or with the clinical picture of DTs. All have gone through detoxification in clinic "Emergency toxicology" of the MHAT-Sofia, MMA between the years 2009 and 2012. The mean age is 43, 50 ± 13, 74 years, in the range between 19 and 69 years.

The distribution by sex is—3 women and 25 men and according to the type of alcohol use. Seven of all patients had binge type of drinking and 21—daily type, as described in Figure 1. The duration of alcohol dependence among patients and the distribution according to the type of drinking is described in Figures 2–4.

The methods used are: documentary, clinical, statistical—all analyzes have been carried out with the program IBM SPSS Statistics 22.0. In cases in which subgroups with a smaller number of persons (<8) have been analyzed, non-parametrical statistical tests were applied.

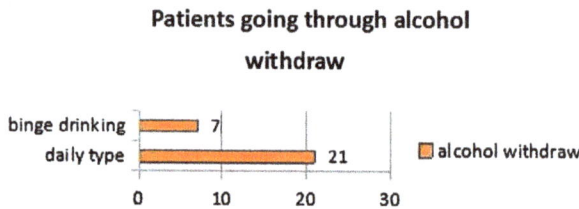

Figure 1. Distribution according to the type of alcohol use.

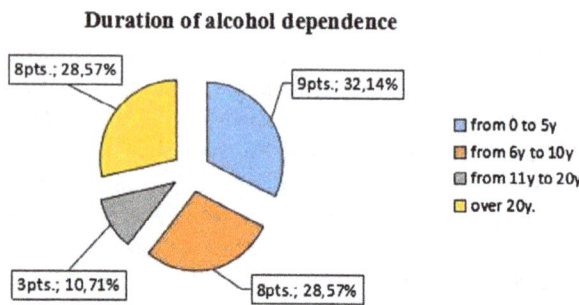

Figure 2. Duration of the dependence.

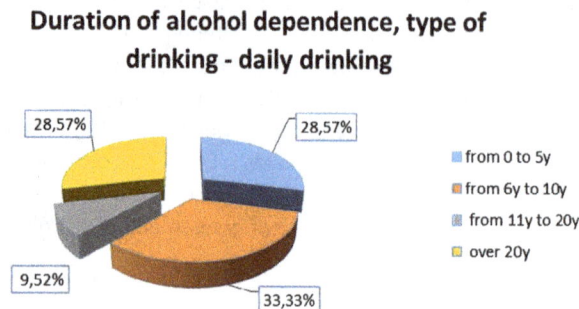

Figure 3. Duration of AD— "daily" type.

Duration of alcohol dependence, type of drinking - binge drinking

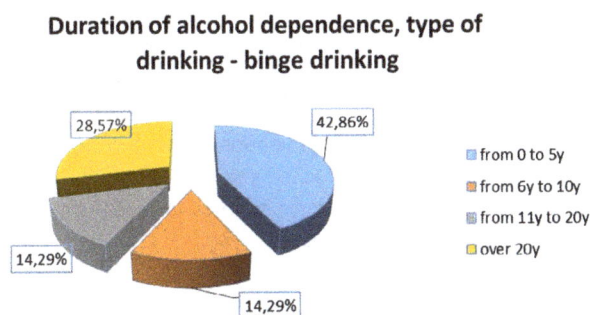

Figure 4. Duration of AD— "binge drinking" type.

4. Results

The main criterion for the selection of patients is the presence of alcohol withdraw—pre-delirium or delirium (DTs). Their development usually is preceded by a long-term alcohol consumption—"daily" or "binge drinking" type. The diagnosis is made by the consultant—psychiatrist.

In the target group of patients with AD, we have observed withdraw symptoms more or less pronounced, of pre-delirium or delirium tremens, after the termination or reduction of alcohol consumption. With respect to differential diagnosis hypoglycemia was ruled-out as possible etiology.

Predominantly affected were patients in the age range of 30–39 years (32%), which raises the hypothesis that alcohol use has started at a young age.

The long-term information for the systematic use of alcohol is the attribute "duration". According to it predilection (33%) have patients with abuse of between 6 and 10 years of "Daily" beer drinkers and the most small is of these between 11 and 20 years old (around 10%). In the subgroup with "binge drinking" type—most of the patients have duration of drinking between 0 and 5 years (almost 43%).

The distribution by sex is 8:1 men toward women. The three female patients have exhibited pre-delirious symptoms.

5. Discussion

The syndrome of alcoholic withdraw with pre-delirium in AD patients is characterized by polymorphic clinical picture. It occurs most frequently between the 10th and 30th hour, but may also start after 40–50 h after the last alcoholic drink. The clinical manifestations observed by our team were: Hyperactivity, fear affect or depressivity, slightly increased temperature, tachycardia, increased diaphoresis, nausea with or without vomiting (in 25 of the patients—89.28%). A remarkable feature of neurological symptoms in patients in pre- delirious state is the absence or very low intensity of the tremor of the upper extremities. Three patients (10.72%) were with a heavy expressed form of AWS—delirium tremens. In their case, in addition to the above stated symptoms combativeness, desorientation, visual and auditory hallucinations were also observed. The psychosis in patients with DTs passed with formally preserved autopsychic (for own personality), but with impaired alopsychichic (for time and place) orientation. The average duration was 15 days.

The structure of the psychopathologic disorders during detoxification varied in relation to the level of anxiety and depressive disorders.

At the time of the study, the patients had laboratory tests done for the assessment of liver pathology. In 75% of them there was an increase in the serum aminotransferase levels, with the predominance of the ratio ASAT > ALAT—indicator of impaired liver function. The levels of gamma glutamyl-transpeptidase (GGT–cholestatic enzyme)—are with increased average value several times above the upper limit of normal. Increase has been noted also in the average values of total cholesterol, TG and LDL-cholesterol.

Patients often complain of withdraw symptoms, but there are no objective evidence of this. This is why close monitoring (hourly in severe cases) for signs of pre-delirium or DTs through observation, BP and pulse, and by applying scales such as CIWA-Ar, DTS and bCAM is of great importance.

This gives the following advantages: individualized medical intervention in accordance with changes in the severity of the AWS (symptom-dependent); facilitates the selection of proper medication regimen and reduces the risk of somatic and neurological complications; shortens the period of detoxification; economic effect—reduces unnecessary treatment costs.

6. Implications

(1) All patients with AD develop withdraw symptoms. AWC is often short lasting, but quickly transforms in pre-delirious or delirious state. In the pre-crisis period somatic manifestations are discreet—a slight tremor, tendency toward tachycardia, small increase in BP, sweating. Careful monitoring and working in liaison with the psychiatric department unit are essential for recognizing those early signs. Beginning treatment in a timely manner often prevents the occurrence of delirium.

(2) Number of factors are important for the development of alcohol delirium:

    (a) Gender–delirium occurs much rarely in women even in severe withdraw. Our research confirms this correlation—male patients are mainly affected (8:1).

    (b) Age–our analysis shows that patients in the age group between 30 and 39 years were predominantly affected—32% of the target group;

    (c) Duration of abuse—with its prolongation and the increase tolerance clinical signs of withdraw tend to get worse. Patients with daily consumption of ethanol—33% of them—have "experience with alcohol" between 6 and 10 years. Individual reactivity against ethanol should be taken into account—this is confirmed by "binge" drinkers—the most numerous group are those with history of abuse between 0 and 5 years (almost 43%). In a large prescription AD, DTs can occur at high levels of alcohol concentration in the blood.

    (d) Type of alcohol use—patients with daily alcohol abuse are three times more compa red to those with "binge drinking" type of abuse.

(3) According to various literature sources pre-delirium or DTs occur in individuals abusing mainly concentrated alcoholic drinks and more frequently after age 30. Our research shows that most often delirium occurs in patients abusing concentrated alcoholic drinks or concentrated along with "light" alcoholic beverages (96.42%).

(4) The intensity of autonomic disorders correlates with the severity of anxiety and depressive disorders. Patients with psychotic symptoms during the AWS, rarely develop depressive symptoms. The severity of psychopathologic signs is determined by a number of factors (including the characteristics of the AD; severity of AWS).

(5) Of great importance for the duration of delirious states is the degree of damage to the liver functioning.

(6) Quick and competent correction of the dysfunction of the liver often significantly reduces the duration and intensity of delirious events.

(7) Standardized assessments i.e. the use of approbated clinical scales for the diagnosis and monitoring of patients going through alcohol withdraw and delirium could be beneficial for the guidance of early interventions in the course of the detoxification and a symptom triggered approach of treatment.

7. Conclusion
Our analysis of the development and the manifestations of pre-delirium and DTs, during detoxification of AD patients, underlines the protracted course of the illness and the need for multidisciplinary intensive care. Dynamic monitoring of the patients and timely correction of therapeutic program are a guarantee for the success of the detoxification.

Funding
The authors received no direct funding for this research.

Competing Interests
The authors have no competing interests.

Author details
Lyudmila Neykova-Vasileva[1]
E-mail: lussi66@abv.bg
Tony Donchev[2]
E-mail: tonyd@abv.bg
Krasimir Kostadinov[2]
E-mail: kkostadinov@ymail.com
[1] Department of Emergency Toxicology, Military Medical Academy, Georgi Sofiiski 3, Sofia 1606, Bulgaria.
[2] Psychiatry Clinic, Military Medical Academy, Georgi Sofiiski 3, Sofia 1606, Bulgaria.

References
American Psychiatric Association. (2000). *Diagnostic and statistical manual of mental disorders*, Fourth Edition, Text Revision (DSM-IV-TR). doi:10.1176/appi.books.9780890423349

Bayard, M., McIntyre, J., Hill, K. R., & Woodside, Jr., J. (2004). Alcohol withdrawal syndrome. *American Family Physician, 69*, 1443–1450.

Collins, M. A. (1988). Acetaldehyde and its condensation products as markers in alcoholism. *Recent Developments in Alcoholism, 6*, 387–403.

Elholm, B., Larsen, K., Hornnes, N., Zierau, F., & Becker, U. (2010). A psychometric validation of the short alcohol withdrawal scale (SAWS). *Alcohol and Alcoholism, 45*, 361–365. doi:10.1093/alcalc/agq033

Elholm, B., Larsen, K., Hornnes, N., Zierau, F., & Becker, U. (2011). Alcohol withdrawal syndrome: symptom-triggered versus fixed-schedule treatment in an outpatient setting. *Alcohol And Alcoholism, 46*, 318–323. doi:10.1093/alcalc/agr020

Entin, M. (1990). *Treamtent of acloholism* (pp. 28–31). Moscow: Moscow Medicine.

Esel, E. (2006). Neurobiology of alcohol withdrawal inhibitory and excitatory neurotransmitters. *Turk Psikiyatri Derg, 17*, 129–137.

Gerevich, J. (2007). Drug and alcohol abuse: A clinical guide to diagnosis and treatment. *Journal of Epidemiology and Community Health, 61*, 173–174. Retrieved from http://doi.org/10.1136/jech.2006.047704 http://dx.doi.org/10.1136/jech.2006.047704

Ivanetz, N. N., & Igonin, A. L. (1987). *Withdraw syndrome in alcoholism (a physician manual)* (pp. 89–97). Moscow: Moscow Medicine.

Khanna, J. M., & Israel, Y. (1980). Ethanol metabolism. *International Review of Physiology, 21*, 275–315.

Koob, G. F., & Le Moal, M. (2006). *Neurobiology of Addiction.* London: Academic Press.

Licata, S. C., & Renshaw, P. F. (2010). Neurochemistry of drug action. *Annals of the New York Academy of Sciences, 1187*, 148–171. http://dx.doi.org/10.1111/j.1749-6632.2009.05143.x

Lofwall, M. R., Schuster, A., & Strain, E. C. (2008). Changing profile of abused substances by older persons entering treatment. *The Journal of Nervous and Mental Disease, 196*, 898–905. doi:10.1097/NMD.0b013e31818ec7ee

Mander, A. J., Smith, M. A., Kean, D. M., Chick, J., Douglas, R. H. B., Rehman, A. U., Best, J. J. K. (1985). Brain water measured in volunteers after alcohol and vasopressin. *The Lancet, 326*, 1075. Retrieved from http://www.thelancet.com/journals/lancet/article/PIIS0140-6736(85)90950-X/abstract http://dx.doi.org/10.1016/S0140-6736(85)90950-X

Naoi, M., Maruyama, W., & Nagy, G. M. (2004, January). Dopamine-derived salsolinol derivatives as endogenous monoamine oxidase inhibitors: occurrence, metabolism and function in human brains. *Neurotoxicology, 25*, 193–204.

Ng, K., Dahri, K., Chow, I., & Legal, M. (2011). Evaluation of an alcohol withdrawal protocol and a preprinted order set at a tertiary care hospital. *The Canadian Journal of Hospital Pharmacy, 64*, 436–445.

Seeto, R. K., Fenn, B., & Rockey, D. C. (2000). Ischemic hepatitis: clinical presentation and pathogenesis. *The American Journal of Medicine, 109*, 109–113. http://dx.doi.org/10.1016/S0002-9343(00)00461-7

Sullivan, J., Sykora, K., Schneiderman, J., Naranjo, C., & Sellers, E. (1989). Assessment of alcohol withdrawal: The revised clinical institute withdrawal assessment for alcohol scale (CIWA-Ar). *Addiction, 84*, 1353–1357. http://dx.doi.org/10.1111/add.1989.84.issue-11

Tatebayashi, H., Motomura, H., & Narahashi, T. (1998). Alcohol modulation of single GABAA receptor-channel kinetics. *NeuroReport, 9*, 1769–1775. http://dx.doi.org/10.1097/00001756-199806010-00018

Thurman, R. G., Glassman, E. B., Handler, J. A., & Forman, D. T. (1989). The swift increase in alcohol metabolism (SIAM): A commentary on the regulation of alcohol metabolism in mammals. In K. E. Crow & R. D. Batt (Eds.), *Human metabolism of alcohol* (Vol. 2, pp. 17–30). Boca Raton, FL: CRC Press.

Westerhoff, H., Groen, A., & Wanders, R. (1984). Modern theories of metabolic control and their applications. *Bioscience Reports, 4*, 1–22. http://dx.doi.org/10.1007/BF01120819

Wetterling, T. (2001). The severity of alcohol withdrawal is not age dependent. *Alcohol And Alcoholism, 36*, 75–78. doi:10.1093/alcalc/36.1.75

4

The physiological role of the brain GLP-1 system in stress

Marie K. Holt[1] and Stefan Trapp[1]*

*Corresponding author: Stefan Trapp, Centre for Cardiovascular and Metabolic Neuroscience, Department of Neuroscience, Physiology & Pharmacology, University College London, WC1E 6BT London, UK
E-mail: s.trapp@ucl.ac.uk

Reviewing editor: William Wisden, Imperial College London, UK

Additional information is available at the end of the article

Abstract: Glucagon-like peptide-1 (GLP-1) within the brain is a potent regulator of food intake and most studies have investigated the anorexic effects of central GLP-1. A range of brain regions have now been found to be involved in GLP-1 mediated anorexia, including some which are not traditionally associated with appetite regulation. However, a change in food intake can be indicative of not only reduced energy demand, but also changes in the organism's motivation to eat following stressful stimuli. In fact, acute stress is well-known to reduce food intake. Recently, more research has focused on the role of GLP-1 in stress and the central GLP-1 system has been found to be activated in response to stressful stimuli. The source of GLP-1 within the brain, the preproglucagon (PPG) neurons, are ideally situated in the brainstem to receive and relay signals of stress and our recent data on the projection pattern of the PPG neurons to the spinal cord suggest a potential strong link with the sympathetic nervous system. We review here the role of central GLP-1 in the regulation of stress responses and discuss the potential involvement of the endogenous source of GLP-1 within the brain, the PPG neurons.

Subjects: Cardiovascular; Endocrinology; Neuroscience

Keywords: glucagon-like peptide-1; preproglucagon neurons; stress; food intake; hypothalamus-pituitary-adrenal axis; sympathetic nervous system

1. GLP-1 is a regulator of homeostasis

Glucagon-like peptide-1 (GLP-1) is an incretin and neuropeptide best known for its role in glucose homeostasis and appetite regulation (Holst, 2007; Kreymann, Ghatei, Williams, & Bloom, 1987; Tang-Christensen et al., 1996; Turton et al., 1996; Wang et al., 1995). In the periphery, GLP-1 is released from L cells in the gut following ingestion of food (Vilsbøll et al., 2003). From the blood it reaches the pancreas where it acts on β-cells to enhance the secretion of insulin in response to

ABOUT THE AUTHORS

Marie K. Holt is pursuing her PhD in Stefan Trapp's group at University College London, UK. The focus of her research is autonomic and metabolic neuroscience with a special interest in the brain GLP-1 system. Her research aims to further our understanding of the physiological role of GLP-1 in the brain and in particular the source of GLP-1 in the brain, the preproglucagon (PPG) neurons. The main focus is on the role of the PPG neurons in both food intake and cardiovascular control. These studies are performed both *in vitro* and *in vivo* on a cellular level and in the context of the whole animal using a combination of viral gene transfer and transgenic mouse strains.

PUBLIC INTEREST STATEMENT

Glucagon-like peptide-1 analogues are arguably the most promising new drugs in the fight against diabetes and obesity. In order to fully understand the therapeutic potential as well as the associated risks for these drugs, a thorough understanding of the effects of GLP-1 in the brain has to be obtained. This review focuses on one important aspect, the potential role of central GLP-1 in stress responses.

glucose while inhibiting the release of glucagon from α-cells (de Heer, Rasmussen, Coy, & Holst, 2008; Holst, 2007; Ørskov, Holst, & Nielsen, 1988; Vilsbøll, Krarup, Madsbad, & Holst, 2003). In addition, peripheral GLP-1 has proliferative and protective effects on islet cells and inhibits gastric emptying (Egan, Bulotta, Hui, & Perfetti, 2003; Farilla et al., 2002; Nauck et al., 1997). Central GLP-1, here defined as GLP-1 acting within the central nervous system (CNS), is well-established as a potent regulator of food intake (Barrera et al., 2011; Larsen, Tang-Christensen, & Jessop, 1997; Turton et al., 1996; Williams, Baskin, & Schwartz, 2009). Within the brain, GLP-1 is produced in a subset of granule cells or short axon cells of the olfactory bulb, some pyramidal cells of the piriform cortex and a few neurons in the lumbar-sacral spinal cord (Larsen, Tang-Christensen, Holst, & Ørskov, 1997; Merchenthaler, Lane, & Shughrue, 1999; Thiebaud et al., 2016; Zheng, Cai, & Rinaman, 2015). However, the primary source of GLP-1 in the brain is in preproglucagon (PPG) neurons in the nucleus tractus solitarii (NTS) and the intermediate reticular nucleus in the lower brainstem (Merchenthaler et al., 1999). PPG neurons, also referred to as GLP-1 neurons, particularly in studies on rat, project throughout the brain to autonomic control centres and this projection pattern largely matches the expression of GLP-1 receptors in the brain (Larsen, Tang-Christensen, Holst, et al., 1997; Llewellyn-Smith, Gnanamanickam, Reimann, Gribble, & Trapp, 2013; Llewellyn-Smith, Reimann, Gribble, & Trapp, 2011; Merchenthaler et al., 1999; Trapp & Cork, 2015; Vrang, Hansen, Larsen, & Tang-Christensen, 2007). PPG neurons are activated *in vitro* and *in vivo* by a range of satiety hormones and peripheral signals relating to food intake and general homeostasis (Hisadome, Reimann, Gribble, & Trapp, 2010, 2011; Merchenthaler et al., 1999; Rinaman, 1999b; Trapp & Richards, 2013).

It is clear that central GLP-1 reduces food intake. What is less clear is the physiological purpose and cause of this anorexic response. Importantly, a reduction in food intake can be a response to not only decreased energy demand, but also to changes in the emotional state that reduces the motivation to eat, or to visceral malaise leading to reduced appetite. Early studies addressed the role of taste aversion and nausea in the regulation of food intake. NTS GLP-1 neurons were activated by intraperitoneal injection of LiCl, a compound which is well known to cause malaise and taste aversion (Rinaman, 1999b). Furthermore, blockade of the GLP-1 receptor using the antagonist Exendin (9-39) reversed the LiCl-induced suppression of appetite in rat, suggesting a role for central GLP-1 in the response to malaise (Rinaman, 1999a). However, these results could not be reproduced in mouse (Lachey et al., 2005), suggesting that subtle, but important species differences may exist in the central GLP-1 system. In another early attempt to anatomically dissect different GLP-1 actions in the brain, van Dijk and Thiele (1999) demonstrated that bilateral lesions in the PVN prevented the induction of satiety by GLP-1, but not the conditioned taste aversion observed with GLP-1. They also showed that lesions in the amygdala prevented GLP-1 induced taste aversion, but rats retained the GLP-1 induced reduction in food intake. These experiments clearly demonstrated the existence of two separate pathways for GLP-1 effects on satiety and malaise. As further evidence for a role of GLP-1 in the response to general malaise, there is now data supporting a link between inflammation and GLP-1 mediated reduction in food intake and the cytokine interleukin-6 (IL-6) has been shown to activate PPG neurons in the NTS (Anesten et al., 2016; Shirazi et al., 2013). More recent studies have revealed that GLP-1 receptor signalling in the mesolimbic system affects food intake by modulating reward pathways (Dickson et al., 2012; Mietlicki-Baase et al., 2013, 2014) and that activation of GLP-1 receptors in the hippocampus, a region not traditionally associated with appetite regulation, reduces food intake (During et al., 2003; Hsu, Hahn, Konanur, Lam, & Kanoski, 2015).

From all these emerging targets it is becoming evident that GLP-1 in the brain does not simply inhibit metabolically driven food intake, but that the reduced appetite may be part of a wider response to emotional stress or visceral malaise (Ghosal, Myers, & Herman, 2013; Kinzig et al., 2003; Kreisler & Rinaman, 2016; Maniscalco, Kreisler, & Rinaman, 2012; Maniscalco, Zheng, Gordon, & Rinaman, 2015; Rinaman, 1999b). In this review we discuss the evidence for the involvement of central GLP-1 in the regulation of stress and consider the potential role of the central source of GLP-1, the PPG neurons. Most evidence has been gathered in mouse and rat and since few anatomical and functional differences have been observed between species (including human and non-human primates), we assume here that most findings are relevant across species (Vrang & Grove, 2011;

Zheng et al., 2015), though any paradigms where conflicting evidence exists between species, will be highlighted. We begin by briefly describing the organism's response to stress. In the next section, we review the evidence for a role for GLP-1 in the regulation of stress responses and finally we discuss the potential role of the central source of GLP-1, the PPG neurons, in the stress response.

2. Stress activates two parallel coping systems, the hypothalamic-pituitary axis and the sympathetic nervous system

Stress is defined as the collection of physiological responses to homeostatic (physical) and psychological (perceived) challenges (Dayas, Buller, Crane, Xu, & Day, 2001; Sawchenko, Li, & Ericsson, 2000; Ulrich-Lai & Herman, 2009). Stress allows the organism to cope with aversive stimuli and appropriate stress responses are essential to the survival of the organism. On the other hand, inappropriate chronic stress responses can lead to long-term damaging disorders such as anxiety and depression. Neural control of stress responses is complex and involves forebrain, brainstem and spinal cord circuits that ultimately converge to activate two important effectors in the body's response to stress: the sympathetic part of the autonomic nervous system (SNS) and the hypothalamus-pituitary-adrenal (HPA) axis (Ulrich-Lai & Herman, 2009) producing the so-called "fight-or-flight" response (Figure 1). A perceived threat to homeostasis and the well-being of the organism leads to rapid activation of the SNS. Heart and blood pressure are increased via recruitment of catecholaminergic neurons in the rostral ventrolateral medulla (RVLM), the raphe pallidus and sympathetic preganglionic neurons in the spinal cord as well as higher order autonomic control sites in the hypothalamus and amygdala (Dampney, 1994; Dayas et al., 2001; Dimicco & Zaretsky, 2007). Increased sympathetic activity leads to release of adrenaline from the adrenal medulla and adrenaline in turn increases cardiac output and respiratory rate while redirecting blood flow to skeletal muscle and mobilising glucose from liver and skeletal muscle (Ulrich-Lai & Herman, 2009). The parallel recruitment of the HPA axis involves the activation of parvocellular neurons in the paraventricular nucleus of the hypothalamus (PVN). These neurons release corticotrophin-releasing hormone (CRH) onto adrenocorticotropic hormone (ACTH) expressing neurons in the anterior pituitary. From the pituitary, ACTH is released into the bloodstream through which it reaches the cortex of the adrenal gland. In the adrenal cortex ACTH elicits release of corticosterone, which works to mobilise glucose by stimulating gluconeogenesis securing the body's demand for glucose during homeostatic challenges (Ulrich-Lai & Herman, 2009). Corticosterone provides negative feedback at the level of the pituitary and the hypothalamus to limit HPA activity (Ulrich-Lai & Herman, 2009).

3. GLP-1 activates both the HPA axis and the sympathetic nervous system

The majority of evidence for a role of GLP-1 in stress has been gathered using supraphysiological doses of GLP-1 or GLP-1 analogues (typically Exendin-4) to activate GLP-1 receptors both peripherally and centrally. These studies have shown that recruitment of central GLP-1 receptors potently activates the HPA axis in both humans and rodents with a resulting increase in both ACTH and corticosterone/cortisol concentrations in blood (Gil-Lozano et al., 2010; Kinzig et al., 2003). The adrenal cortex does not express GLP-1 receptors and isolated cells from the adrenal glands do not release corticosterone in response to GLP-1 ruling out the possibility that GLP-1 acts directly on the adrenal cortex (Dunphy, Taylor, & Fuller, 1998; Gil-Lozano et al., 2014). Importantly, central administration of Exendin-4 leads to an increase in corticosterone in rodents, suggesting an involvement of central GLP-1 receptors possibly expressed on CRH expressing neurons in the hypothalamus (Gil-Lozano et al., 2014; Larsen, Tang-Christensen, & Jessop, 1997; Sarkar, Fekete, Légrádi, & Lechan, 2003). In fact, central blockade of CRH receptors blocks Exendin-4-induced increases in ACTH and corticosterone, establishing a role for central GLP-1 receptors in HPA axis regulation (Gil-Lozano et al., 2014).

In a study investigating the involvement of central GLP-1 in both homeostatic and psychogenic stress, Kinzig et al. (2003) found that injections of GLP-1 into the PVN increased blood ACTH and corticosterone concentrations. Targeting the amygdala increased anxiety-like behaviour with animals spending significantly less time in the open arms of an elevated plus maze after infusion of GLP-1. Different types of stressors, i.e. homeostatic vs. psychogenic, are known to activate distinct neural pathways and the data described above suggests that GLP-1 regulates both homeostatic and psychogenic stress responses through distinct neural pathways (Dayas et al., 2001; Kinzig et al., 2003).

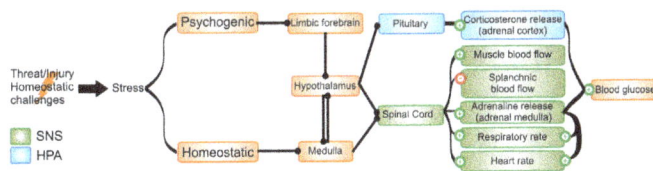

Figure 1. Stressors activate forebrain and brainstem regions to prepare for "fight-or-flight".

Notes: Acute injury or a perceived threat leads to activation of psychogenic and/or homeostatic stress. Psychogenic stress recruits areas in the limbic areas of the forebrain, including the hippocampus and the amygdala, which influence the activity of hypothalamic nuclei. A homeostatic challenge activates brainstem nuclei, which either relay this information to hypothalamic regions or directly to sympathetic preganglionic neurons in the spinal cord. The hypothalamus receives modulatory input from forebrain and medullary nuclei and is the main driver of the HPA axis (in blue) and the sympathetic nervous system (SNS, in green). Release of corticotropin-releasing hormone (CRH) from parvocellular neurons in the paraventricular nucleus of the hypothalamus on cells expressing adrenocorticotropic hormone (ACTH) leads to release of ACTH into the bloodstream via the blood, ACTH reaches the adrenal cortex where it stimulates the release of corticosterone (cortisol in humans). Corticosterone mobilises glucose by increasing gluconeogenesis and by counteracting the effects of insulin. Parallel activation of the SNS (green) leads to release of adrenalin from postganglionic sympathetic neurons onto target tissues, including blood vessels, the adrenal medulla, heart and the respiratory system. This leads to increased muscle blood flow and decreased blood flow to the organs in the abdominal cavity (splanchnic blood flow) ensuring sufficient oxygen and energy supply to the muscles. The secretion of adrenaline from the adrenal medulla into the blood further increases heart and respiratory rate and the mobilisation of glucose through glycogenolysis and lipolysis.

Intriguingly, activation of central GLP-1 receptors not only stimulates the HPA axis, but also appears to increase sympathetic activity, the other important pathway for the physiological response to stress (Smits et al., 2016; Yamamoto et al., 2002). This is measured as an increase in heart rate in both rodents and humans following GLP-1 receptor activation (Gil-Lozano et al., 2014; Griffioen et al., 2011; Robinson, Holt, Rees, Randeva, & O'Hare, 2013; Smits et al., 2016; Yamamoto et al., 2002). Central GLP-1 receptor activation was found to stimulate both autonomic regulatory neurons, neurons in the spinal cord and cells in the adrenal medulla signifying clear recruitment of the sympathetic nervous system (Yamamoto et al., 2002). This suggests that central GLP-1 could activate both arms of the stress response, the HPA axis and the sympathetic nervous system.

4. NTS GLP-1 neurons are ideally situated to integrate signals of stress

Studies using supraphysiological activation of GLP-1 receptors provide little information about the physiological role of the endogenous GLP-1 system in stress regulation. Exogenous activation of central GLP-1 receptors clearly triggers release of ACTH and corticosterone, but the question remains, whether there is an endogenous source of GLP-1 eliciting these responses under physiological conditions? Further, which signals trigger release of GLP-1 from that source? GLP-1 released from L cells in the gut is rapidly degraded in the liver and bloodstream making it unlikely that GLP-1 reaches receptors in the brain in large quantities (Deacon, 2004; Hansen, Deacon, Ørskov, & Holst, 1999; Holst & Deacon, 2005; Kieffer, McIntosh, & Pederson, 1995; Vilsboll, Krarup, Deacon, Madsbad, & Holst, 2001). In contrast, the central source of GLP-1, the PPG neurons, are ideally situated in the NTS in the caudal brainstem to receive and process signals of stress from the rest of the body (Kreisler & Rinaman, 2016; Maniscalco et al., 2012; Merchenthaler et al., 1999; Rinaman, 1999b; Vrang, Phifer, Corkern, & Berthoud, 2003). The NTS is a well-established central site of integration of visceral afferent signals concerning general homeostasis, which are relayed to higher brain centres (Grill & Hayes, 2012). From the NTS, PPG neurons project to autonomic control sites throughout the brain including the PVN, the dorsomedial hypothalamus and the RVLM, which are all involved in the control of the HPA axis and/or sympathetic activity (Larsen, Tang-Christensen, Holst, et al., 1997; Llewellyn-Smith et al., 2011, 2013; Vrang et al., 2007). In the PVN, there is dense expression of GLP-1 receptors and GLP-1 immunoreactive axons make contact with parvocellular CRH producing neurons, supporting a role for PPG neurons in the regulation of CRH secretion from the hypothalamus (Cork et al., 2015; Larsen, Tang-Christensen, Holst, et al., 1997; Sarkar et al., 2003).

NTS neurons are generally thought to regulate sympathetic activity indirectly through ascending projections to either the RVLM or hypothalamic nuclei. Recent findings in our laboratory have demonstrated that PPG neurons in the brainstem not only send ascending projections to autonomic control sites mainly in the hypothalamus, but also directly innervate sympathetic preganglionic neurons in the spinal cord (Llewellyn-Smith et al., 2015). These data highlight the possibility that the central GLP-1 system may integrate incoming stress signals and relay them via both ascending projections to CRH neurons in the PVN to elicit HPA activity and descending projections to preganglionic sympathetic neurons in the intermediolateral column (IML) and central autonomic area (CAA) of the spinal cord to increase sympathetic outflow.

In a thorough dissection of HPA-GLP-1 crosstalk, Lee et al. (2016) explored the neural pathways underlying GLP-1 receptor initiated increases in corticosterone in rat. Systemic (intraperitoneal; i.p.) exendin-4 activated catecholaminergic (CA), non-PPG neurons in the NTS and RVLM. Most of these neurons were found to project to the parvocellular and magnocellular PVN. Selective ablation of this CA-PVN connection using DBH-saporin prevented the i.p. exendin-4 induced increase in blood concentrations of corticosterone. This demonstrated that activation of the HPA axis by systemic exendin-4 is dependent on CA input to the PVN. On first sight this seems surprising given that PPG neurons project heavily to the PVN and make contacts to CRH neurons as discussed above. However, keeping in mind that PPG neurons do not express GLP-1 receptors, these findings might just indicate that the peripheral and central GLP-1 systems are more separate than widely thought. It also emphasises that still more studies are needed that explore which exact peripheral signals activate PPG neurons and which do not.

5. The central GLP-1 system is activated in response to both homeostatic and psychogenic stress

An early study suggesting a link between central GLP-1 and stress was conducted by Rinaman (1999b). Interoceptive stress was induced through intraperitoneal injection of LiCl. LiCl is considered a nauseogenic agent and is known to reduce food intake while increasing the concentration of stress hormones ACTH and corticosterone in the blood (Kinzig, Hargrave, & Honors, 2008). LiCl was found to activate GLP-1 neurons in the NTS which were found to project to the HPA-regulating parvocellular region of the PVN (Rinaman, 1999b). Kinzig et al. later demonstrated that the LiCl-induced increase in stress hormones is dependent on central GLP-1 signalling (Kinzig et al., 2003). Third ventricular infusion of a GLP-1 receptor antagonist abolished the increase in both ACTH and corticosterone following systemic LiCl injections. The discovery of close appositions from GLP-1 immunoreactive axon terminals on CRH producing PVN neurons in rat further substantiate these findings (Tauchi, Zhang, D'Alessio, Stern, & Herman, 2008). Similarly, axons of mouse PPG neurons have close appositions on CRH producing PVN neurons (personal communication, Ida Llewellyn-Smith). These data suggest that homeostatic stress following a toxic challenge activates the central GLP-1 system to recruit systemic stress pathways.

Similarly, acute psychogenic stress induced by physical restraint reduces food intake and activates the HPA axis (Kinzig et al., 2008). Maniscalco et al. recently demonstrated that the number of cFOS positive GLP-1 neurons increases following 30mins restraint stress or 5mins exposure on an elevated platform, suggesting that GLP-1 neurons are activated by psychogenic stress (Maniscalco et al., 2015). Furthermore, they found that 30 mins restraint stress reduced food intake at the onset of dark phase and that this hypophagic response was blocked by central infusion of GLP-1 receptor antagonist, suggesting that psychogenic stress recruits the central GLP-1 system to reduce food intake (Maniscalco et al., 2015).

Finally, in a study focusing on the role of GLP-1 in cocaine addiction, GLP-1 neurons were found to be activated by an injection of corticosterone into the fourth ventricle (Schmidt et al., 2016). Fourth ventricle corticosterone reduced cocaine self-administration and this reduction was blocked by GLP-1 receptor antagonism in the ventral tegmental area (Schmidt et al., 2016). These data suggest that not only does central GLP-1 activate the HPA axis, but corticosterone in turn activates the central GLP-1 system.

6. Conclusions

It is clear that the central GLP-1 system plays a role in the regulation of food intake. However, increasing numbers of studies report effects of GLP-1 in brain regions not classically associated with appetite control and it is becoming increasingly clear that central GLP-1 may be responsible for much wider homeostatic control. In particular, the anorexic effects of GLP-1 may in some cases be secondary to responses to homeostatic and psychogenic stress.

We have discussed here evidence for a role of central GLP-1 in the regulation of the body's stress response. It is clear that overactivation of brain GLP-1 receptors enhances secretion of stress hormones and activity of the SNS. In contrast, the role of the central source of GLP-1, the PPG neurons, is less explored and the neural pathways underlying the GLP-1 mediated modulation of stress are largely unknown. The evidence discussed here suggests a model in which peripheral signals of homeostatic and psychogenic stress activate PPG neurons (Figure 2). The PPG neurons are ideally situated in the NTS to integrate signals of stress and relay that signal on to parvocellular neurons in the PVN. Furthermore, in this model psychogenic or homeostatic stress would lead to release of GLP-1 from PPG neurons. Downstream activation of GLP-1 receptors then increases sympathetic activity via direct and indirect pathways. Directly, PPG neurons modulate the activity of sympathetic preganglionic neurons in the spinal cord. Indirectly, GLP-1 from PPG neurons activates presympathetic RVLM and PVN neurons which in turn project to spinal sympathetic preganglionic neurons.

The evidence discussed here suggests that central GLP-1 does not simply regulate food intake in response to changes in energy demand, but that the PPG neurons are activated by stress and that central GLP-1 modulates acute stress, homeostatic or psychogenic, by increasing corticosterone, mobilising glucose and increasing heart rate, allowing the organism to cope with potential threats.

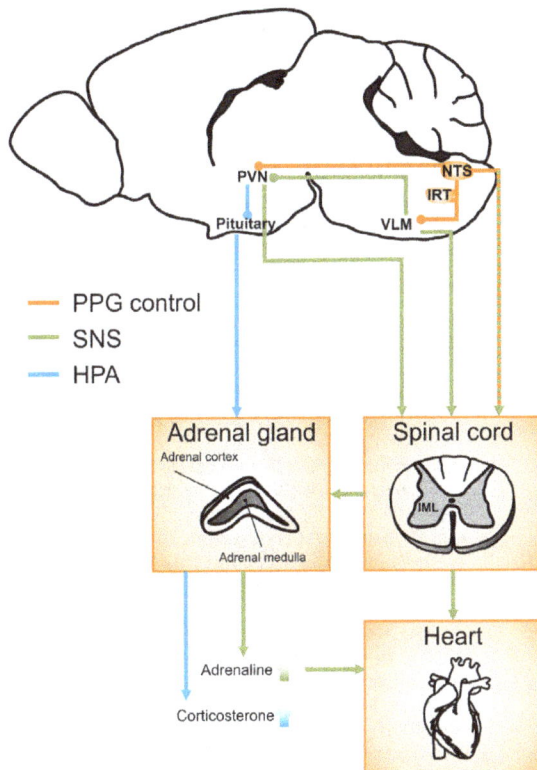

Figure 2. PPG pathways to activate both HPA axis and sympathetic nervous system in the control of stress responses.

Notes: PPG neurons in the nucleus of the solitary tract (NTS) and the intermediate reticular nucleus (IRT) are activated by stressors. Activation of PPG neurons leads to release of GLP-1 (orange) onto parvocellular cells in the PVN, which in turn activate the HPA axis (blue arrows) by stimulating cells in the pituitary to release adrenocorticotropic hormone (ACTH). ACTH acts on the adrenal cortex to increase the secretion of corticosterone. In parallel, PPG neurons send descending axons directly to spinal sympathetic preganglionic neurons in the intermediolateral column (IML) and ascending axons to presympathetic neurons in VLM and PVN. Recruitment of PVN and VLM neurons also leads to activation of the sympathetic nervous system (green arrows) via descending fibres to the sympathetic preganglionic neurons in the IML in the spinal cord, which in turn increase heart rate and stimulate the release of adrenaline from the adrenal medulla.

Funding
Work in our laboratory is funded by the Medical Research Council, UK [grant number MR/N02589X/1]; the British Heart Foundation [grant number FS/14/43/30960]. MKH holds a University College London Graduate Research Scholarship.

Competing Interests
The authors declare no competing interest.

Author details
Marie K. Holt[1]
E-mail: m.holt@ucl.ac.uk
Stefan Trapp[1]
E-mail: s.trapp@ucl.ac.uk
ORCID ID: http://orcid.org/0000-0003-0665-4948
[1] Centre for Cardiovascular and Metabolic Neuroscience, Department of Neuroscience, Physiology & Pharmacology, University College London, WC1E 6BT London, UK.

References

Anesten, F., Holt, M. K., Schéle, E., Pálsdóttir, V., Reimann, F., Gribble, F. M., ... Jansson, J. O. (2016). Preproglucagon neurons in the hindbrain have IL-6 receptor-α and show Ca 2+ influx in response to IL-6. *American Journal of Physiology - Regulatory, Integrative and Comparative Physiology, 311*, R115–R123.
http://dx.doi.org/10.1152/ajpregu.00383.2015

Barrera, J. G., Jones, K. R., Herman, J. P., D'Alessio, D. A., Woods, S. C., & Seeley, R. J. (2011). Hyperphagia and increased fat accumulation in two models of chronic cns glucagon-like peptide-1 loss of function. *Journal of Neuroscience, 31*, 3904–3913. http://dx.doi.org/10.1523/JNEUROSCI.2212-10.2011

Cork, S. C., Richards, J. E., Holt, M. K., Gribble, F. M., Reimann, F., & Trapp, S. (2015). Distribution and characterisation of glucagon-like peptide-1 receptor expressing cells in the mouse brain. *Molecular Metabolism, 4*, 718–731.
http://dx.doi.org/10.1016/j.molmet.2015.07.008

Dampney, R. A. (1994). Functional organization of central pathways regulating the cardiovascular system. *Physiological Reviews, 74*, 323–364.

Dayas, C. V., Buller, K. M., Crane, J. W., Xu, Y., & Day, T. A. (2001). Stressor categorization: Acute physical and psychological stressors elicit distinctive recruitment patterns in the amygdala and in medullary noradrenergic cell groups. *European Journal of Neuroscience, 14*, 1143–1152.
http://dx.doi.org/10.1046/j.0953-816x.2001.01733.x

Deacon, C. F. (2004). Circulation and degradation of GIP and GLP-1. *Hormone and Metabolic Research, 36*, 761–765. http://dx.doi.org/10.1055/s-2004-826160

de Heer, J., Rasmussen, C., Coy, D. H., & Holst, J. J. (2008). Glucagon-like peptide-1, but not glucose-dependent insulinotropic peptide, inhibits glucagon secretion via somatostatin (receptor subtype 2) in the perfused rat pancreas. *Diabetologia, 51*, 2263–2270.
http://dx.doi.org/10.1007/s00125-008-1149-y

Dickson, S. L., Shirazi, R. H., Hansson, C., Bergquist, F., Nissbrandt, H., & Skibicka, K. P. (2012). The glucagon-like peptide 1 (glp-1) analogue, exendin-4, decreases the rewarding value of food: A new role for mesolimbic glp-1 receptors. *Journal of Neuroscience, 32*, 4812–4820.
http://dx.doi.org/10.1523/JNEUROSCI.6326-11.2012

Dimicco, J. A., & Zaretsky, D. V. (2007). The dorsomedial hypothalamus: A new player in thermoregulation. *American Journal of Physiology - Regulatory, Integrative and Comparative Physiology, 292*, R47–63.

Dunphy, J. L., Taylor, R. G., & Fuller, P. J. (1998). Tissue distribution of rat glucagon receptor and GLP-1 receptor gene expression. *Molecular and Cellular Endocrinology, 141*, 179–186.
http://dx.doi.org/10.1016/S0303-7207(98)00096-3

During, M. J., Cao, L., Zuzga, D. S., Francis, J. S., Fitzsimons, H. L., Jiao, X., ... Haile, C. N. (2003). Glucagon-like peptide-1 receptor is involved in learning and neuroprotection. *Nature Medicine, 9*, 1173–1179.
http://dx.doi.org/10.1038/nm919

Egan, J. M., Bulotta, A., Hui, H., & Perfetti, R. (2003). GLP-1 receptor agonists are growth and differentiation factors for pancreatic islet beta cells. *Diabetes/Metabolism Research and Reviews, 19*, 115–123.
http://dx.doi.org/10.1002/(ISSN)1520-7560

Farilla, L., Hui, H., Bertolotto, C., Kang, E., Bulotta, A., Di Mario, U., & Perfetti, R. (2002). Glucagon-like peptide-1 promotes islet cell growth and inhibits apoptosis in zucker diabetic rats. *Endocrinology, 143*, 4397–4408.
http://dx.doi.org/10.1210/en.2002-220405

Ghosal, S., Myers, B., & Herman, J. P. (2013). Role of central glucagon-like peptide-1 in stress regulation. *Physiology & Behavior, 122*, 201–207.
http://dx.doi.org/10.1016/j.physbeh.2013.04.003

Gil-Lozano, M., Pérez-Tilve, D., Alvarez-Crespo, M., Martís, A., Fernandez, A. M., Catalina, P. A. F., ... Mallo, F. (2010). GLP-1(7-36)-amide and exendin-4 stimulate the HPA axis in rodents and humans. *Endocrinology, 151*, 2629–2640.
http://dx.doi.org/10.1210/en.2009-0915

Gil-Lozano, M., Romaní-Pérez, M., Outeiriño-Iglesias, V., Vigo, E., González-Matías, L. C., Brubaker, P. L., & Mallo, F. (2014). Corticotropin-releasing hormone and the sympathoadrenal system are major mediators in the effects of peripherally administered exendin-4 on the hypothalamic-pituitary-adrenal axis of male rats. *Endocrinology, 155*, 2511–2523.
http://dx.doi.org/10.1210/en.2013-1718

Griffioen, K. J., Wan, R., Okun, E., Wang, X., Lovett-Barr, M. R., Li, Y., ... Mattson, M. P. (2011). GLP-1 receptor stimulation depresses heart rate variability and inhibits neurotransmission to cardiac vagal neurons. *Cardiovascular Research, 89*, 72–78.
http://dx.doi.org/10.1093/cvr/cvq271

Grill, H. J., & Hayes, M. R. (2012). Hindbrain neurons as an essential hub in the neuroanatomically distributed control of energy balance. *Cell Metabolism, 16*, 296–309.
http://dx.doi.org/10.1016/j.cmet.2012.06.015

Hansen, L., Deacon, C. F., Ørskov, C., & Holst, J. J. (1999). Glucagon-like peptide-1-(7-36)amide is transformed to glucagon-like peptide-1-(9-36)amide by dipeptidyl peptidase IV in the capillaries supplying the L cells of the porcine intestine. *Endocrinology, 140*, 5356–5363.

Hisadome, K., Reimann, F., Gribble, F. M., & Trapp, S. (2010). Leptin directly depolarizes preproglucagon neurons in the nucleus tractus solitarius: Electrical properties of glucagon-like peptide 1 neurons. *Diabetes, 59*, 1890–1898.
http://dx.doi.org/10.2337/db10-0128

Hisadome, K., Reimann, F., Gribble, F. M., & Trapp, S. (2011). CCK stimulation of GLP-1 neurons involves α1-adrenoceptor-mediated increase in glutamatergic synaptic inputs. *Diabetes, 60*, 2701–2709.
http://dx.doi.org/10.2337/db11-0489

Holst, J. J. (2007). The physiology of glucagon-like peptide 1. *Physiological Reviews, 87*, 1409–1439.
http://dx.doi.org/10.1152/physrev.00034.2006

Holst, J. J., & Deacon, C. F. (2005). Glucagon-like peptide-1 mediates the therapeutic actions of DPP-IV inhibitors. *Diabetologia, 48*, 612–615.
http://dx.doi.org/10.1007/s00125-005-1705-7

Hsu, T. M., Hahn, J. D., Konanur, V. R., Lam, A., & Kanoski, S. E. (2015). Hippocampal GLP-1 receptors influence food intake, meal size, and effort-based responding for food through volume transmission. *Neuropsychopharmacology, 40*, 327–337. http://dx.doi.org/10.1038/npp.2014.175

Kieffer, T. J., McIntosh, C. H., & Pederson, R. A. (1995). Degradation of glucose-dependent insulinotropic polypeptide and truncated glucagon-like peptide 1 *in vitro* and *in vivo* by dipeptidyl peptidase IV. *Endocrinology, 136*, 3585–3596.

Kinzig, K. P., D'Alessio, D. A., Herman, J. P., Sakai, R. R., Vahl, T. P., Figueiredo, H. F., ... Seeley, R. J. (2003). CNS glucagon-like peptide-1 receptors mediate endocrine and anxiety responses to interoceptive and psychogenic stressors. *The Journal of Neuroscience, 23*, 6163–6170.

Kinzig, K. P., Hargrave, S. L., & Honors, M. A. (2008). Binge-type eating attenuates corticosterone and hypophagic responses to restraint stress. *Physiology & Behavior, 95*, 108–113. http://dx.doi.org/10.1016/j.physbeh.2008.04.026

Kreisler, A. D., & Rinaman, L. (2016). Hindbrain glucagon-like peptide-1 neurons track intake volume and contribute to injection stress-induced hypophagia in meal-entrained rats. *American Journal of Physiology - Regulatory, Integrative and Comparative Physiology, 310*, R906–R916. http://dx.doi.org/10.1152/ajpregu.00243.2015

Kreymann, B., Ghatei, M.A., Williams, G., & Bloom, S. R. (1987). Glucagon-like peptide-1 7-36: A physiological incretin in man. *The Lancet, 330*, 1300–1304. http://dx.doi.org/10.1016/S0140-6736(87)91194-9

Lachey, J. L., D'Alessio, D. A., Rinaman, L., Elmquist, J. K., Drucker, D. J., & Seeley, R. J. (2005). The role of central glucagon-like peptide-1 in mediating the effects of visceral illness: Differential effects in rats and mice. *Endocrinology, 146*, 458–462. http://dx.doi.org/10.1210/en.2004-0419

Larsen, P. J., Tang-Christensen, M., Holst, J. J., & Ørskov, C. (1997). Distribution of glucagon-like peptide-1 and other preproglucagon-derived peptides in the rat hypothalamus and brainstem. *Neuroscience, 77*, 257–270. http://dx.doi.org/10.1016/S0306-4522(96)00434-4

Larsen, P. J., Tang-Christensen, M., & Jessop, D. S. (1997). Central administration of glucagon-like peptide-1 activates hypothalamic neuroendocrine neurons in the rat. *Endocrinology, 138*, 4445–4455.

Lee, S. J., Diener, K., Kaufman, S., Krieger, J.-P., Pettersen, K. G., Jejelava, N., ... Langhans, W. (2016). Limiting glucocorticoid secretion increases the anorexigenic property of Exendin-4. *Molecular Metabolism, 5*, 552–565. http://dx.doi.org/10.1016/j.molmet.2016.04.008

Llewellyn-Smith, I. J., Gnanamanickam, G. J., Reimann, F., Gribble, F. M., & Trapp, S. (2013). Preproglucagon (PPG) neurons innervate neurochemicallyidentified autonomic neurons in the mouse brainstem. *Neuroscience, 229*, 130–143. http://dx.doi.org/10.1016/j.neuroscience.2012.09.071

Llewellyn-Smith, I. J., Marina, N., Manton, R. N., Reimann, F., Gribble, F. M., & Trapp, S. (2015). Spinally projecting preproglucagon axons preferentially innervate sympathetic preganglionic neurons. *Neuroscience, 284*, 872–887. http://dx.doi.org/10.1016/j.neuroscience.2014.10.043

Llewellyn-Smith, I. J., Reimann, F., Gribble, F. M., & Trapp, S. (2011). Preproglucagon neurons project widely to autonomic control areas in the mouse brain. *Neuroscience, 180*, 111–121. http://dx.doi.org/10.1016/j.neuroscience.2011.02.023

Maniscalco, J. W., Kreisler, A. D., & Rinaman, L. (2012). Satiation and stress-induced hypophagia: Examining the role of hindbrain neurons expressing prolactin-releasing peptide or glucagon-like peptide 1. *Frontiers in Neuroscience, 6*, 199.

Maniscalco, J. W., Zheng, H., Gordon, P. J., & Rinaman, L. (2015). Negative energy balance blocks neural and behavioral responses to acute stress by "silencing" central glucagon-like peptide 1 signaling in rats. *Journal of Neuroscience, 35*, 10701–10714. http://dx.doi.org/10.1523/JNEUROSCI.3464-14.2015

Merchenthaler, I., Lane, M., & Shughrue, P. (1999). Distribution of pre-pro-glucagon and glucagon-like peptide-1 receptor messenger RNAs in the rat central nervous system. *The Journal of Comparative Neurology, 403*, 261–280. http://dx.doi.org/10.1002/(ISSN)1096-9861

Mietlicki-Baase, E. G., Ortinski, P. I., Reiner, D. J., Sinon, C. G., McCutcheon, J. E., Pierce, R. C., ... Hayes, M. R. (2014). Glucagon-like peptide-1 receptor activation in the nucleus accumbens core suppresses feeding by increasing glutamatergic ampa/kainate signaling. *Journal of Neuroscience, 34*, 6985–6992. http://dx.doi.org/10.1523/JNEUROSCI.0115-14.2014

Mietlicki-Baase, E. G., Ortinski, P. I., Rupprecht, L. E., Olivos, D. R., Alhadeff, A. L., Pierce, R. C., & Hayes, M. R. (2013). The food intake-suppressive effects of glucagon-like peptide-1 receptor signaling in the ventral tegmental area are mediated by AMPA/kainate receptors. *AJP: Endocrinology and Metabolism, 305*, E1367–E1374. http://dx.doi.org/10.1152/ajpendo.00413.2013

Nauck, M. A., Niedereichholz, U., Ettler, R., Holst, J. J., Ørskov, C., Ritzel, R., & Schmiegel, W. H. (1997). Glucagon-like peptide 1 inhibition of gastric emptying outweighs its insulinotropic effects in healthy humans. *American Journal of Physiology, 273*, E981–988.

Ørskov, C., Holst, J. J., & Nielsen, O. V. (1988). Effect of truncated glucagon-like peptide-1 [proglucagon-(78–107) amide] on endocrine secretion from pig pancreas, antrum, and nonantral stomach. *Endocrinology, 123*, 2009–2013. http://dx.doi.org/10.1210/endo-123-4-2009

Rinaman, L. (1999a). A functional role for central glucagon-like peptide-1 receptors in lithium chloride-induced anorexia. *American Journal of Physiology, 277*, R1537–1540.

Rinaman, L. (1999b). Interoceptive stress activates glucagon-like peptide-1 neurons that project to the hypothalamus. *American Journal of Physiology, 277*, R582–590.

Robinson, L. E., Holt, T. A., Rees, K., Randeva, H. S., & O'Hare, J. P. (2013). Effects of exenatide and liraglutide on heart rate, blood pressure and body weight: Systematic review and meta-analysis. *BMJ Open, 3*. doi:10.1136/bmjopen-2012-001986

Sarkar, S., Fekete, C., Légrádi, G., & Lechan, R. M. (2003). Glucagon like peptide-1 (7-36) amide (GLP-1) nerve terminals densely innervate corticotropin-releasing hormone neurons in the hypothalamic paraventricular nucleus. *Brain Research, 985*, 163–168. http://dx.doi.org/10.1016/S0006-8993(03)03117-2

Sawchenko, P. E., Li, H. Y., & Ericsson, A. (2000). Circuits and mechanisms governing hypothalamic responses to stress: A tale of two paradigms. *Progress in Brain Research, 122*, 61–78. http://dx.doi.org/10.1016/S0079-6123(08)62131-7

Schmidt, H. D., Mietlicki-Baase, E. G., Ige, K. Y., Maurer, J. J., Reiner, D. J., Zimmer, D. J., ... Hayes, M. R. (2016). Glucagon-like peptide-1 receptor activation in the ventral tegmental area decreases the reinforcing efficacy of cocaine. *Neuropsychopharmacology, 41*, 1917–1928. http://dx.doi.org/10.1038/npp.2015.362

Shirazi, R., Palsdottir, V., Collander, J., Anesten, F., Vogel, H., Langlet, F., ... Skibicka, K. P. (2013). Glucagon-like peptide 1 receptor induced suppression of food intake, and body weight is mediated by central IL-1 and IL-6. *Proceedings of the National Academy of Sciences, 110*, 16199–16204. http://dx.doi.org/10.1073/pnas.1306799110

Smits, M. M., Muskiet, M. H., Tonneijck, L., Hoekstra, T., Kramer, M. H., Diamant, M., & van Raalte, D. H. (2016). Exenatide acutely increases heart rate in parallel with augmented sympathetic nervous system activation in healthy overweight males. *British Journal of Clinical Pharmacology, 81*, 613–620.
http://dx.doi.org/10.1111/bcp.v81.4

Tang-Christensen, M., Larsen, P. J., Goke, R., Fink-Jensen, A., Jessop, D. S., Moller, M., & Sheikh, S. P. (1996). Central administration of GLP-1-(7-36) amide inhibits food and water intake in rats. *American Journal of Physiology, 271*, R848–856.

Tauchi, M., Zhang, R., D'Alessio, D. A., Stern, J. E., & Herman, J. P. (2008). Distribution of glucagon-like peptide-1 immunoreactivity in the hypothalamic paraventricular and supraoptic nuclei. *Journal of Chemical Neuroanatomy, 36*, 144–149.
http://dx.doi.org/10.1016/j.jchemneu.2008.07.009

Thiebaud, N., Llewellyn-Smith, I. J., Gribble, F., Reimann, F., Trapp, S., & Fadool, D. A. (2016). The incretin hormone glucagon-like peptide 1 increases mitral cell excitability by decreasing conductance of a voltage-dependent potassium channel. *The Journal of Physiology, 594*, 2607–2628.
http://dx.doi.org/10.1113/tjp.2016.594.issue-10

Trapp, S., & Cork, S. C. (2015). PPG neurons of the lower brain stem and their role in brain GLP-1 receptor activation. *American Journal of Physiology - Regulatory, Integrative and Comparative Physiology, 309*, R795–R804.
http://dx.doi.org/10.1152/ajpregu.00333.2015

Trapp, S., & Richards, J. E. (2013). The gut hormone glucagon-like peptide-1 produced in brain: Is this physiologically relevant? *Current Opinion in Pharmacology, 13*, 964–969.
http://dx.doi.org/10.1016/j.coph.2013.09.006

Turton, M. D., O'Shea, D., Gunn, I., Beak, S. A., Edwards, C. M., Meeran, K., ... Bloom, S. R. (1996). A role for glucagon-like peptide-1 in the central regulation of feeding. *Nature, 379*, 69–72.
http://dx.doi.org/10.1038/379069a0

Ulrich-Lai, Y. M., & Herman, J. P. (2009). Neural regulation of endocrine and autonomic stress responses. *Nature Reviews Neuroscience, 10*, 397–409.
http://dx.doi.org/10.1038/nrn2647

van Dijk, G., & Thiele, T. E. (1999). Glucagon-like peptide-1 (7-36) amide: A central regulator of satiety and interoceptive stress. *Neuropeptides, 33*, 406–414.
http://dx.doi.org/10.1054/npep.1999.0053

Vilsboll, T., Krarup, T., Deacon, C. F., Madsbad, S., & Holst, J. J. (2001). Reduced postprandial concentrations of intact biologically active glucagon-like peptide 1 in type 2 diabetic patients. *Diabetes, 50*, 609–613.
http://dx.doi.org/10.2337/diabetes.50.3.609

Vilsbøll, T., Krarup, T., Madsbad, S., & Holst, J. J. (2003). Both GLP-1 and GIP are insulinotropic at basal and postprandial glucose levels and contribute nearly equally to the incretin effect of a meal in healthy subjects. *Regulatory Peptides, 114*, 115–121.
http://dx.doi.org/10.1016/S0167-0115(03)00111-3

Vilsbøll, T., Krarup, T., Sonne, J., Madsbad, S., Vølund, A., Juul, A. G., & Holst, J. J. (2003). Incretin secretion in relation to meal size and body weight in healthy subjects and people with type 1 and type 2 diabetes mellitus. *The Journal of Clinical Endocrinology & Metabolism, 88*, 2706–2713.
http://dx.doi.org/10.1210/jc.2002-021873

Vrang, N., & Grove, K. (2011). The brainstem preproglucagon system in a non-human primate (Macaca mulatta). *Brain Research, 1397*, 28–37.
http://dx.doi.org/10.1016/j.brainres.2011.05.002

Vrang, N., Hansen, M., Larsen, P. J., & Tang-Christensen, M. (2007). Characterization of brainstem preproglucagon projections to the paraventricular and dorsomedial hypothalamic nuclei. *Brain Research, 1149*, 118–126.
http://dx.doi.org/10.1016/j.brainres.2007.02.043

Vrang, N., Phifer, C. B., Corkern, M. M., & Berthoud, H. R. (2003). Gastric distension induces c-Fos in medullary GLP-1/2-containing neurons. *American Journal of Physiology - Regulatory, Integrative and Comparative Physiology, 285*, R470–R478.
http://dx.doi.org/10.1152/ajpregu.00732.2002

Wang, Z., Wang, R. M., Owji, A. A., Smith, D. M., Ghatei, M. A., & Bloom, S. R. (1995). Glucagon-like peptide-1 is a physiological incretin in rat. *Journal of Clinical Investigation, 95*, 417–421.
http://dx.doi.org/10.1172/JCI117671

Williams, D. L., Baskin, D. G., & Schwartz, M. W. (2009). Evidence that intestinal glucagon-like peptide-1 plays a physiological role in satiety. *Endocrinology, 150*, 1680–1687.
http://dx.doi.org/10.1210/en.2008-1045

Yamamoto, H., Lee, C. E., Marcus, J. N., Williams, T. D., Overton, J. M., Lopez, M. E., ... Elmquist, J. K. (2002). Glucagon-like peptide-1 receptor stimulation increases blood pressure and heart rate and activates autonomic regulatory neurons. *Journal of Clinical Investigation, 110*, 43–52.
http://dx.doi.org/10.1172/JCI0215595

Zheng, H., Cai, L., & Rinaman, L. (2015). Distribution of glucagon-like peptide 1-immunopositive neurons in human caudal medulla. *Brain Structure and Function, 220*, 1213–1219.
http://dx.doi.org/10.1007/s00429-014-0714-z

5

Estimation of *in vivo* neuropharmacological and *in vitro* antioxidant effects of *Tetracera sarmentosa*

Muhammad Moin Uddin Mazumdar[1], Md. Ariful Islam[1], Mohammad Tanvir Hosen[1], Mohammad Shahin Alam[1], Mohammad Nazmul Alam[1]*, Md. Faruk[1], Md. Mominur Rahman[1], Mohammed Abu Sayeed[1], Md. Masudur Rahman[1] and Shaikh Bokhtear Uddin[2]

*Corresponding author: Mohammad Nazmul Alam, Department of Pharmacy, International Islamic University Chittagong, Chittagong, Bangladesh
E-mail: nazmul_pharmacy@yahoo.com

Reviewing editor: Tsai-Ching Hsu, Chung Shan Medical University, Taiwan
Additional information is available at the end of the article

Abstract: To determine the *in vivo* neuropharmacological and *in vitro* antioxidant activities of methanolic extract of *Tetracera sarmentosa*. Open field (OFT), hole cross (HCT), thiopental-induced sleeping time (TIST), elevated plus-maze (EPMT) tests were used to determine the neuropharmacological activity and 1,1-diphenyl-2-picrylhydrazyl (DPPH) free radical scavenging assay, reducing power assay, total phenolic content tests were used to evaluate the *in vitro* antioxidant activity of *T. sarmentosa*. In the case of OFT and HCT, the extract showed a decrease in exploratory and locomotion activities at both dose levels (200 and 400 mg/kg body weight). In the thiopental-induced hypnosis test, 400 mg/kg dose of *T. sarmentosa* produced quick onset of sleep and prolonged duration of sleep than that of 200 mg/kg dose. *T. sarmentosa* extract showed the lessening percentage of entries of mice into the open arm and decreased percentage of time spent in open arm compared to the standard drug diazepam. In the case of DPPH scavenging activity, IC_{50} value of methanolic plant extract of *T. sarmentosa* is 151.56 µg/ml whereas the value of ascorbic acid is 23.53 µg/ml. In this current study, the phenolic content of *T. sarmentosa* was found to be 140.34 ± 1.56 GAE mg/gm dry extract. Results of this study revealed that methanolic extract of *T. sarmentosa* contains significant neuropharmacological and antioxidant activities.

ABOUT THE AUTHOR

Mohammad Nazmul Alam is the corresponding author of this experiment. He has completed his Master of Pharmacy degree with Thesis from the Department of Pharmaceutical Sciences, North South University and Bachelor of Pharmacy (Honors) degree from the Department of Pharmacy, International Islamic University Chittagong. His research interest is the exploration of Pharmacological activities, such as *in vivo* neuropharmacological effect, *in vitro* antioxidant effect of different plant extracts. He is also doing research work on Alzheimer's disease, Isoprenaline-induced cardiovascular dysfunction in aged rats, high fat diet-induced obesity in rats, CCl4-induced cirrhotic liver in rats. His work includes prevention of inflammation, oxidative stress, cardiac remodeling, obesity, dyslipidemia, fibrosis using various natural products and safer chemical compounds for living creatures.

PUBLIC INTEREST STATEMENT

Day by day the use of synthetic drugs is increasing which is responsible for various side effects as well as sometimes for hazardous adverse effects. Synthetic drugs are also expensive. So it is the high time to explore indigenous medicinal plants for the discovery of natural drug molecules which will be safer for human being and less expensive than synthetic drugs. Our approach is to discover new medicinal plant which will be a great alternative way for the treatment of various diseases. In this experiment, we are evaluating the neuropharmacological and antioxidant effects of *Tetracera sarmentosa*.

Subjects: Pharmacology; Medicine; Pharmaceutical Medicine

Keywords: *Tetracera sarmentosa*; neuropharmacology; OFT; HCT; TIST; EPMT; antioxidant; DPPH; reducing power; TPC

1. Introduction

Medicinal plants are the prominent source of secondary metabolites as well as active drug compounds. It plays a pivotal role for the discovery of new drug molecules. Additionally, medicinal plants contain different essential bioactive compounds such as antioxidants, alkaloids, flavonoids, saponins, steroids, terpenoids, polysaccharides, and so on which are the important part of modern and traditional medicines (Doughari, Ndakidemi, Human, & Benade, 2012; Mahboubi, Haghi, Kazempour, & Hatemi, 2013; Ming, Khang, Sai, & Fatt, 2003). *Tetracera sarmentosa* (Family: Dilleniaceae) is a scandent shrub which is generally found in Chittagong, Chittagong Hill Tracts, and Cox's Bazar (Bangladesh Ethnobotany Online Database, n.d.). It is also widely distributed through Southern China, India, Sri Lanka, Mayanmar, Thailand, Malaysia, and Indonesia (Flora of China, n.d.). Traditionally, the root of this plant extract is taken for the treatment of rheumatism (Bangladesh Ethnobotany Online Database, n.d.). In Sri Lanka, it is also used as a healing agent for the treatment of bone fracture. Antinociceptive activity of this plant was also found when aqueous leaf extract was tested on rats (Fernando, Ratnasooriya, & Deraniyagala, 2009). According to a research of chemical components of this plant by GC-MS, few essential oils were isolated from leaves of this plant. Among them hexadecanoic acid (41.59%), phytol (11.10%), and linoleic acid (5.08%) are notable chemical components (Da, Zhu, Zhao, Teng, & Gan, 2014). In this modern era, anxiety, stress, and depression are the most common form of mental disorder. Anxiety is a normal response to stress. But, when it becomes excessive, turns to a disastrous psychiatric disorder. Depression is an effect which shows reduction of mental interests and enjoyments through the absence of proper control. It may also show emotional and cognitive difficulties. It includes the central nervous system depression activity by showing sedative, hypnotic, and tranquilizer, and anxiolytic properties. Various methods like hole cross, open field, thiopental sodium-induced sleep, elevated plus-maze (EPM) tests are used to obtain the actual analytical result. Anxiety disorders affect about 40 million American adults age 18 years and older (about 18%). It is also the most common emotional disorders affecting people in all countries worldwide. It is reported that more than 20% of the adult population suffer from these conditions at some stage during their life (Abid, Hrishikeshavan, & Asad, 2006; Buller & Legrand, 2001; Titov, Andrews, Kemp, & Robinson, 2010; Yadav, Kawale, & Nade, 2008). Antioxidants provides great protection to the physiological system of our body through protecting cells from Reactive Oxygen Species (ROS), such as superoxide, peroxide, peroxynitrite, lipid peroxidation. On the contrary, antioxidants protect the cell from oxidative stress and notorious ROS that can cause great harm to the cell, destroy the cell, and causes various diseases, such as cardiovascular diseases, renal, hepatic diseases (Díaz-Muñoz, Álvarez-Pérez, & Yáñez, 2006; Rodrigo, Libuy, Feliú, & Hasson, 2013; Takimoto & Kass, 2007; Zweier & Talukder, 2006). In this study, methanolic plant extract of *T. sarmentosa* was evaluated to find its *in vivo* neuropharmacological and *in vitro* antioxidant activities.

2. Materials and methods

2.1. Plant material

The leaves of *T. sarmentosa* were collected from Hathazari area of Chittagong district, Bangladesh and authenticated by Dr Shaikh Bokhtear Uddin, associate professor, Department of Botany, University of Chittagong. A voucher specimen has been deposited at the Department of Pharmacy, International Islamic University, Chittagong, Bangladesh.

2.2. Preparation of extract

The leaves were dried for a period of two weeks under shade and ground, followed by grounding to course powder. The ground leaves (250 gm) were soaked in sufficient amount of methanol (1:3) for one week at room temperature with an occasional shaking and stirring then filtered through a cotton plug followed by Whitman filter paper (NO. 1). The solvent was evaporated under reduced pressure at room temperature to yield semisolid. The extract was then preserved in a refrigerator till further use.

2.3. In-vivo neuropharmacological activity

2.3.1. Experimental animals
Swiss Albino mice weighing 25–30 gm of both male and female were collected from International Center for Diarrheal Diseases Research, Bangladesh (ICDDRB) and housed in polypropylene cages under controlled conditions. The animals were exposed to alternative 12:12 h light and dark cycle at an ambient temperature of 26 ± 2°C. Animals were allowed free access to drinking water and pellet diet, collected from ICDDRB, Dhaka. Mice were acclimatized for seven days in the laboratory environment prior to the study. The set of rules followed for animal experiment were approved by the institutional animal ethics committee, Department of Pharmacy, International Islamic University Chittagong, Bangladesh according to governmental guidelines (Zimmermann, 1983).

2.3.2. Experimental design
The animals were randomly divided into four groups and each group consisting of five mice. The test groups received methanolic leaf extract of *T. sarmentosa* at the doses of 200 and 400 mg/kg while positive control was treated with diazepam (1 mg/kg) and control with vehicle (1% Tween 80 in water).

2.3.3. Open field test
The method described by Gupta was slightly modified and used for screening depressive action of the test drugs on CNS in mice. The animals were divided into control, positive control, and test groups. The test groups received *T. sermentosa* methanolic leaf extracts at the doses of 200 and 400 mg/kg body weight orally whereas the control group received vehicle (1% Tween 80 in water) and reference standard drug Diazepam at the dose of 1 mg/kg (i.p). The floor of an OFT of half square meter was divided into a series of squares each alternatively colored black and white. The apparatus had a 40 cm height wall. The number of squares visited by the animals was counted for 3 min, at 0, 30, 60 and 90 min during the study period (Gupta, Dandiya, & Gupta, 1971).

2.3.4. Hole cross test (HCT)
The test was observed by the method described by Takagi et al. for screening CNS depressant activity in mice. The animals were divided into three groups—control, positive control, and test. The test groups received methanol extract of *T. sarmentosa* at the doses of 200 and 400 mg/kg body weight orally whereas the control group received vehicle (1% Tween 80 in water). A steel partition was fixed in the middle of a cage having a size of 30 × 20 × 14 cm. A hole of 3 cm diameter was made at a height of 7.5 cm in the center of the cage. The number of passages of a mouse through the hole from one chamber to other was counted for a period of 3 min on 0, 30, 60, 90, and 120 min after the oral treatment with test drugs. Diazepam was used in the positive control group as reference standard at the dose of 1 mg/kg (i.p) (Takagi, Watanabe, & Saito, 1971).

2.3.5. Thiopental sodium-induced sleeping time test
The experiment was conducted following the method described by Ferrini et al. The animals were randomly divided into three groups consisting of five mice each. The test groups received methanol extract from the leaves of *T. sarmentosa* at a dose of 200 mg/kg and 400 mg/kg (p.o) body weight while the standard group was treated with diazepam (1 mg/kg, p.o) and control group with vehicle (1% Tween 80 in water). Twenty minutes later, thiopental sodium (40 mg/kg, i.p) were administered to each mice to induce sleep. The animals were observed for the latent period (time between thiopental administrations to loss of righting reflex) and duration of sleep i.e. time between the loss and recovery of righting reflex (Ferrini, Miragoli, & Taccardi, 1974).

2.3.6. Elevated plus-maze (EPM test)
The elevated plus maze (EPM) is a rodent model of anxiety that is used as a screening test for puta-tive anxiolytic or anxiogenic compounds and as a general research tool in neurobiological anxiety research. The EPM apparatus consists of two open arms (5 × 10 cm) and two closed arms (5 × 10 × l5 cm) radiating from a platform (5 × 5 cm) to form a plus sign figure. The apparatus was situated 40 cm above the floor (Lister, 1987). The open arms edges were 0.5 cm in height to keep the mice from falling and the closed-arms edges were 15 cm in height. Sixty minutes after administra-tion of the test drugs, each animal was placed at the center of the maze facing one of the enclosed arms. During the 5 min test period, the number of open and enclosed arms entries was recorded (Pellow & File, 1986). Entry into an arm was defined as the point when the animal places all four paws onto the arm. The procedure was conducted in a sound attenuated room; observations made from an adjacent corner (Braida, Prana, & Hiberty, 2009; Braida et al., 2008).

$$\% \text{ of entries in open arm} = \frac{\text{Number of entries in open arm}}{\text{Number of entries in open arm} + \text{Number of entries in closed arm}}$$

2.3.7. Acute oral toxicity test
An acute oral toxicity test was performed according to the "Organization for Environmental Control Development" guidelines (OECD: Guidelines 420; Fixed Dose Method). Swiss Albino mice ($n = 5$) (both male and female) overnight fasted for 18 h were used for the study. Different doses of methanolic plant extract were administered orally into the mice. The maximum given dose was 600 mg/kg body weight. Then the animals were observed for the first three hours of administration and mortality recorded within 48 h. No mortality was observed at maximum dose which is 600 mg/kg.

2.3.8. Evaluation of antioxidant effect
A number of assays such as 1,1-diphenyl-2-picrylhydrazyl (DPPH) free radical scavenging assay, fer-ric reducing power assay, and total phenolic content (TPC) tests were performed for the evaluation of antioxidant properties of the plant extract.

2.3.9. DPPH free radical scavenging assay
Free radical scavenging activity of methanolic extract of *T. sarmentosa* was determined according to the Brand-Willium (Brand-Williams, Cuvelier, & Berset, 1995) method with a slight modification. This activity was determined spectrophotometrically by taking absorbance of DPPH at 517 nm. Then the % of free radical inhibition was calculated using following equation:

$$\% \text{ of inhibition} = [(\text{Abs of control} - \text{Abs of sample}) \div \text{Abs of control}] \times 100$$

Lower the absorbance with high concentration of extract indicates potential antioxidant activity of test sample. Ascorbic acid was used as a reference standard.

2.3.10. Reducing power assay
Reducing power capacity of the methanolic extract of *T. sarmentosa* was determined according to Oyaizu (1986) method. The amount of ferrous complex was determined spectrophotometrically by taking absorbance at 700 nm where plant extracts having excellent antioxidant property show greater absorbance with the higher concentration of extract solution. Ascorbic acid was used as a reference standard.

2.3.11. Total phenolic content (TPC) determination
TPC in the plant extract was determined by following the Folin-ciocalteu method (Slinkard & Singleton, 1977). TPC was determined spectrophotometrically by taking absorbance at 760 nm. Standard Gallic acid solution of different concentrations with same procedure were used to prepare a calibration curve by plotting the absorbance against their respective concentrations from where a standard equation was formulated to determine the unknown concentration of Gallic acid equiva-lent (GAE) phenolic concentration of the sample by putting the value of absorbance in the equation. Then the TPC was determined by following equation:

$$TPC = (C \times V)/m \text{ (mg GAE/gm)}$$

2.3.12. Statistical analysis

The data were expressed as mean ± standard error of mean (S.E.M.). Statistical comparisons were performed using one-way ANOVA followed by Dunnett's multiple comparison test. The values obtained were compared with the vehicle control group and were considered statistically significant when $p < 0.05$.

3. Results

3.1. In-vivo neuropharmacological activity

3.1.1. Open field test
In the OFT, the extract showed a decrease in locomotion in the test animals at both dose levels (200 and 400 mg/kg body weight). The depressant activity was slowly reduced with time. The results were dose-dependent & statistically significant (Table 1).

3.1.2. HCT
In the animal treated with methanol extract at 400 mg/kg dose showed a dose-dependent reduction in the locomotor activity and at higher dose, it was comparable with that of standard drug diazepam. Diazepam was used as the standard drug in the experimental animals to evaluate the CNS depressant effect of the plant extract. The extract produced reduction in spontaneous motor activity, and this effect may be attributed to CNS depression, as depression of locomotor activity is common to most neuroleptics. The CNS was depressed till observation and the results were statistically significant (Table 2).

3.1.3. Thiopental sodium-induced sleeping time test
In the thiopental-induced hypnosis test, the extract at doses, 200 and 400 mg/kg showed a significant reduction in the time of onset of sleep in a dose-dependent manner. The effect of the extract (200 and 400 mg/kg) on the onset of sleep were comparable to that of the standard. In our study, 400 mg/kg dose of TSME produced quick onset of sleep and prolonged duration of sleep than that of 200 mg/kg dose (Table 3).

3.1.4. Elevated plus-maze (EPM) test
The EPM test is probably the most widely used model of animal anxiety. A substance which has anxiolytic effect generally increases time and proportion of entrance into the open arms when treated animals are exposed to EPM. Our present results showed that the treatment with T. sarmentosa extract decreased the percentage of entries of mice into the open arm and percentage of time spent in open arm compared to standard drug diazepam. Reference drug diazepam showed significant anxiolytic effect than T. sarmentosa (Table 4).

Table 1. CNS depressant activity of TSME in OFT test

Group	Treatment	Dose	Numbers of movements				
			0 min	30 min	60 min	90 min	120 min
Control	1% tween 80 in water	10 ml/kg	74.00 ± 1.458	59.20 ± 0.962	45.80 ± 1.710	46.80 ± 0.962	49.40 ± 0.908
Standard	Diazepam	1 mg/kg	68.60 ± 1.151	54.00 ± 1.275*	28.00 ± 0.79*	20.20 ± 0.962*	18.80 ± 1.140*
Test	TSME	400 mg/kg	65.80 ± 1.98*	57.00 ± 2.574	41.00 ± 1.225	24.80 ± 2.485*	19.40 ± 1.304*
		200 mg/kg	79.80 ± 1.55	61.60 ± 3.962	49.80 ± 3.362	30.40 ± 1.204*	23.80 ± 1.673*

Notes: Number of squares traveled by the mice of different groups in the OFT. All values are expressed as mean ± SEM ($n = 5$); One-way Analysis of Variance (ANOVA) followed by Dunnett's test. TSME = T. sermentosa methanolic extract.

*$p < 0.05$, significant compared to control.

Table 2. CNS depressant activity of TSME in hole cross test

Group	Treatment	Dose	Numbers of movements				
			0 min	30 min	60 min	90 min	120 min
Control	1% tween 80 in water	10 ml/kg	20.40 ± 1.037	16.20 ± 1.387	13.00 ± 1.118	9.40 ± 0.570	7.00 ± 0.791
Standard	Diazepam	1 mg/kg	17.00 ± 0.791	8.00 ± 0.79*	7.00 ± 0.79*	4.60 ± 0.57*	3.20 ± 0.652*
Test	TSME	400 mg/kg	19.20 ± 1.432	10.20 ± 0.65*	8.20 ± 0.65*	6.20 ± 0.65*	5.20 ± 0.652
		200 mg/kg	21.00 ± 1.620	13.00 ± 0.79	11.00 ± 0.79	7.60 ± 0.57	7.60 ± 0.510

Notes: Number of hole crossed by the mice of different groups in the hole cross test. All values are expressed as mean ± SEM ($n = 5$); One-way Analysis of Variance (ANOVA) followed by Dunnett's test. TSME = T. sermentosa methanolic extract.
*$p < 0.05$, significant compared to control.

Table 3. Thiopental sodium-induced hypnosis test

Group	Treatment	Dose	Onset of sleep (min)	Duration of sleep (min)
Control	1% tween 80 in water	10 ml/kg	42.60 ± 0.908	47.80 ± 0.82
Standard	Diazepam	1 mg/kg	14.60 ± 0.570*	147.80 ± 2.945*
Test	TSME	400 mg/kg	16.00 ± 0.791*	97.40 ± 2.864*
		200 mg/kg	19.80 ± 0.822*	60.40 ± 2.657*

Notes: CNS depressant activity of methanolic extract of leaves of T. sermentosa on thiopental sodium-induced sleeping time test in mice. All values are expressed as mean ± SEM ($n = 5$); One-way Analysis of Variance (ANOVA) followed by Dunnett's test. TSME = T. sermentosa methanolic extract.
*$p < 0.05$, significant compared to control.

Table 4. EPM test for the evaluation of anxiety

Group	Treatment	Dose	Percent of entry into open arm	Percent of time spent in open arm
Control	1% tween 80 in water	10 ml/kg	55.18 ± 1.891	58.14 ± 2.280
Standard	Diazepam	1 mg/kg	78.11 ± 1.963*	80.47 ± 2.806*
Test	TSME	400 mg/kg	41.43 ± 0.834*	56.91 ± 1.821
		200 mg/kg	30.95 ± 1.003*	45.80 ± 1.982*

Notes: CNS depressant activity of methanolic extract of leaves of T. sermentosa on elevated plus-maze test in mice. All values are expressed as mean ± SEM ($n = 5$); One-way Analysis of Variance (ANOVA) followed by Dunnett's test. TSME = T. sermentosa methanolic extract.
*$p < 0.05$, significant compared to control.

3.2. Evaluation of in vitro antioxidant effect

3.2.1. DPPH free radical scavenging assay
The DPPH radical scavenging activity of T. sarmentosa is shown in Table 5. This activity was found to increase with increasing concentration of the extract. DPPH antioxidant assay is based on the ability of DPPH, a stable free radical, to decolorize in the presence of antioxidants. In Table 5, IC_{50} value of the extract was 151.56 μg/ml as compared to that of ascorbic acid (IC_{50} 23.53 μg/ml) which is a well-known antioxidant.

Table 5. DPPH scavenging activity of Ascorbic acid and TSME

Sample	IC_{50} (μg/ml)
Ascorbic acid	23.53
TSME	151.56

Note: TSME = T. sermentosa methanolic extract.

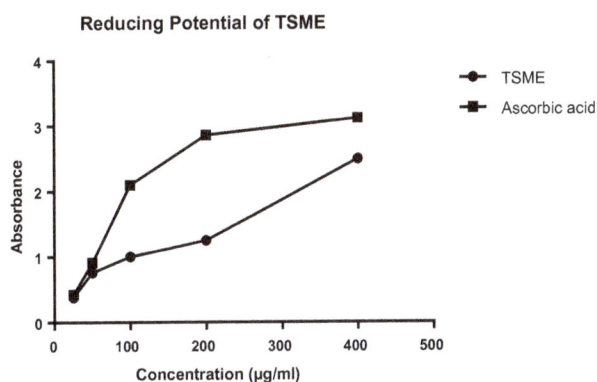

Figure 1. Reducing power capacity of ascorbic acid and methanolic extract of *T. sarmentosa*.

Table 6. Determination of total phenolic content as GAE mg/gm extract

Sample solution of TSME (μg/ml)	Weight of dry extract per ml, m (gm)	Absorbance At 760 nm	GAE conc. C (μg/ml)**	GAE conc. C (mg/ml)	TPC as GAE, $A = \frac{C \times V}{m}$ (mg/gm)	Mean ± SD
500	0.0005	0.304	69.487	0.069	138.974	140.34 ± 1.56
500	0.0005	0.301	71.025	0.071	142.05	
500	0.0005	0.306	70	0.070	140	

Note: TSME = *T. sermentosa* methanolic extract

**$Y = 0.0039x + 0.033$.

3.2.2. Reducing power capacity

Reducing power is related with the antioxidant effect. Components containing reducing power decreases the oxidized intermediated of lipid peroxidation process. The amount of ferrous complex was determined spectrophotometrically by taking absorbance at 700 nm while the TSME displayed increased absorbance with the increasing concentrations of the extract solution demonstrating that the plant had significant reducing power capacity and the results have been shown in Figure 1. The extract increased the absorbance significantly, demonstrating the antioxidant potential of TSME. The plant extract showed activity with an absorbance of 2.49 and 1.00 at 400 and 100 μg/ml concentration, respectively, which was appeared to be comparable to the activity of the reference standard ascorbic acid that gave an absorbance of 3.12 and 2.10 at the same concentration.

3.2.3. Determination of TPCs

Phenolic compounds are important plant constituents, having excellent antioxidant properties due to their hydroxyl groups that act as free radical terminator. The TPC in the methanolic extract was expressed as the number of GAE. Final result of TPC in extract was expressed as mg GAE/gm of dry extract. In our current study, the TPC of TSME was found to be 140.34 ± 1.56 GAE mg/gm dry extract (Table 6).

4. Discussion

Anxiety and depression-related disorders are increasing day by day all over the world. In 1990, people suffered from anxiety or depression-related disorder were 416 million (World Health Organization, 2016). In 2013, it is almost increase by closely 50% which is 615 million (World Health Organization, 2016). Humanitarian emergencies and ongoing conflict add further to the need for scale up of treatment options. WHO estimates that, during emergencies, as many as 1 in 5 people are affected by depression and anxiety (World Health Organization, 2016). The CNS depressant activity obtained for extract was evidenced from the suppression of the number of squares traveled by the mice in the test group throughout the study period. For both OFT and HCT, reduced movements of mice were observed in the treatment group especially at 400 mg/kg dose, which indicates the sedative action of the test extract. In TIST, the extract at both doses showed a significant reduction in the time of onset of sleep in a dose-dependent manner. EPM is a popular test for anxiolytic-related behavioral assessment. Compared to the reference drug diazepam test extract showed minor anxiolytic action at highest dose which is 400 mg/kg dose. On experimenting, it was found that IC_{50} value of the extract was 151.56 µg/ml as compared to that of ascorbic acid (IC_{50} 23.53 µg/ml) which is a well-known antioxidant in the case of DPPH radical scavenging assay. Reducing power test of this plant extract showed a promising activity with an absorbance of 2.49 at 400 µg/ml compared to reference standard ascorbic acid that gave an absorbance of 3.12 at the same concentration. In our current study, the TPC of TSME was found to be 140.34 ± 1.56 GAE mg/gm dry extract which demonstrated strong reduction capability.

Many research showed that plant containing phenols, flavonoids, saponins, and tannins are useful in many CNS disorders (Bhattacharya & Satyan, 1997). Phytochemical investigations also showed the presence of phenols, flavonoids, saponins, and steroids in this plant (Mungole & Chaturvedi, 2011). So might be this phytoconstituents are responsible for its antioxidant activity. The oxidative imbalance by ROS such as superoxide, peroxide, peroxynitrite, nitric oxide, lipid peroxidation plays a key role in anxiety development. Studies in both humans and animals have shown a strong correlation between anxiety and ROS. It has been suggested that a medicine or therapy specifically focus in reducing ROS production may have a beneficial effect in reducing anxiety. However, the neurobiological pathways underlying the effect of oxidative stress on anxiety symptoms are not fully comprehended. The challenge now is to identify the oxidative stress mechanisms likely to be involved in the induction of anxiety symptoms and produce a therapy or medicine which may prevent this disorder. Antioxidants are great source to neutralize the free radicals in lipid chains by contributing a hydrogen atom usually from a phenolic hydroxyl group, which in turn converts phenolic groups into stable free radicals that do not initiate or propagate further oxidation of lipids (Uttara, Singh, Zamboni, & Mahajan, 2009). These findings suggest that the plant could be useful as a therapeutic agent for anxiety-related disorders. Further investigations are needed to elucidate the prime constituents of this plant.

5. Conclusion
From the above experiments, it could be concluded that methanolic extract of *T. sarmentosa* contains significant neuropharmacological and antioxidant activities. To elucidate the exact mechanism action and bioactive compounds responsible for the neuropharmacological and antioxidant activites of this plant extract, further pharmacological studies must be performed.

Acknowledgments
The authors are grateful to Department of Pharmacy, International Islamic University Chittagong for giving all facilities to carry out the whole project successfully. Authors also convey their gratitude to the committee of Animal Research Branch of the International Centre for Diarrheal Disease and Research, Bangladesh (ICDDRB) for providing experimental rats.

Funding
The authors received no direct funding for this research.

Competing Interests
The authors declare no competing interest.

Author details
Muhammad Moin Uddin Mazumdar[1]
E-mail: saddam.moin@gmail.com
Md. Ariful Islam[1]
E-mail: arif_iiuc2013@yahoo.com
Mohammad Tanvir Hosen[1]
E-mail: pharmacist.tanvir@gmail.com
Mohammad Shahin Alam[1]
E-mail: shahiniiuc@gmail.com
Mohammad Nazmul Alam[1]
E-mail: nazmul_pharmacy@yahoo.com
Md. Faruk[1]
E-mail: mdfaruk.pharm@gmail.com
Md. Mominur Rahman[1]
E-mail: momin.rahman@gmail.com
Mohammed Abu Sayeed[1]
E-mail: ssmas2003@yahoo.com
Md. Masudur Rahman[1]
E-mail: mamun2001@hotmail.com
Shaikh Bokhtear Uddin[2]
E-mail: bokhtear@cu.ac.bd
[1] Department of Pharmacy, International Islamic University Chittagong, Chittagong, Bangladesh.
[2] Department of Botany, University of Chittagong, Chittagong, Bangladesh.

References

Abid, M., Hrishikeshavan, H. J., & Asad, M. (2006). Pharmacological evaluation of *Pachyrrhizus Erosus* (L.) seeds for central nervous system depressant activity. *Indian Journal of Physiology and Pharmacology, 50*, 143–151.

Bangladesh Ethnobotany Online Database. (n.d.). *Tetracera sarmentosa*. Retrieved from February 21, 2016, from http://www.ebbd.info/tetracera-sarmentosa.html

Bhattacharya, S. K., & Satyan, K. S. (1997). Experimental methods for evaluation of psychotropic agents in rodents: I-Anti-anxiety agents. *Indian Journal of Experimental Biology, 35*, 565–575.

Braida, B., Prana, V., & Hiberty, P. C. (2009). The physical origin of Saytzeff's rule. *Angewandte Chemie International Edition, 48*, 5724–5728. http://dx.doi.org/10.1002/anie.v48:31

Braida, D., Limonta, V., Capurro, V., Fadda, P., Rubino, T., Mascia, P., Zani, A., Gori, E., Fratta, W., Parolaro, D., & Sala, M. (2008). Involvement of κ-opioid and endocannabinoid system on salvinorin a-induced reward. *Biological Psychiatry, 63*, 286–292. http://dx.doi.org/10.1016/j.biopsych.2007.07.020

Brand-Williams, W., Cuvelier, M. E., & Berset, C. (1995). Use of a free radical method to evaluate antioxidant activity. *LWT - Food Science and Technology., 28*, 25–30. http://dx.doi.org/10.1016/S0023-6438(95)80008-5

Buller, R., & Legrand, V. (2001). Novel treatments for anxiety and depression: Hurdles in bringing them to the market. *Drug Discovery Today, 6*, 1220–1230. http://dx.doi.org/10.1016/S1359-6446(01)02043-8

Da, F., Zhu, H., Zhao, X., Teng, J., & Gan, J. (2014). Analysis of the chemical components of essential oils from leaves of *Tetracera sarmentosa* by GC-MS. *Medicinal Plant, 5*, 27–28.

Díaz-Muñoz, M., Álvarez-Pérez, M. A., & Yáñez, L. (2006). Correlation between oxidative stress and alteration of intracellular calcium handling in isoproterenol-induced myocardial infarction. *Molecular and Cellular Biochemistry, 289*, 125–136. http://dx.doi.org/10.1007/s11010-006-9155-1

Doughari, J. H., Ndakidemi, P. A., Human, I. S., & Benade, S. (2012). Antioxidant, antimicrobial and antiverotoxic potentials of extracts of *Curtisia dentata*. *Journal of Ethnopharmacology, 141*, 1041–1050. http://dx.doi.org/10.1016/j.jep.2012.03.051

Fernando, T., Ratnasooriya, W. D., & Deraniyagala, S. A. (2009). Antinociceptive activity of aqueous leaf extract of *Tetracera sarmentosa* L. in rats. *Pharmacognosy Research, 1*, 381–386.

Ferrini, R., Miragoli, G., & Taccardi, B. (1974). Neuro-pharmacological studies on SB 5833, a new psychotherapeutic agent of the benzodiazepine class. *Arzneimittel Forschung, 24*, 2029–2032.

Flora of China [FOC]. (n.d.). *Tetracera sarmentosa* (Vol. 12, page 331). Author. Retrieved February 22, 2016, from http://www.efloras.org/florataxon.aspx?flora_id=2&taxon_id=242351569

Gupta, B. D., Dandiya, P. C., & Gupta, M. L. (1971). A psycho-pharmacological analysis of behaviour in rats. *The Japanese Journal of Pharmacology, 21*, 293. http://dx.doi.org/10.1254/jjp.21.293

Lister, R. G. (1987). The use of a plus-maze to measure anxiety in the mouse. *Psychopharmacology (Berl), 92*, 180–185.

Mahboubi, M., Haghi, G., Kazempour, N., & Hatemi, A. R. (2013). Total phenolic content, antioxidant and antimicrobial activities of Blepharis edulis extracts. *Songklanakarin Journal of Science and Technology, 35*, 11–16.

Ming, K. J., Khang, G. N., Sai, C. L., & Fatt, C. T. (2003). Recent advances in traditional plant drugs and orchids. *Acta Pharmacologica Sinica, 24*, 7–21.

Mungole, A., & Chaturvedi, A. (2011). *Hibiscus sabdariffa* L. a rich source of secondary metabolites. *International Journal of Pharmaceutical Sciences Review and Research, 6*, 83–87.

Oyaizu, M. (1986). Studies on product of browning reaction prepared from glucose amine. *The Japanese Journal of Nutrition and Dietetics, 44*, 307–315. http://dx.doi.org/10.5264/eiyogakuzashi.44.307

Pellow, S., & File, S. E. (1986). Anxiolytic and anxiogenic drug effects on exploratory activity in an elevated plus-maze: A novel test of anxiety in the rat. *Pharmacology Biochemistry and Behavior, 24*, 525–529. http://dx.doi.org/10.1016/0091-3057(86)90552-6

Rodrigo, R., Libuy, M., Feliú, F., & Hasson, D. (2013). Oxidative stress-related biomarkers in essential hypertension and ischemia-reperfusion myocardial damage. *Disease Markers, 35*, 773–790. http://dx.doi.org/10.1155/2013/974358

Slinkard, K., & Singleton, V. L. (1977). Total phenol analyses: Automation and comparison with manual methods. *American Journal of Enology and Viticulture, 28*, 49–55.

Takagi, K., Watanabe, M., & Saito, H. (1971). Studies of the spontaneous movement of animals by the hole cross test; effect of 2-dimethyl-aminoethanol and its acyl esters on the central nervous system. *The Japanese Journal of Pharmacology, 21*, 797–810. http://dx.doi.org/10.1254/jjp.21.797

Takimoto, E., & Kass, D. A. (2007). Role of oxidative stress in cardiac hypertrophy and remodeling. *Hypertension, 49*, 241–248.

Titov, N., Andrews, G., Kemp, A., & Robinson, E. (2010). Characteristics of adults with anxiety or depression treated at an internet clinic: Comparison with a national survey and an outpatient Clinic. *PLoS ONE, 5*(5), e10885. http://dx.doi.org/10.1371/journal.pone.0010885

Uttara, B., Singh, A. V., Zamboni, P., & Mahajan, R. (2009). Oxidative stress and neurodegenerative diseases: A review of upstream and downstream antioxidant therapeutic options. *Current Neuropharmacology, 7*, 65–74. http://dx.doi.org/10.2174/157015909787602823

World Health Organization. (2016, April). *Investing in treatment for depression and anxiety leads to fourfold return*. Washington, DC: Lancet Psychiatry.

Yadav, A. V., Kawale, L. A., & Nade, V. S. (2008). Effect of *Morus alba* L. (mulberry) leaves on anxiety in mice. *Indian Journal of Pharmacology, 40*, 32–36.

Zimmermann, M. (1983). Ethical guidelines for investigations of experimental pain in conscious animals. *Pain, 16*, 109. http://dx.doi.org/10.1016/0304-3959(83)90201-4

Zweier, J. L., & Talukder, M. A. H. (2006). The role of oxidants and free radicals in reperfusion injury. *Cardiovascular Research, 70*, 181–190. http://dx.doi.org/10.1016/j.cardiores.2006.02.025

6

Investigation of intrinsic dynamics of enzymes involved in metabolic pathways using coarse-grained normal mode analysis

Sarah M. Meeuwsen[1], An N. Hodac[1], Lauren M. Adams[1], Ryan D. McMunn[1], Maxwell S. Anschutz[1], Kari J. Carothers[1], Rachel E. Egdorf[1], Peter M. Hanneman[1], Jonathan P. Kitzrow[1], Cynthia K. Keonigsberg[1], Oscar Lopez-Martinez[1], Paul A. Matthew[1], Ethan H. Richter[1], Jonathan E. Schenk[1], Heidi L. Schmit[1], Matthew A. Scott[1], Eva M. Volenec[1] and Sanchita Hati[1]*

*Corresponding author: Sanchita Hati, Chemistry Department, University of Wisconsin – Eau Claire, Eau Claire, WI 54702, USA

E-mail: hatis@uwec.edu

Reviewing editor: Yusuf Akhter, Central University of Himachal Pradesh, India

Additional information is available at the end of the article

Abstract: Intrinsic dynamics of proteins are known to play important roles in their function. In particular, collective dynamics of a protein, which are defined by the protein's overall architecture, are important in promoting the active site conformation that favors substrate binding and effective catalysis. The primary sequence of a protein, which determines its three-dimensional structure, encodes unique dynamics. The intrinsic dynamics of a protein actually link protein structure to its function. In the present study, coarse-grained normal mode analysis was performed to examine the intrinsic dynamic patterns of 24 different enzymes involved in primary metabolic pathways. We observed that each metabolic enzyme exhibits unique patterns of motions, which are conserved across multiple species and functionally relevant. Dynamic cross-correlation matrices (DCCMs) are visibly identical for a given enzyme family but significantly different from DCCMs of other protein families, reinforcing that proteins with similar function exhibit a similar pattern of motions. The present work also reasserted that correct identification of unknown proteins is possible based on their intrinsic mobility patterns.

ABOUT THE AUTHORS

Our research is focused on the simple question - how protein motions influence substrate binding and catalysis. The relationship between protein dynamics and its role in function (molecular recognition, catalysis, and allostery) is an evolving perspective in enzymology. For the past several years, we have been investigating the interplay of protein dynamics and enzymatic processes using computations and experiments. In particular, we are exploring how dynamics of different domains modulate substrate recognition, catalysis, and allosteric communication in multi-domain enzymes. Our overall goal is to advance the drug design and screening process, by taking into consideration proteins' intrinsic flexibility.

PUBLIC INTEREST STATEMENT

In the present work, the intrinsic dynamics of enzymes from 24 different families, which are involved in key metabolic pathways, were analyzed. The goal of this study is to better understand the connection between intrinsic dynamics and enzymatic function. As each enzyme exhibits a unique pattern of motions, these intrinsic dynamic patterns can be used to characterize the function of unknown proteins. This work demonstrates that (i) dynamic-based functional characterization is a highly reliable method and (ii) computationally less expensive coarse-grained normal mode analysis (NMA) is extremely useful to explore intrinsic dynamics of a vast number of proteins in relatively short duration of time. Overall, this work reasserted the importance of dynamics in protein function and suggested that coarse-grained NMA could be used for protein identification and classification. These findings can be applied by researchers working in the fields of protein chemistry and enzymology.

Subjects: Biochemistry; Bioinformatics; Biophysics; Chemistry

Keywords: normal mode analysis; protein dynamics; dynamic cross-correlation; metabolic enzymes

1. Introduction

Proteins play crucial roles in nearly all biological processes. Each protein has a primary amino acid sequence that dictates the tertiary structure of the protein, and ultimately determines the protein's overall function. In other words, the amino acid sequence → structure → function relationship exists for proteins and enzymes. Although the static structure of proteins is commonly correlated with their function, proteins are intrinsically dynamic in nature, and their internal motions, over a wide-range of timescales, assist their function. In fact, it has been suggested that the protein's dynamics is the fundamental link between its structure and function (Hensen et al., 2012). However, the local dynamical character of an enzyme active site does not alone produce the functionally/catalytically competent state necessary to perform the protein's biological function. In fact, the whole protein's dynamism is essential for its overall function. Several studies revealed the influence of protein dynamics on enzyme catalysis, substrate binding, allosteric signal propagation, and protein–protein interactions (Agarwal, 2005; Agarwal, Billeter, Rajagopalan, Benkovic, & Hammes-Schiffer, 2002; Bhabha et al., 2011; Chennubhotla & Bahar, 2007; Chennubhotla, Yang, & Bahar, 2008; Eisenmesser et al., 2005; Hammes-Schiffer & Benkovic, 2006; Henzler-Wildman & Kern, 2007; Henzler-Wildman et al., 2007; Klinman & Kohen, 2013; Lisi & Patrick Loria, 2016; Liu et al., 2016; Singh, Francis, & Kohen, 2015; Vöhringer-Martinez & Dörner, 2016; Zen, Micheletti, Keskin, & Nussinov, 2010). In addition, it has been observed that the dynamics of residues that are not directly involved in catalysis are also important for enzyme function (Roston, Kohen, Doron, & Major, 2014; Tousignant & Pelletier, 2004).

For a majority of known protein sequences and structures, functional information is lacking (Erdin, Lisewski, & Lichtarge, 2011). To date, the functional classification of proteins has been largely dominated by sequence- or structure-based methods. However, both methods have strict limitations. Sequence-based methods are limited due to evolutionary divergence, as a protein structure is more likely to remain conserved than its sequence (Chothia & Lesk, 1986), and effective sequence homology requires a minimum of 40% identity over significant portions of the sequence (Devos & Valencia, 2000; Hensen et al., 2012; Park, Teichmann, Hubbard, & Chothia, 1997; Whisstock & Lesk, 2003). Structure-based methods have revealed more direct functional relationships when coupled with sequence data. Unfortunately, structure-based prediction methods are limited due to diverse structures possessing similar functions or similar structures carrying out a diverse range of functions. Structure-based methods alone do not produce reliable functional predictions (reliability ≤ 30%) (Hensen et al., 2012). Hensen et al. (2012) found that the intrinsic dynamics of a protein provide a more accurate depiction of protein function due to the large correlation between protein dynamics and function. They observed that functionally similar proteins tend to exhibit similar dynamic patterns and thereby, protein functional annotations could be accomplished based on dynamics (Hensen et al., 2012). Their work demonstrated that the functional classification of proteins based solely upon protein dynamics has a success rate of 46% within the broad functional classes, which is much higher than the structure-based prediction rate of 32%. Other researchers have also explored protein dynamics-function relationships by performing dynamics-based comparisons of proteins (Micheletti, 2013; Munz, Lyngsø, Hein, & Biggin, 2010; Pandini, Mauri, Bordogna, & Bonati, 2007; Tobi, 2013). Here, we have explored the dynamics of enzymes involved in some important metabolic pathways in an attempt to further our understanding of the relationships between protein structure, dynamics, and function.

Earlier, Hensen et al. have performed long timescale (100 ns) atomistic molecular dynamic (MD) simulations and derived 34 dynamic descriptors to characterize dynamics of proteins (Hensen et al., 2012). Picosecond to nanosecond timescale atomic fluctuations can be simulated using all-atom MD simulations, and important information about functional dynamics could be obtained. Unfortunately, it is computationally expensive to investigate functionally significant displacements occurring in microsecond to millisecond timescale using MD simulations. Alternatively, normal mode analysis (NMA) can be utilized to examine protein dynamics at a slower timescale (Bahar & Rader, 2005; Hinsen, 1998; Hinsen, Thomas, & Field, 1999). Both low-frequency vibrations that represent the protein's collective motions as well as the higher frequency vibrations, which represent local deformations, can be studied by NMA. Earlier studies have shown that low-frequency vibrational modes or normal modes (large fluctuations) of proteins are functionally relevant (Brooks & Karplus, 1983; Go, Noguti, & Nishikawa, 1983; Marques & Sanejouand, 1995). Despite a few limitations such as local fluctuations upon substrate binding being poorly captured by the coarse-grained NMA simulations (Strom, Fehling, Bhattacharyya, & Hati, 2014), NMA is not limited by the system size and can be used to analyze functionally relevant conformational dynamics from protein structure, making it an incredibly powerful method of analysis. Additionally, the NMA method that employs course-grained elastic network model (ENM) offers a dependable yet computationally inexpensive way to evaluate large displacements (conformational changes) of proteins comprised of thousands of residues (Frank & Agrawal, 2000; Strom et al., 2014; Tama & Brooks, 2005; Weimer, Shane, Brunetto, Bhattacharyya, & Hati, 2009).

In the present study, ENM-based coarse-grained NMA was employed to characterize intrinsic dynamics of metabolically relevant enzymes and formulate a protein functional characterization method emphasizing dynamical importance. The dynamics of 24 enzymes involved in primary metabolic pathways were studied and sequence, structure, and dynamic similarities were calculated across six species to determine the degree of conservation of sequence, structure, and dynamics within a protein group. We examined the relationship among these three properties, as well as developed a catalog of dynamic cross-correlation matrices (DCCMs). We then attempted to match a set of proteins to the catalog of DCCMs for their functional identification. The present study demonstrated that each metabolic enzyme exhibits unique patterns of motions, which are conserved across multiple species; proteins with similar function exhibit a similar pattern of motions. Moreover, this work also demonstrated that the coarse-grained NMA is a faster method to develop protein-specific DCCMs, which could be used for functional characterization. Also, it is established from our study that dynamic-based functional characterization is highly predictable. This study could be extended to include all known protein families and generate a database of DCCMs of proteins for functional annotation.

2. Methods

2.1. Protein sequences and structures

Twenty-four proteins from common metabolic pathways, as listed in Table 1, were selected from the Protein Data Bank (PDB) (Berman et al., 2000) based upon structural data availability. We attempted to select proteins from six different species for which crystal structures are known. Also, proteins were screened for similar amino acid chain length. A total of 141 proteins were considered in this study because only 5 crystal structures were available for 3 of the 24 protein families. The symmetry and stoichiometry analyses revealed that all but one of the selected protein families are homopolymers. Therefore, a single polypeptide chain (chain A) was chosen to compare the sequence and structure, as well as intrinsic dynamics between proteins (from different species) within a given family. For the heterodimer protein, succinyl-CoA synthetase, both chains were used. In the present

Table 1. Comprehensive list of the 24 protein groups and the six selected proteins for each group with the organism name, PDB ID, and number of amino acid residues (sequence length) in chain A of the protein

Protein group	Organism	PDB ID	Length
Citrate synthase (citric acid cycle)	Vibrio vulnificus	4E6Y	426
	Escherichia coli	4G6B	432
	Acetobacter aceti	2H12	436
	Francisella tularensis	3MSU	427
	Pyrobaculum aerophilum	2IBP	409
	Salmonella typhimurium	3O8J	404
Fumarase (citric acid cycle)	Escherichia coli	1YFE	467
	Mycobacterium tuberculosis	4APA	474
	Sinorhizobium meliloti	4HGV	495
	Rickettsia prowazekii	3GTD	482
	Thermus thermophilus	1VDK	466
	Homo sapiens	3E04	490
Isocitrate dehydrogenase (citric acid cycle)	Ruminiclostridium thermocellum	4AOU	402
	Escherichia coli	9ICD	416
	Sulfolobus tokodaii	2DHT	409
	Desulfotalea psychrophila	2UXR	402
	Burkholderia pseudomallei	3DMS	427
	Aeropyrum pernix	1V94	435
Malate dehydrogenase (citric acid cycle)	Escherichia coli	1EMD	312
	Thermus thermophilus	2CVQ	327
	Haloarcula marismortui	1O6Z	303
	Chloroflexus aurantiacus	4CL3	309
	Picrophilus torridus	4BGV	323
	Archaeoglobus fulgidus	2XOI	294
Succinyl CoA synthetase (citric acid cycle)	Methanocaldococcus jannaschii	2YV1	294
	Thermus thermophilus	1OI7	288
	Sus scrofa	2FP4	305
	Escherichia coli	1CQI	286
	Aeropyrum pernix	2YV2	297
	Thermus aquaticus	3UFX	296
Alcohol dehydrogenase (alcohol metabolism)	Mus musculus	1E3L	376
	Homo sapiens	1HTB	374
	Equus caballus	6ADH	374
	Thermus sp	4CPD	347
	Arabidopsis thaliana	4RQT	375
	Thermotoga maritima	3IP1	404
Lactate dehydrogenase (lactate metabolism)	Plasmodium falciparum	1CET	316
	Squalus acanthias	3LDH	330
	Plasmodium falciparum	1T25	322
	Staphylococcus aureus	3D0O	317
	Cryptosporidium parvum	4ND5	321
	Enterococcus mundtii	3WSV	322

(Continued)

Table 1. (Continued)			
Protein group	**Organism**	**PDB ID**	**Length**
Enolase (glycolysis)	*Homarus gammarus*	1PDZ	434
	Homo sapiens	3B97	433
	Saccharomyces cerevisiae	1ELS	436
	Bacillus subtilis	4A3R	430
	Enterococcus hirae	1IYX	432
	Trypanosoma brucei	2PTX	436
Fructose bisphosphate aldolase (glycolysis)	*Encephalitozoon cuniculi*	3MBF	342
	Thermus caldophilus	2FJK	305
	Mycobacterium tuberculosis	4LV4	349
	Babesia bovis	3KX6	379
	Toxoplasma gondii	4TU1	355
	Oryctolagus cuniculus	3DFN	353
Glyceraldehyde phosphate dehydrogenase (glycolysis)	*Geobacillus stearothermophilus*	1NQO	334
	Streptococcus agalactiae	4QX6	356
	Toxoplasma gondii	3STH	361
	Borrelia burgdorferi	3HJA	356
	Bacillus anthracis	4DIB	345
	Bartonella henselae	3LOD	356
Hexokinase (glycolysis)	*Schistosoma mansoni*	1BDG	451
	Arabidopsis thaliana	4QS8	474
	Xenopus laevis	3W0L	458
	Homo sapiens	4IWV	456
	Kluyveromyces lactis	4JAX	485
	Saccharomyces cerevisiae	3B8A	485
Phosphoglycerate kinase (glycolysis)	*Escherichia coli*	1ZMR	387
	Francisella tularensis	4FEY	395
	Thermus thermophilus	1V6S	390
	Geobacillus stearothermophilus	1PHP	394
	Thermotoga maritima	1VPE	398
Phosphoglycerate mutase (glycolysis)	*Saccharomyces cerevisiae*	3PGM	244
	Bacillus anthracis	3R7A	237
	Toxoplasma gondii	4ODI	281
	Plasmodium falciparum	1XQ9	258
	Homo sapiens	1YFK	262
	Saccharomyces cerevisiae	1BQ4	246
Pyruvate decarboxylase (glycolysis)	*Saccharomyces cerevisiae*	1PYD	556
	Vibrio parahaemolyticus	2QMA	497
	Saccharomyces pastorianus	1QPB	563
	Azospirillum brasilense	2NXW	565
	Enterobacter cloacae	1OVM	552

(Continued)

Table 1. (Continued)

Protein group	Organism	PDB ID	Length
Pyruvate kinase (glycolysis)	Homo sapiens	3SRF	551
	Felis catus	1PKM	530
	Oryctolagus cuniculus	1AQF	530
	Toxoplasma gondii	3EOE	511
	Cryptosporidium parvum	4DRS	526
	Leishmania mexicana	3E0 V	539
Triose phosphate isomerase (glycolysis)	Escherichia coli	4IOT	255
	Tenebrio molitor	2I9E	259
	Gallus gallus	8TIM	247
	Saccharomyces cerevisiae	1YPI	248
	Homo sapiens	1WYI	250
	Rhipicephalus microplus	3TH6	249
Phosphoenolpyruvate carboxykinase (gluconeo-genesis)	Thermus thermophilus	2PC9	529
	Escherichia coli	1OS1	540
	Trypanosoma cruzi	1II2	524
	Escherichia coli	1AYL	541
	Anaerobiospirillum succiniciproducens	1YTM	532
	Actinobacillus succinogenes	1YGG	560
Arginase (urea cycle)	Thermus thermophilus	2EF4	290
	Homo sapiens	4IXU	306
	Leishmania Mexicana	4ITY	330
	Rattus norvegicus	1ZPE	314
	Bacillus caldovelox	1CEV	299
	Deinococcus peraridilitor	XXXX	294
Argininosuccinate synthetase (urea cycle)	Campylobacter jejuni	4NZP	409
	Thermotoga maritima	1VL2	421
	Mycobacterium thermoresistibile	4XFJ	408
	Escherichia coli	1KP2	455
	Homo sapiens	2NZ2	413
	Thermus thermophilus	1J1Z	400
Ornithine transcarbamoylase (urea cycle)	Pseudomonas Aeruginosa	1ORT	335
	Escherichia coli	1AKM	333
	Pyrococcus Furiosus	1A1S	314
	Thermus thermophilus	2EF0	301
	Brucella Melitensis	4OH7	320
	Gloeobacter Violaceus	3GD5	323
Amine oxidase (urea cycle, amino acid metabolism)	Escherichia coli	2WO0	727
	Pisum sativum	1KSI	642
	Arthrobacter globiformis	1IVU	638
	Ogataea angusta	3SX1	692
	Aspergillus nidulans	3PGB	797
	Homo sapiens	3HI7	731

(Continued)

Table 1. (Continued)

Protein group	Organism	PDB ID	Length
Methionine adenosyltransferase (biosynthesis of S-adenosylmethionine)	Escherichia coli	1FUG	383
	Burkholderia pseudomallei	3IML	399
	Mycobacterium avium	3S82	407
	Rattus norvegicus	1QM4	396
	Homo sapiens	2P02	396
	Entaoeba histolytica	3SO4	415
Pantothenate synthetase (coenzyme A biosynthesis)	Thermotoga maritima	2EJC	280
	Mycobacterium tuberculosis	2A86	300
	Thermus thermophilus	1UFV	276
	Escherichia coli	1IHO	283
	Staphylococcus aureus	3AG5	283
	Yersinia pestis	3Q10	287
Alkaline phosphatase (signal transduction)	E. coli	3BDF	458
	Homo sapiens	1EW2	513
	Rattus norvegicus	4KJG	488
	Vibrio sp. G15-21	3E2D	502
	Pandalus borealis	1SHN	478
	Halobacterium salinarum	2X98	431

Note: The metabolic pathway(s) in which these enzymes are involved is given in parentheses.

study, structures of proteins were visualized using the Visual Molecular Dynamics (VMD) program (Humphrey, Dalke, & Schulten, 1996).

2.2. Sequence comparison

The sequence identity of proteins in each group was examined. These pairwise sequence alignments were carried out with the Basic Local Alignment Search Tool (BLAST) (Altschul et al., 1997), which aligns two sequences and provides the percentage of identical residues. The sequence identity is the percentage of identical residues between protein sequences being compared. A mean and standard deviation was calculated for each protein group.

2.3. Structure comparison

The pairwise DaliLite server (Holm & Rosenstrom, 2010) was used to determine the degree of structural similarity between two proteins. Pairwise structural alignments were performed between all proteins of a given group. Root-mean-square deviation (RMSD) was used to address structural (protein backbone) comparison, which represents the variation of a particular locus between two superimposed structures. The mean and standard deviation of the sample were calculated for each protein group.

2.4. Single and comparative NMA

In the coarse-grained ENM, developed by Hinsen, Petrescu, Dellerue, Bellissent-Funel, and Kneller (2000), the protein is simplified to a string of beads, where each bead represents a C_α atom. The interactions between C_α atoms are expressed as a harmonic pair potential as follows:

$$U_{ij}(r) = \frac{k_{ij}}{2}\left(\left|r_i - r_j\right| - \left|r_i^0 - r_j^0\right|\right)^2 \tag{1}$$

where r_i and r_j are the positions of residues i and j in the current conformation of the protein, and the superscript 0 denotes the equilibrium conformation; k_{ij} is the force constant for the spring connecting residues i and j. The force constant k_{ij} is distance dependent and expressed as

$$k_{ij}(r) = \left\{ \begin{array}{ll} ar - b, & \text{for } r < d \\ cr^{-6}, & \text{for } r \geq d \end{array} \right\} \qquad (2)$$

where a, b, c, and d are determined by fitting the equation to an all-atom model as discussed by Hinsen et al. (2000) and Tiwari et al. (2014). The overall potential of the system, which is the sum of the pair potential over all pairs, is expressed as follows:

$$U(r) = \sum_{\text{all pairs } i,j} U_{ij}(r) \qquad (3)$$

Eigenvectors and eigenvalues of the mass weighted matrix of the second derivative of $U(r)$ correspond to normal modes and the squares of the frequencies for each normal mode, respectively (Tiwari et al., 2014). In the present study, normal mode calculations were carried out using the WEBnm@ server (Fuglebakk, Tiwari, & Reuter, 2014; Hollup, Salensminde, & Reuter, 2005; Tiwari et al., 2014). WEBnm@ uses a molecular modeling toolkit and employs the coarse-grained ENM to calculate the low frequency normal modes. Individual PDB files were submitted to the online WEBnm@ server (Hollup, Fuglebakk, Taylor, & Reuter, 2011; Hollup et al., 2005; Skjaerven, Hollup, & Reuter, 2009; Tiwari et al., 2014) for single NMA analysis. The correlated/anticorrelated motions between residues (C_α atoms) were identified from the DCCMs. The DCCM was calculated from the normal modes using the method described by Ichiye and Karplus (Ichiye & Karplus, 1991). Each element in the DCCM quantifies the coupling between two C_α atoms i and j as:

$$C_{ij} = \frac{\sum_{m=1}^{M} \frac{1}{\gamma_m} [X_m]_i [X_m]_j}{\left(\sum_{m=1}^{M} \frac{1}{\gamma_m} [X_m]_i [X_m]_i \right)^{1/2} \left(\sum_{m=1}^{M} \frac{1}{\gamma_m} [X_m]_j [X_m]_j \right)^{1/2}} \qquad (4)$$

In Equation (3), X_m and Y_m represent eigenvectors and eigenvalues of the mth normal mode, respectively. In the present study, the DCCMs were obtained using the default setting of WEBnm@ i.e. the combined 200 modes were used to generate each DCCM. The DCCMs of the 24 protein families were compared and a catalog of all DCCMs was assembled for identification of unknown (test) proteins. The single WEBnm@ analysis also provides atomic (C_α atoms) fluctuation profiles (*vide infra*) and vector coordinates files, which allow for the visualization of the protein's dynamics in different modes. In WEBnm@, modes 1–6 correspond to the rotational and translational motions and are ignored. Therefore, mode 7 actually represents the lowest frequency mode of a protein system.

The comparative NMA was performed for each enzyme to examine the conservation of dynamics across species (Fuglebakk, Echave, & Reuter, 2012). In the present study, dynamics of a given enzyme from five to six different species were compared. Aligned FASTA sequence (See Supplementary Table S1) and corresponding PDB coordinate files of each protein group were used for the comparative NMA (Sievers et al., 2011). From the comparative analysis, the dynamic similarity, in terms of Bhattacharyya coefficient (BC) (Bhattacharyya, 1943), as well as the heat map showing the theoretical evolutionary relatedness between proteins solely based upon their intrinsic dynamics were obtained.

2.5. Bhattacharyya coefficient

The BC measures the dynamical similarity between proteins by comparing their covariance matrices, which are obtained from the normal modes of the conserved parts of proteins under consideration (Tiwari et al., 2014). The BC calculations were performed following the method described by Tiwari et al. (2014). The BC ranges from 0 to 1, and represents the amount of overlap between the collective dynamics of the aligned proteins; BC of 1 represents maximum overlap or dynamical similarity between proteins. For a given protein group, the average BC with the sample standard deviation is calculated for comparative purposes.

Table 2. The number of species studied (N), the sequence identity, structural similarity, and dynamic similarity (in terms of BC) of the 24 protein groups studied

Proteins	N	Sequence identity	Structural similarity (RMSD, Å)	Dynamic similarity (BC)
Citrate synthase	6	45 ± 14	2.4 ± 0.4	0.81 ± 0.04
Fumarase	6	56 ± 5	1.6 ± 0.4	0.87 ± 0.02
Isocitrate dehydrogenase	6	38 ± 17	2.9 ± 1.0	0.78 ± 0.06
Malate dehydrogenase	6	31 ± 6	2.5 ± 0.7	0.82 ± 0.03
Succinyl-CoA synthetase	6	59 ± 11	1.1 ± 0.3	0.89 ± 0.02
Alcohol dehydrogenase	6	39 ± 17	2.1 ± 0.7	0.82 ± 0.07
Lactate dehydrogenase	6	32 ± 6	1.8 ± 0.3	0.81 ± 0.03
Enolase	6	58 ± 7	1.3 ± 0.3	0.89 ± 0.03
Fructose bisphosphate aldolase	6	42 ± 14	2.7 ± 1.0	0.77 ± 0.09
Glyceraldehyde phosphate dehydrogenase	6	51 ± 5	1.1 ± 0.1	0.88 ± 0.01
Hexokinase	6	42 ± 14	3.0 ± 1.0	0.82 ± 0.04
Phosphoglycerate kinase	5	51 ± 8	2.5 ± 1.2	0.89 ± 0.02
Phosphoglycerate mutase	6	51 ± 17	1.8 ± 0.7	0.83 ± 0.08
Pyruvate decarboxylase	6	36 ± 8	3.1 ± 1.0	0.75 ± 0.09
Pyruvate kinase	6	56 ± 21	1.9 ± 0.7	0.87 ± 0.03
Triose phosphate isomerase	6	59 ± 13	1.0 ± 0.2	0.88 ± 0.03
PEP carboxylase	6	54 ± 13	1.8 ± 0.5	0.87 ± 0.03
Arginase	5	42 ± 8	1.5 ± 0.3	0.78 ± 0.08
Arginosuccinate synthetase	6	42 ± 11	1.7 ± 0.4	0.78 ± 0.03
Ornithine transcarbamo-ylase	6	44 ± 7	1.6 ± 0.3	0.84 ± 0.03
Amine oxidase	5	24 ± 4	2.6 ± 0.6	0.73 ± 0.05
Methionine adenosyltrans-ferase	6	61 ± 11	1.3 ± 0.3	0.86 ± 0.02
Pantothernate synthetase	6	46 ± 7	2.8 ± 1.1	0.84 ± 0.02
Alkaline phosphatase	6	36 ± 12	2.2 ± 0.8	0.80 ± 0.05

2.6. The C_α atom fluctuations

The normalized squared C_α atom fluctuations for each protein were calculated as the sum of the displacement of each C_α atom along the lowest modes (Tiwari et al., 2014). The fluctuations are the sum of the atomic (C_α atoms only) displacements in each mode weighted by the inverse of their corresponding eigenvalues. In the present study, the default setting of WEBnm@, which includes the first 200 modes, was used. The x-axis of a fluctuation profile represents residue index in the protein sequence, while the y-axis represents the normalized displacement corresponding to each amino acid. Peaks in the fluctuation profile correspond to flexible regions of proteins.

2.7. Root-mean-squared inner product

For quantitative comparison of atomic (C_α atoms) fluctuations between proteins, the RMSIP was computed for the lowest normal modes using the following equation as described in Amadei, Ceruso, and Di Nola (1999), Carnevale, Pontiggia, and Micheletti (2007), Fuglebakk et al. (2014), Tiwari et al. (2014) and Yang, Song, Carriquiry, and Jernigan (2008):

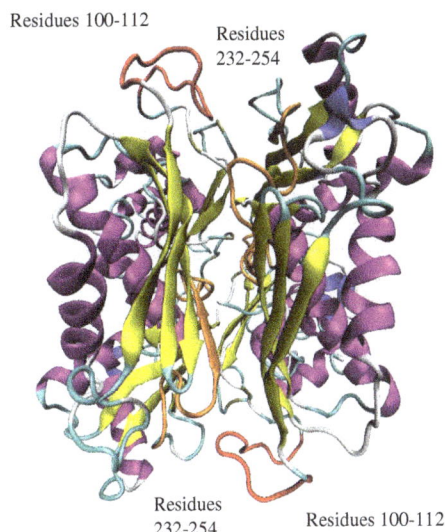

Residues 100-112

Residues 232-254

Residues 232-254   Residues 100-112

Figure 1. The structure of *E. coli* methionine adenosyltransferase homodimer (1FUG) with the flexible gating loops shown in red and the active site residues in orange. The structure visualization was achieved with VMD.

$$\text{RMSIP} = \left(\frac{1}{n} \left[\sum_{i=1}^{n} \sum_{j=1}^{n} \left(X_i Y_j \right)^2 \right] \right)^{\frac{1}{2}} \tag{5}$$

where X_i and Y_j represent the eigenvectors of a pair of proteins being compared and i and j represent the mode numbers. In the present study, the default setting of WEBnm@, which includes the first 10 lowest energy nontrivial modes (Amadei et al., 1999; Tiwari et al., 2014), was used. The RMSIP values range from 0 to 1; RMSIP of 1 represents maximum similarity in C_α atom fluctuations between proteins being compared.

2.8. Identification of test proteins

PDB coordinate files of 12 separate proteins, which either belong to one of the 24 protein groups included in the present study or were randomly chosen proteins, were selected from the PDB. The coordinate files of these proteins were modified to remove any identifying evidence, leaving only the Cartesian coordinates of each atom in the PDB file to run a single NMA analysis. The DCCMs of these test proteins were then compared to the catalog of DCCMs (obtained from the NMA of 24 protein groups) in an attempt to correctly match a test protein to a known protein group for functional identification. Wherever possible, quantitative analyses were performed using the comparative NMA analysis option of WEBnm@. The comparative analysis provides the extent of dynamic similarities between proteins in terms of BC and RMSIP values.

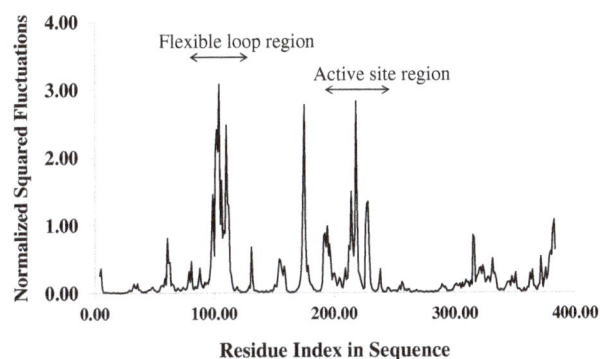

Figure 2. The C_α atom fluctuations of the *E. coli* methionine adenosyltransferase.

Notes: The fluctuations describe the flexibility of C_α atoms of the enzyme. The residue numbers are given on the x-axis, and the amplitude of the fluctuations on the y-axis. The graph showing normalized squared fluctuations of each C_α atom and calculated using the default setting of WEBnm@.

3. Results and discussion

3.1. Protein sequences and structures

A total of 141 protein sequences and structures were collected from the PDB files. The proteins are from metabolic pathways including glycolysis, gluconeogenesis, fermentation, the citric acid cycle, and the urea cycle (Table 1). These 24 protein groups along with the name of organisms, PDB codes, and the number of amino acids in polypeptide chain A are listed in Table 1. Each protein group has six different organisms associated with it, with three exceptions; arginase, phosphoglycerate kinase, and ornithine transcarbamoylase, which only have five structures. Each protein group possessed a similar number of residues per chain.

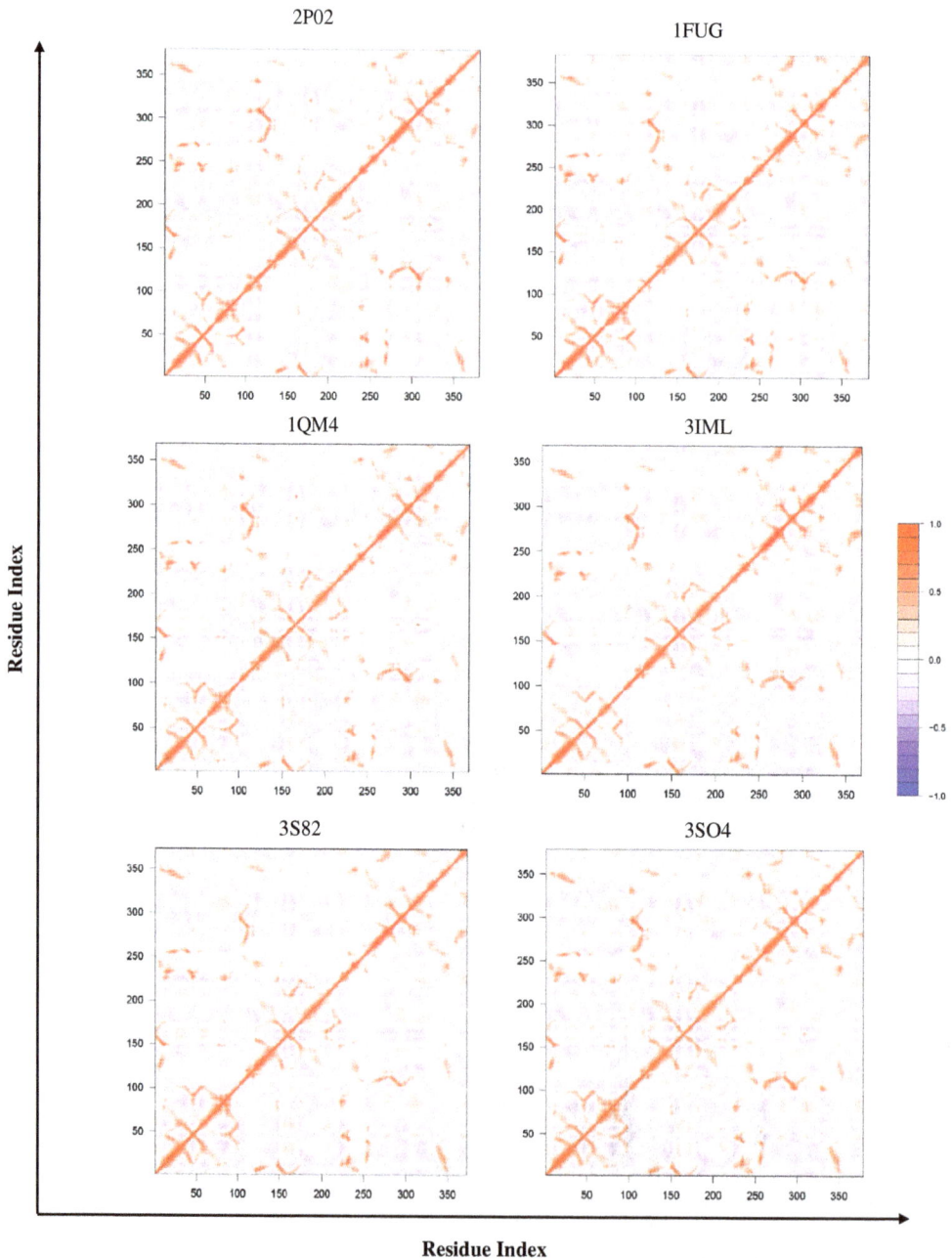

Figure 3. Dynamic cross-correlation matrices of six proteins in the methionine adenosyltransferase group with a color key depicting the correlated motions in red and anticorrelated motions in blue.

Note: The overall pattern remains consistent across all species within the group, apart from minute differences that are barely visible to the naked eye.

Table 3. A list of proteins with PDB codes, DCCM of the protein represented in bold, and the group BC heat map

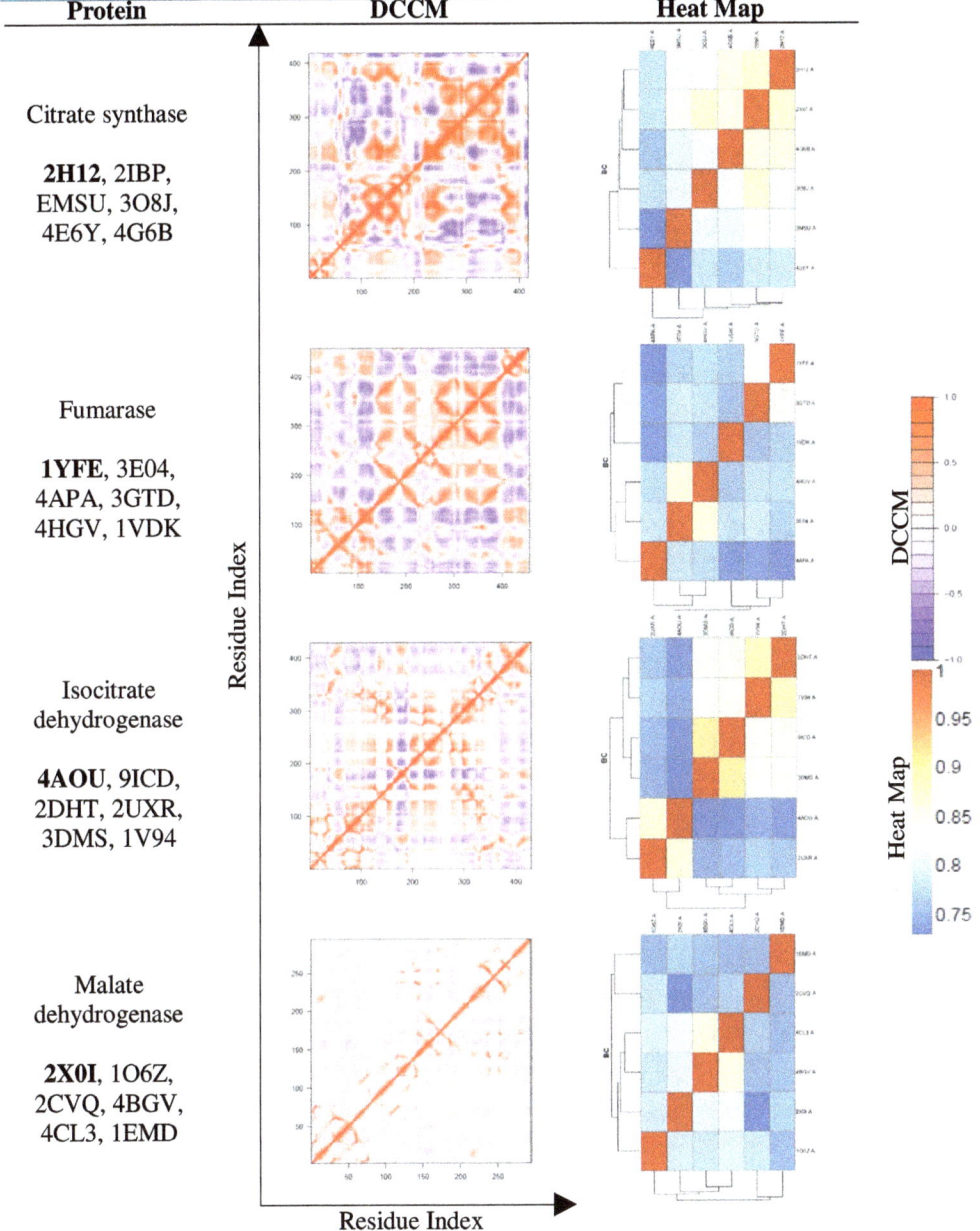

Protein	DCCM	Heat Map
Citrate synthase **2H12**, 2IBP, EMSU, 3O8J, 4E6Y, 4G6B	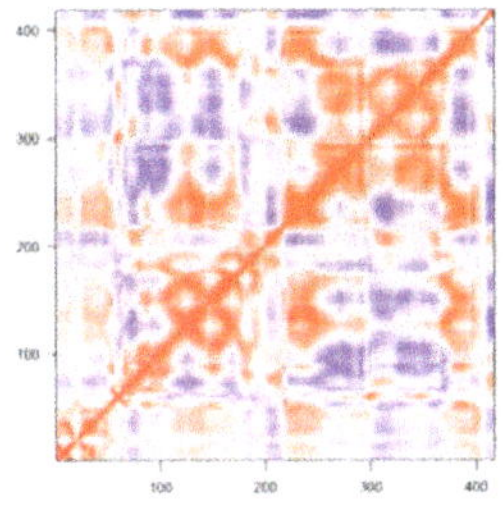	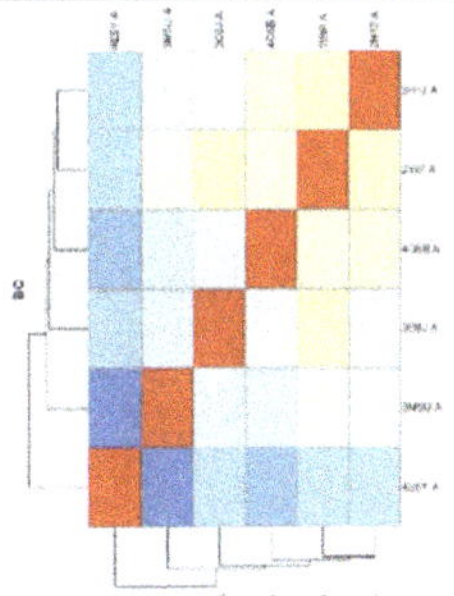
Fumarase **1YFE**, 3E04, 4APA, 3GTD, 4HGV, 1VDK	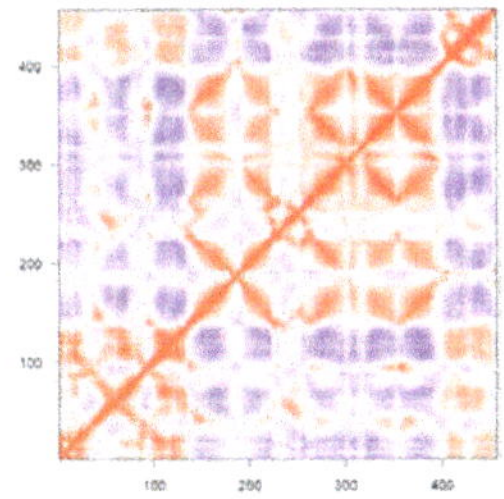	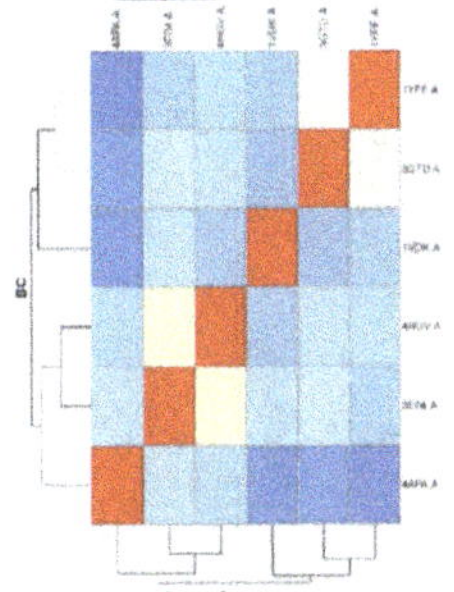
Isocitrate dehydrogenase **4AOU**, 9ICD, 2DHT, 2UXR, 3DMS, 1V94	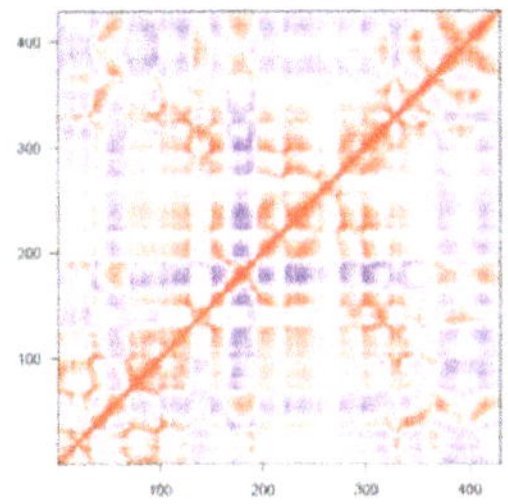	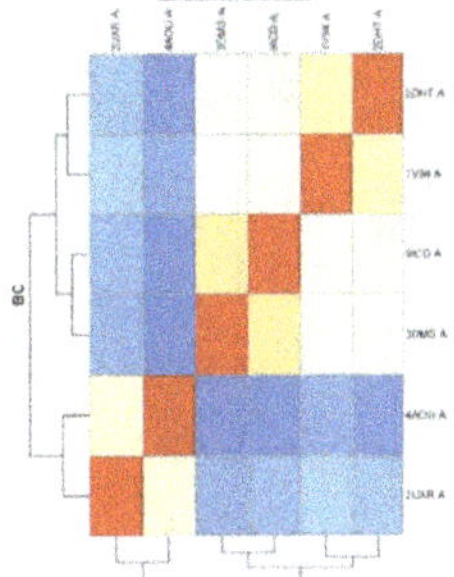
Malate dehydrogenase **2X0I**, 1O6Z, 2CVQ, 4BGV, 4CL3, 1EMD	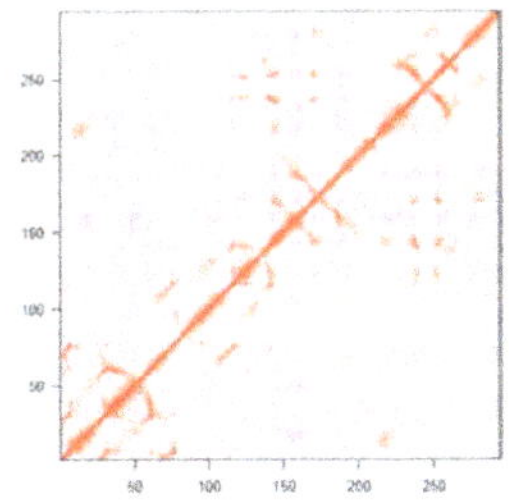	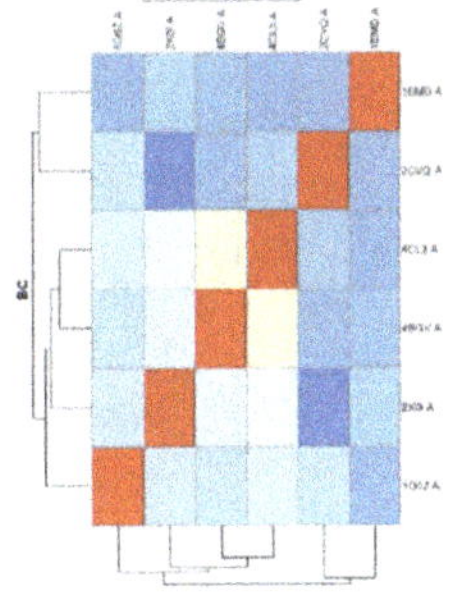

Residue Index (axis, DCCM plots)

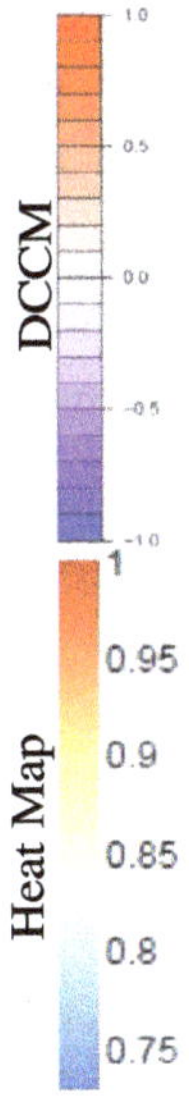
DCCM
1.0
0.5
0.0
-0.5
-1.0
Heat Map
1
0.95
0.9
0.85
0.8
0.75

(*Continued*)

Table 3. (Continued)

Protein	DCCM	Heat Map
Succinyl-CoA synthetase **1CQI,** 2YV1, 1OI7, 2FP4, 2YV1, 2YV2, 3UFX		
Alcohol dehydrogenase **1E3L,** 1HTB, 6ADH, 4CPD, 4RQT, 3IP1		
Lactate dehydrogenase **1CET,** 3LDH, 1T25, 3D0O, 2ND5, 3WSV		
Enolase **1ELS,** 1PDZ, 1IYX, 3B97, 4A3R		

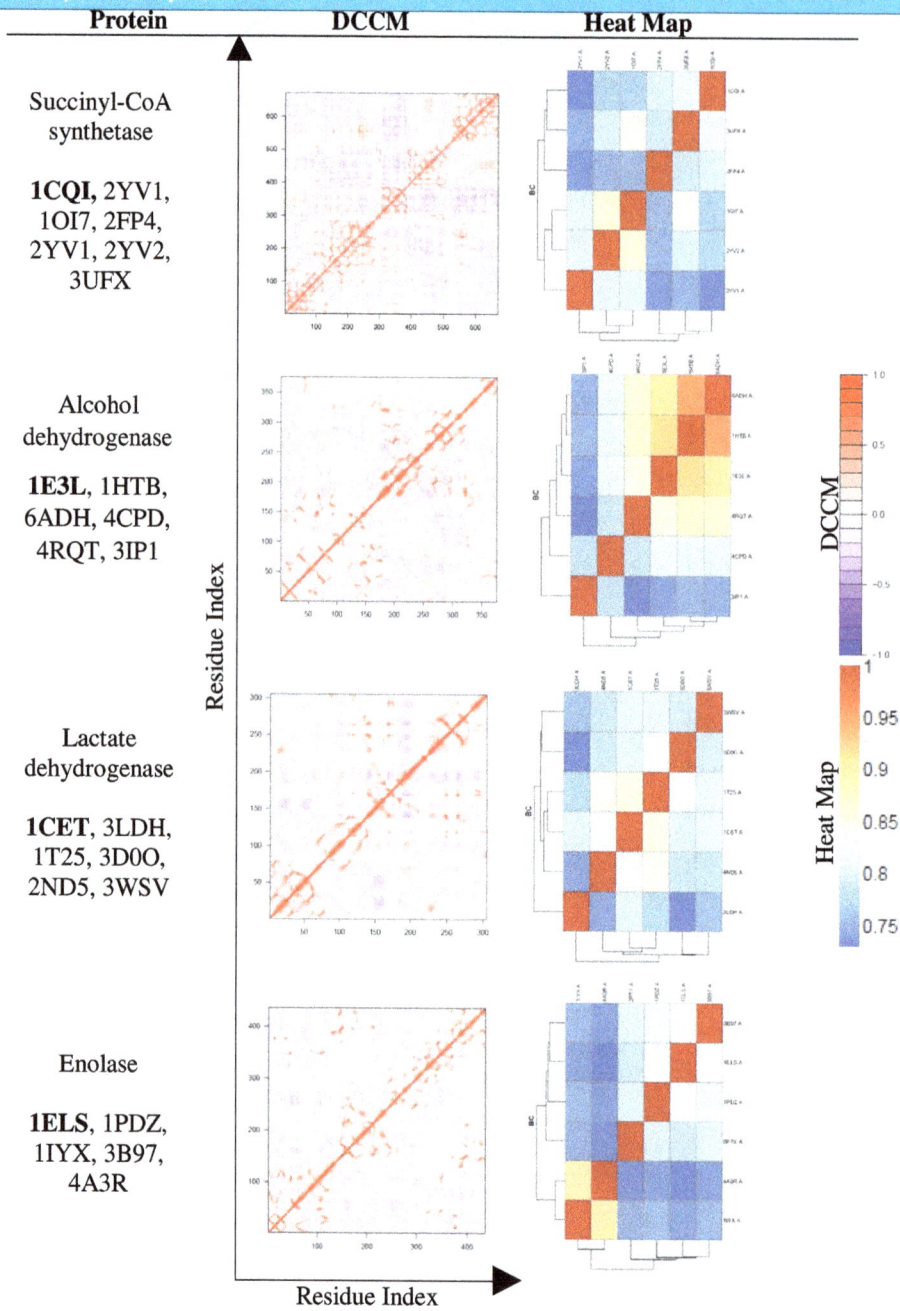

Residue Index

Residue Index

(Continued)

Table 3. (Continued)

Protein	DCCM	Heat Map
Fructose-bisphosphate aldolase **3MBF**, 2FJK, 4LV4, 3KX6, 4TU1, 3DFN	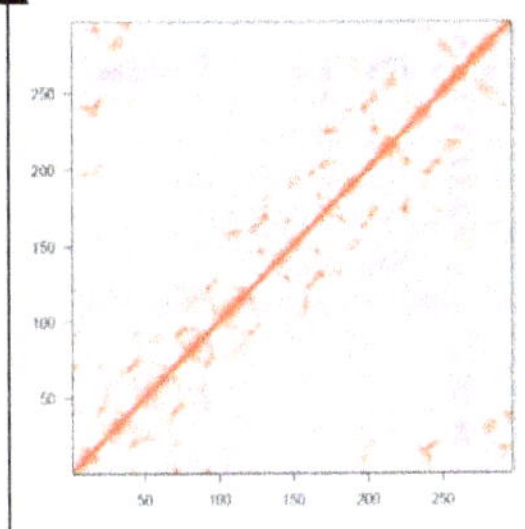	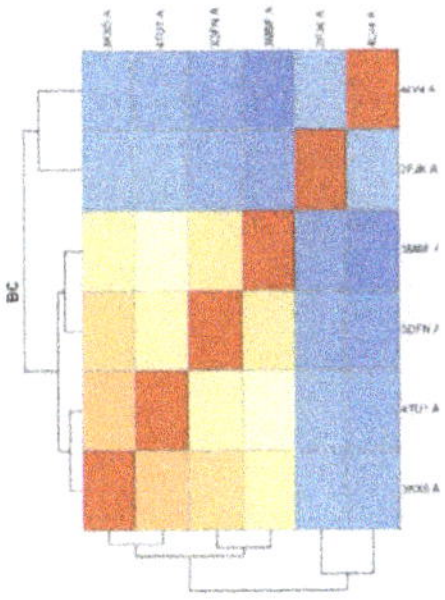
Glyceraldehyde phosphate dehydrogenase **1NQO**, 4QX6, 3STH, 3HJA, 4DIB, 3L0D	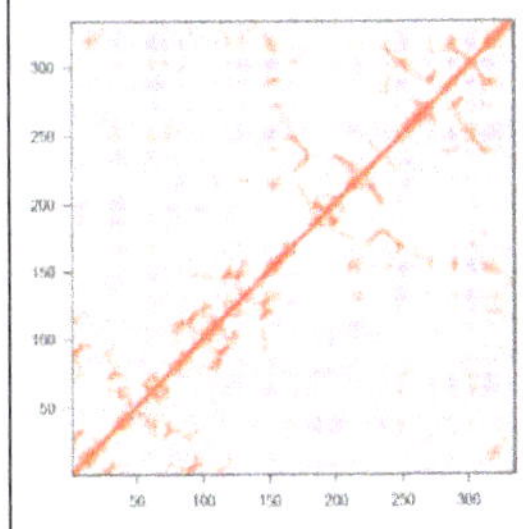	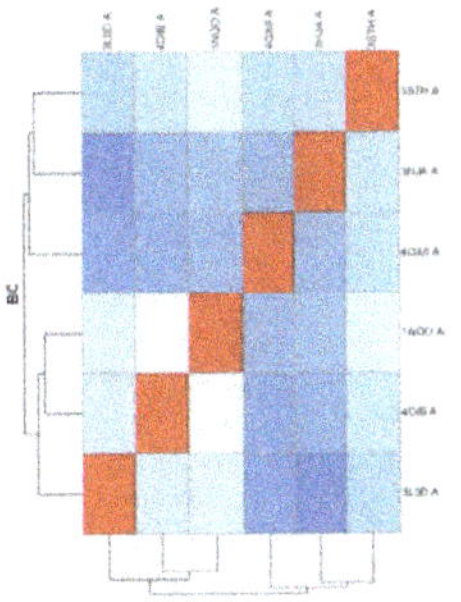
Hexokinase **1BDG**, 3B8A, 3W0L, 4JAX, 4QS8, 4IWV	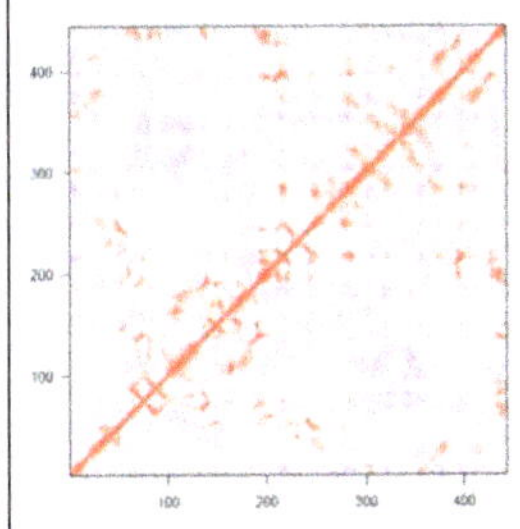	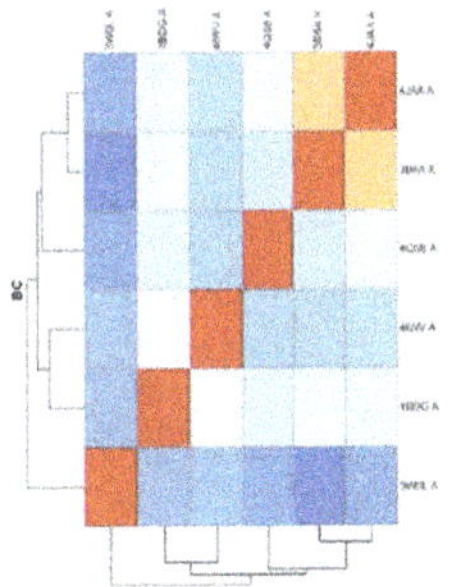
Phospho-glycerate kinase **1PHP**, 1V6S, 1VPE, 1ZMR, 3UWD, 4FEY	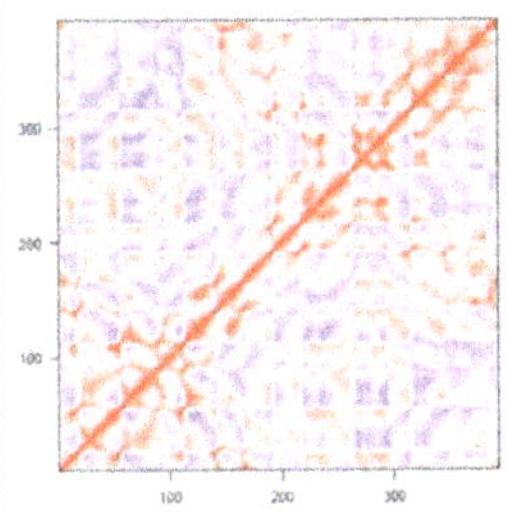	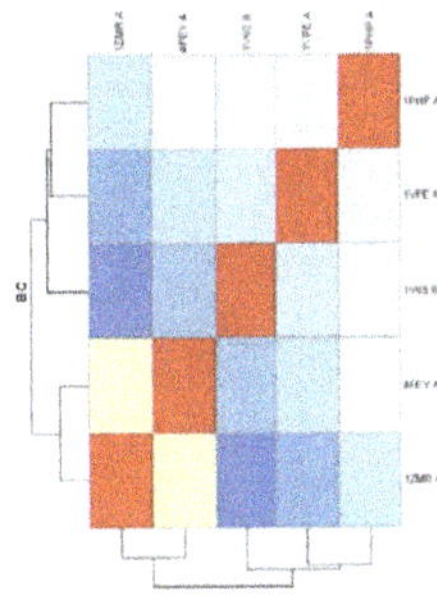

Residue Index

Residue Index

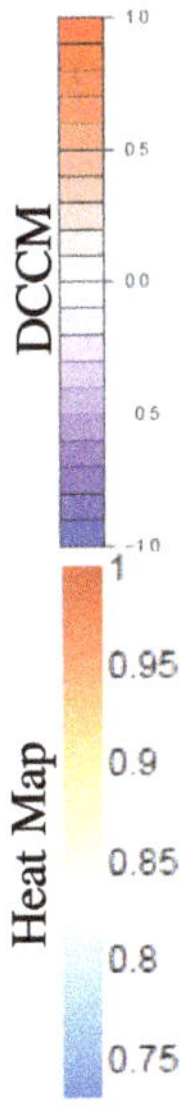

(Continued)

Table 3. (Continued)

Protein	DCCM	Heat Map

Phospho-glycerate mutase

3PGM, 3R7A, 4ODI, 1XQ9, 1YFK, 1BQ4

Pyruvate decarboxylase

1PYD, 2QMA, 1QPB, 2NXW, 1OVM

Pyruvate kinase

1AQF, 1PKM, 3E0V, 3EOE, 3SRF, 4DRS

Triose phosphate isomerase

1WYI, 1YPI, 2I9E, 3TH6, 4IOT, 8TIM

Residue Index

Residue Index

DCCM

Heat Map

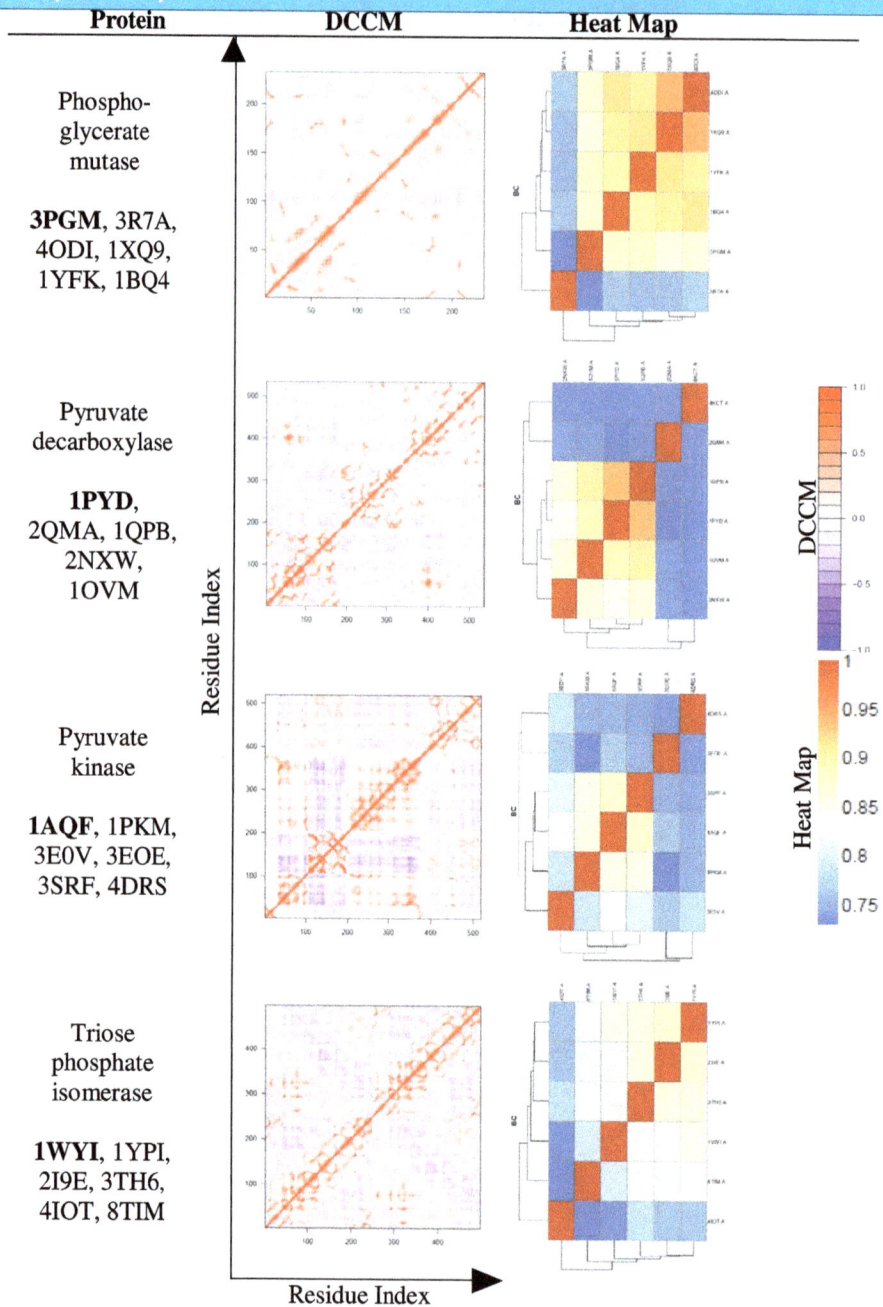

(Continued)

Table 3. (Continued)

Protein	DCCM	Heat Map

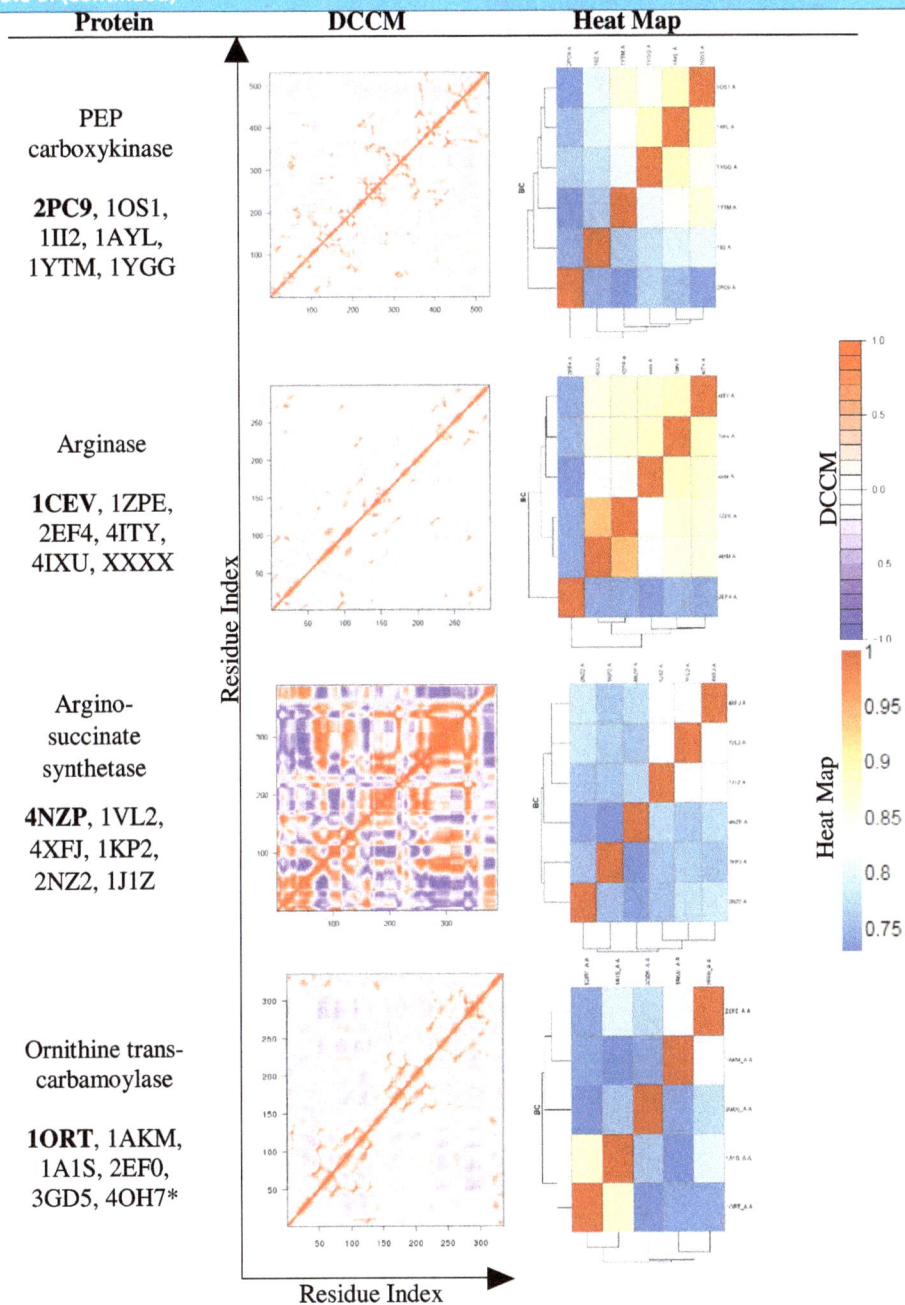

PEP carboxykinase

2PC9, 1OS1, 1II2, 1AYL, 1YTM, 1YGG

Arginase

1CEV, 1ZPE, 2EF4, 4ITY, 4IXU, XXXX

Argino-succinate synthetase

4NZP, 1VL2, 4XFJ, 1KP2, 2NZ2, 1J1Z

Ornithine trans-carbamoylase

1ORT, 1AKM, 1A1S, 2EF0, 3GD5, 4OH7*

Residue Index

Residue Index

(Continued)

Table 3. (*Continued*)

Protein	DCCM	Heat Map

Amine oxidase

1IVU, 1KS1, 2WO0, 3HI7, 3PGB, 3SX1

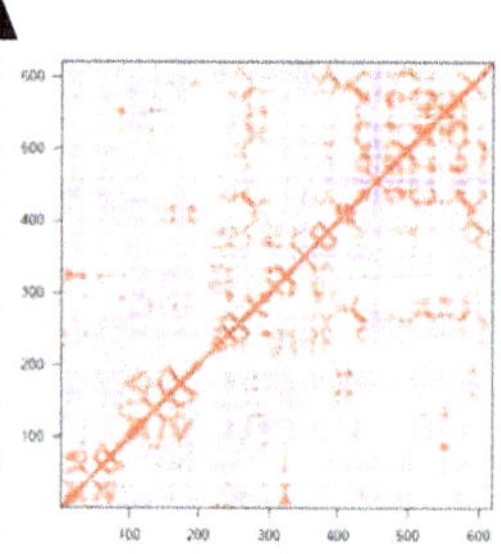

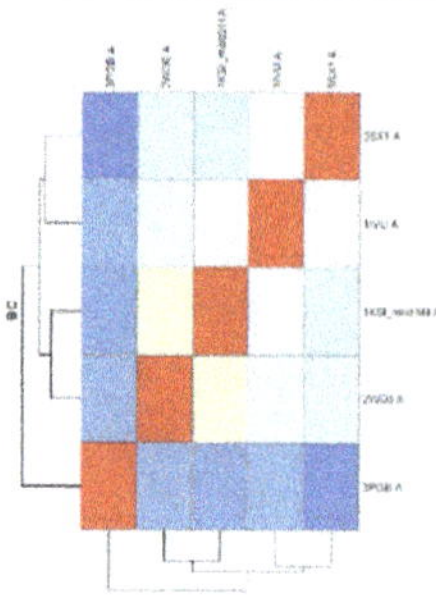

Methionine adenosyl-transferase

2OBV, 1FUG, 1QM4, 3IML, 3S82, 3SO4

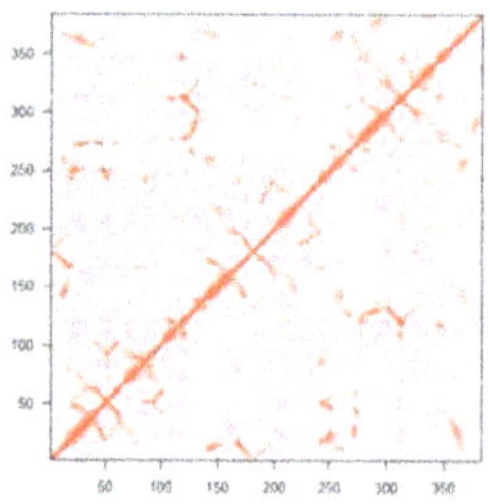

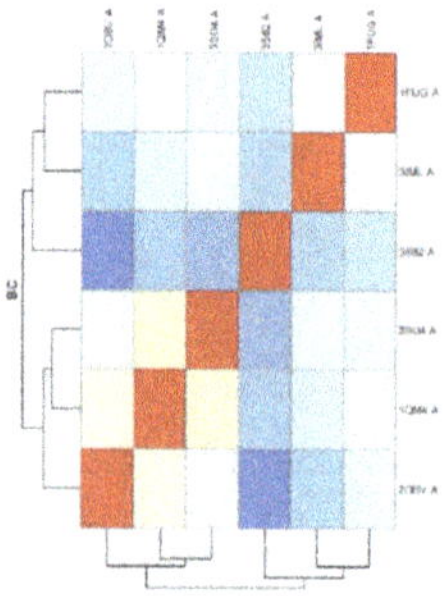

Pantothenate synthetase

2EJC, 2A86, 1UFV, 1IHO, 3AG5, 3Q10

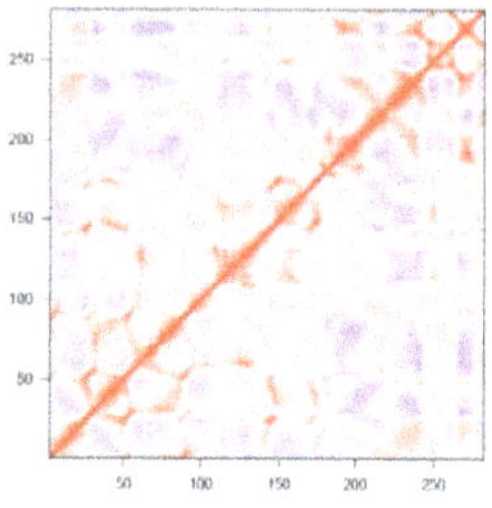

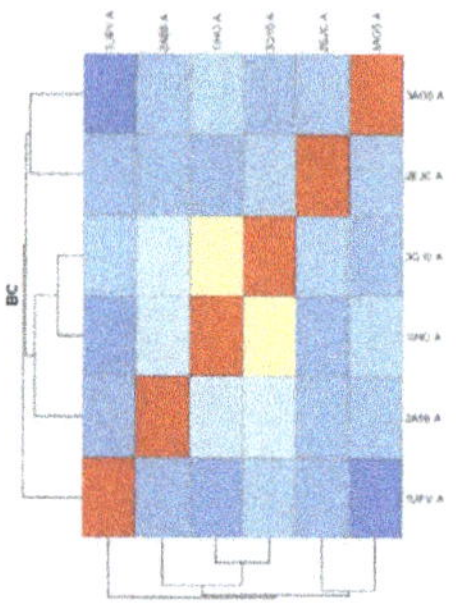

Alkaline phosphatase

1EW2, 1SHN, 2X98, 3BDF, 3E2D, 4KJG

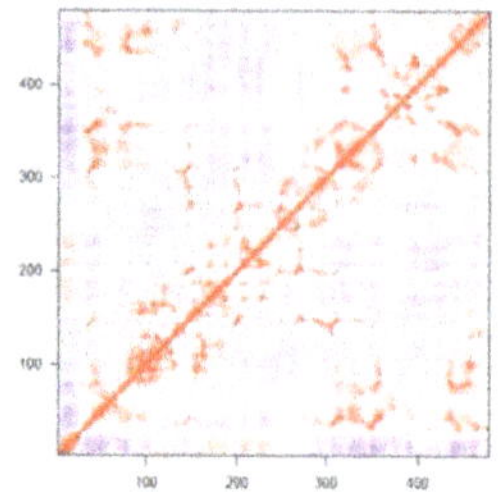

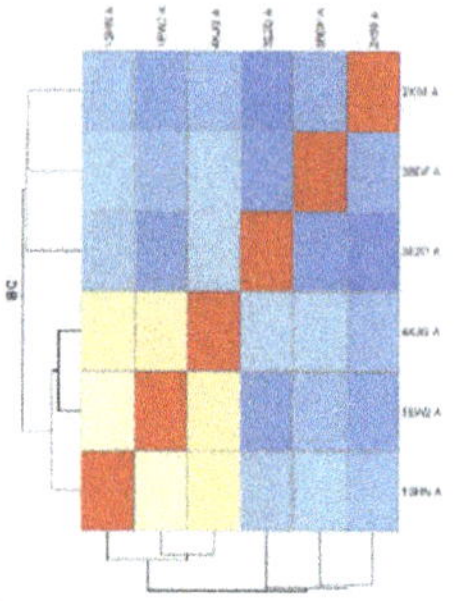

Residue Index

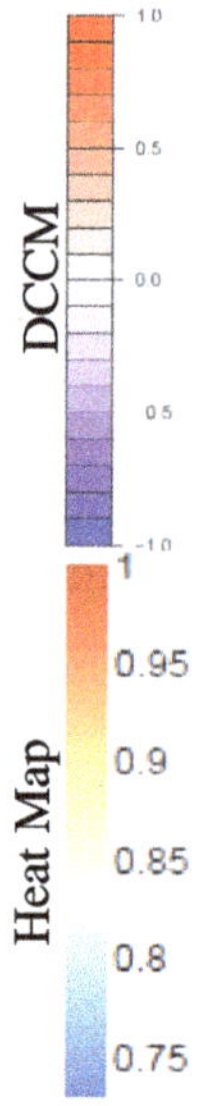

Note: Chain A of all enzymes were used for generating DCCMs and heat maps, except for succinyl-CoA synthetase, which has both chain A and chain B.

3.2. Sequence comparison

The sequence identities between proteins of a given family are computed and listed in Table 2. The sequence identities range from about 24 to 61% with an average of 46% among all protein groups. Nine of the protein families had sequence identities greater than 50%, which means that the individual protein structures have at least 90% of their residues within the common cores (Chothia & Lesk, 1986). For the other 15 protein families with sequence identities between 20 and 50% are found to have 42–98% of individual protein residues within the common cores (Chothia & Lesk, 1986).

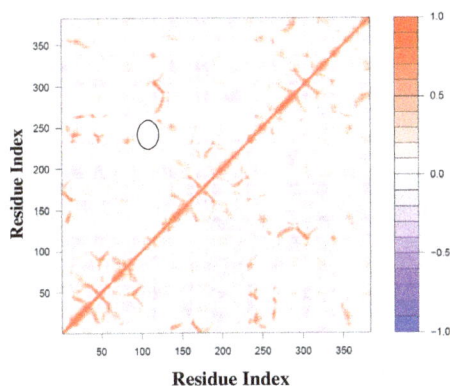

Figure 4. A dynamic cross-correlation matrix for *E. coli* Methionine Adenosyltransferase (1FUG) showing the correlated (red) and anticorrelated (blue) motions between C_α atoms.

Note: The black oval on DCCM represents the anti-correlated motions between the active site (residues 247–258) and the flexible loop (residues 101–109).

3.3. Structure comparison

The RMSD values that represent structural similarities between proteins within an enzyme family are also shown in Table 2. The RMSD represents the deviation between two superimposed structures. The observed RMSD values ranged from 1.0 to 3.1 Å, with the lower numbers representing more similar structures. The RMSD is usually small (<3 Å) for homologous proteins (Chothia & Lesk, 1986; Kosloff & Kolodny, 2008). The RMSD between homologous proteins within an enzyme family studied in this work were observed to be ≤3 Å, which suggested that these homologous proteins possess considerable extent of structural similarities. For higher sequence identities, we expected to see a low RMSD for the protein group, which was satisfied (Table 2).

3.4. The C_α atom fluctuations

The fluctuation profiles for C_α atoms of all 24 groups of proteins were obtained from the NMA study. The calculated normalized squared fluctuations of C_α atoms matched satisfactorily with the experimental results for these enzymes. For example, the structure of the *Escherichia coli* methionine adenosyltransferase (MAT) (Figure 1) revealed a dynamic loop region shown in red (Fu, Hu, Markham, & Takusagawa, 1996). Both the loop region (residues 100 to 112 for *E. coli* MAT) and the adjacent residues displayed high flexibility as observed in the normalized fluctuation (C_α atoms) profile (Figure 2). This observation confirmed that the experimentally observed flexible regions of MAT correlate well

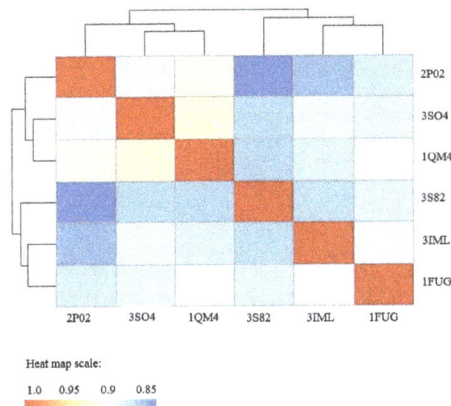

Figure 5. Methionine Adenosyltransferase heat map of BC showing the dynamic relatedness between proteins, with the color key at the bottom.

Notes: Red describes identical dynamics, and blue describes the most unrelated dynamics within the group. Cluster data is shown on the x- and y-axis, grouping 1FUG, 3IML, 3S82 together with 3IML being more similar to 1FUG than to 3S82. Similarly, 2P02, 3SO4, and 1QM4 form another cluster, with 1QM4 and 3SO4 being more related to each other than to 2P02.

with peaks in the fluctuation profile. Also, high flexibility was noticed from the active site motions, which can be seen around residues 232 to 254.

3.5. Correlated and anticorrelated motions

The single NMA for each protein within a particular group revealed strikingly similar correlation matrices. Slight differences were seen with respect to the extent of correlated and anticorrelated motions; however the overall patterns remained consistent. As an example, Figure 3 shows the six DCCM files for the MAT group. All six protein matrices appear identical, as was the case with all other protein groups. Therefore, only one representative DCCM was chosen for each enzyme group and listed in Table 3. Because DCCMs of homomultimeric proteins revealed that each polypeptide chain displays identical patterns of motions (Figure S1), only one polypeptide chain of multimeric proteins was used to represent the intrinsic dynamic patterns (DCCMs) of all protein families, except for succinyl-CoA synthetase. For succinyl-CoA synthetase, a heteromultimeric protein, both polypeptide

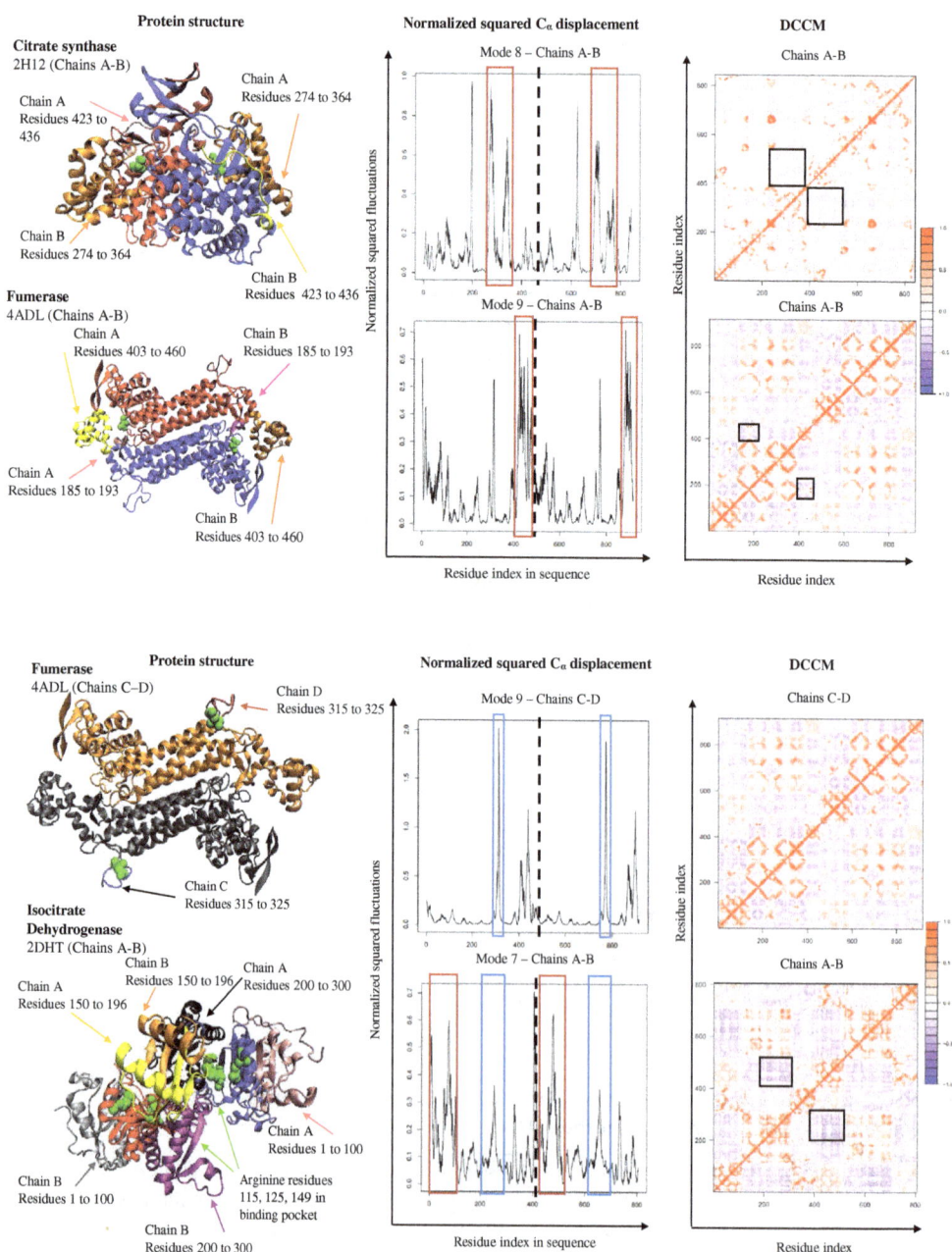

Figure 6. (Continued).

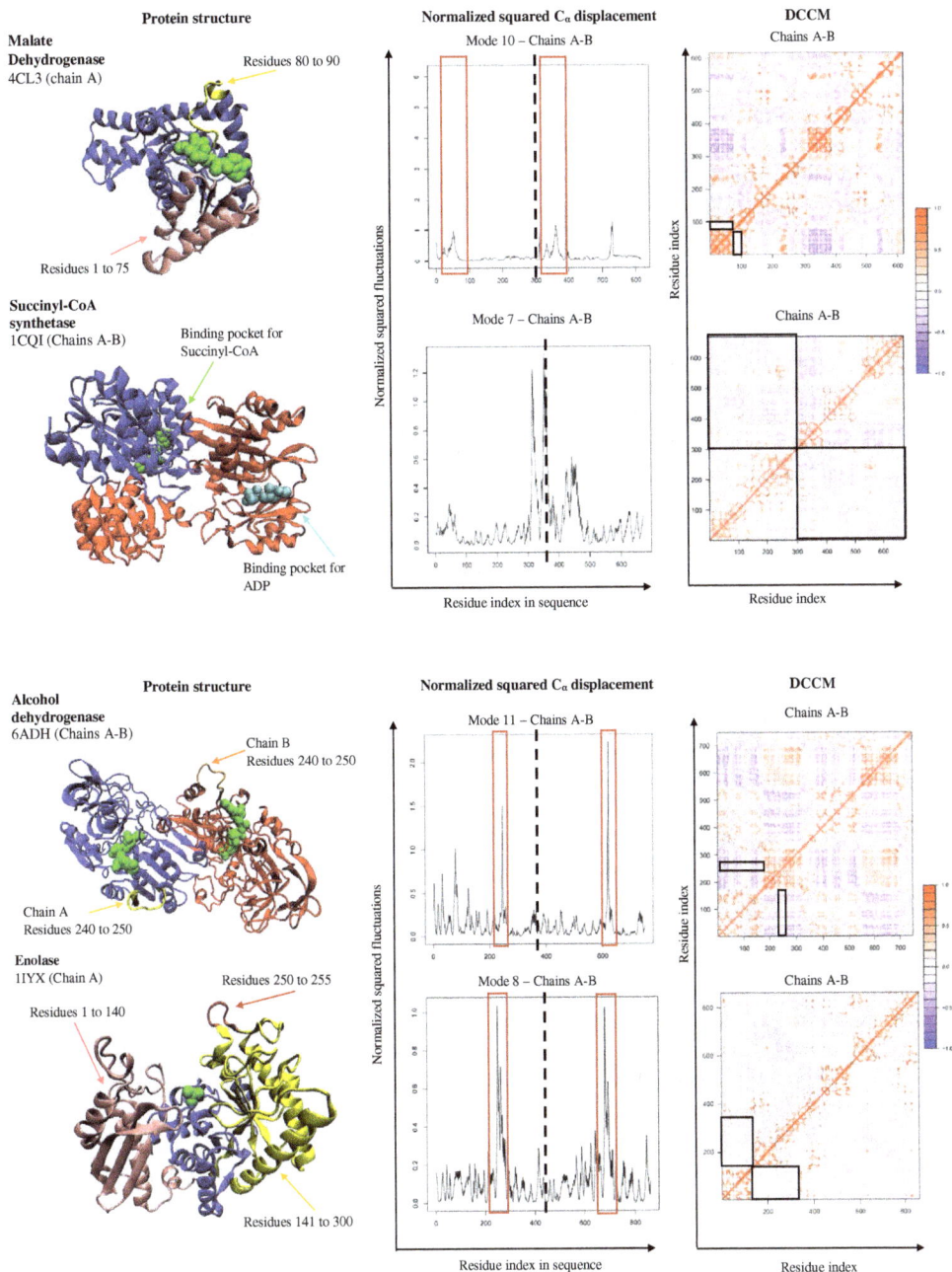

Figure 6. (Continued).

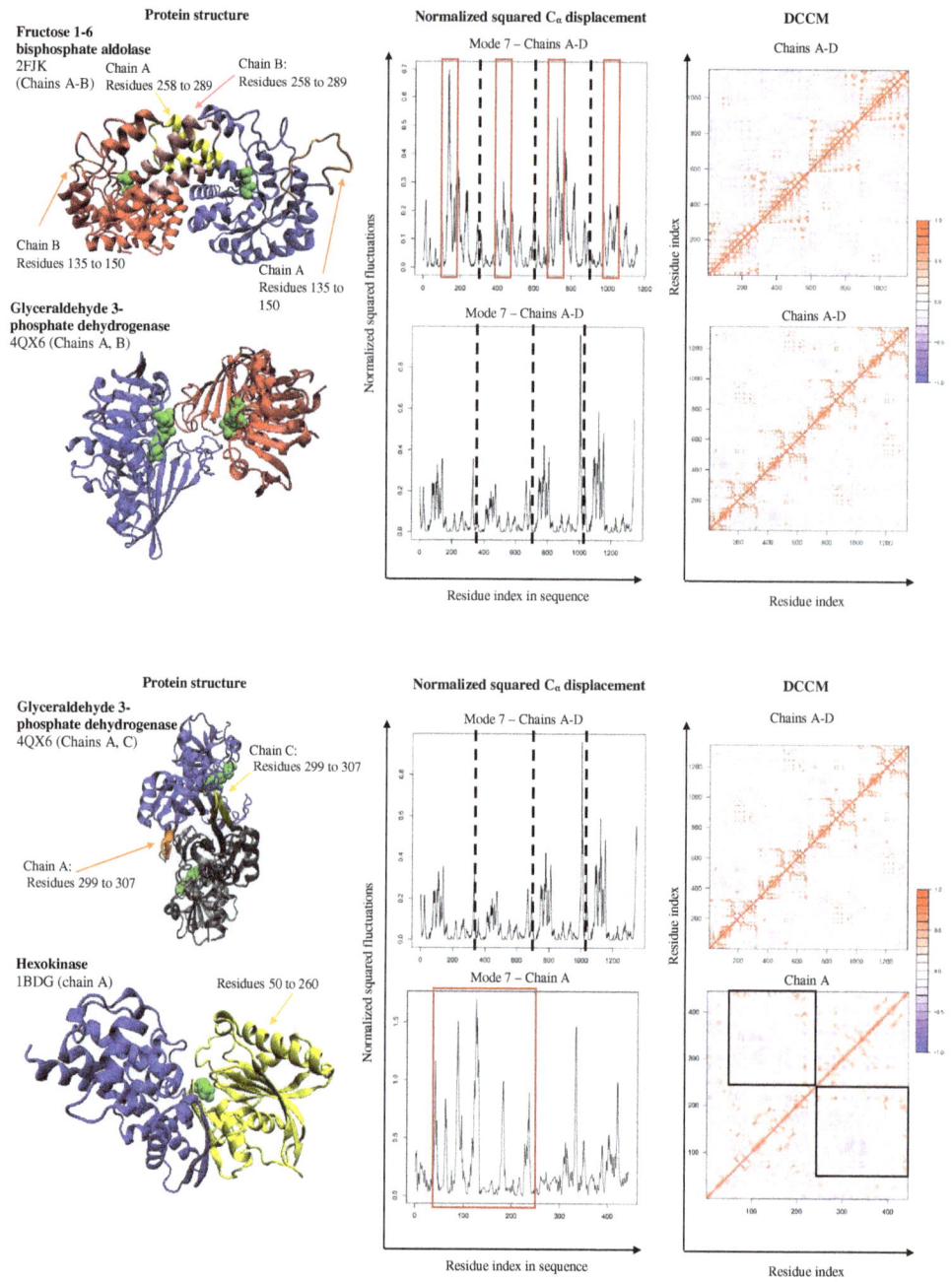

Figure 6. (Continued).

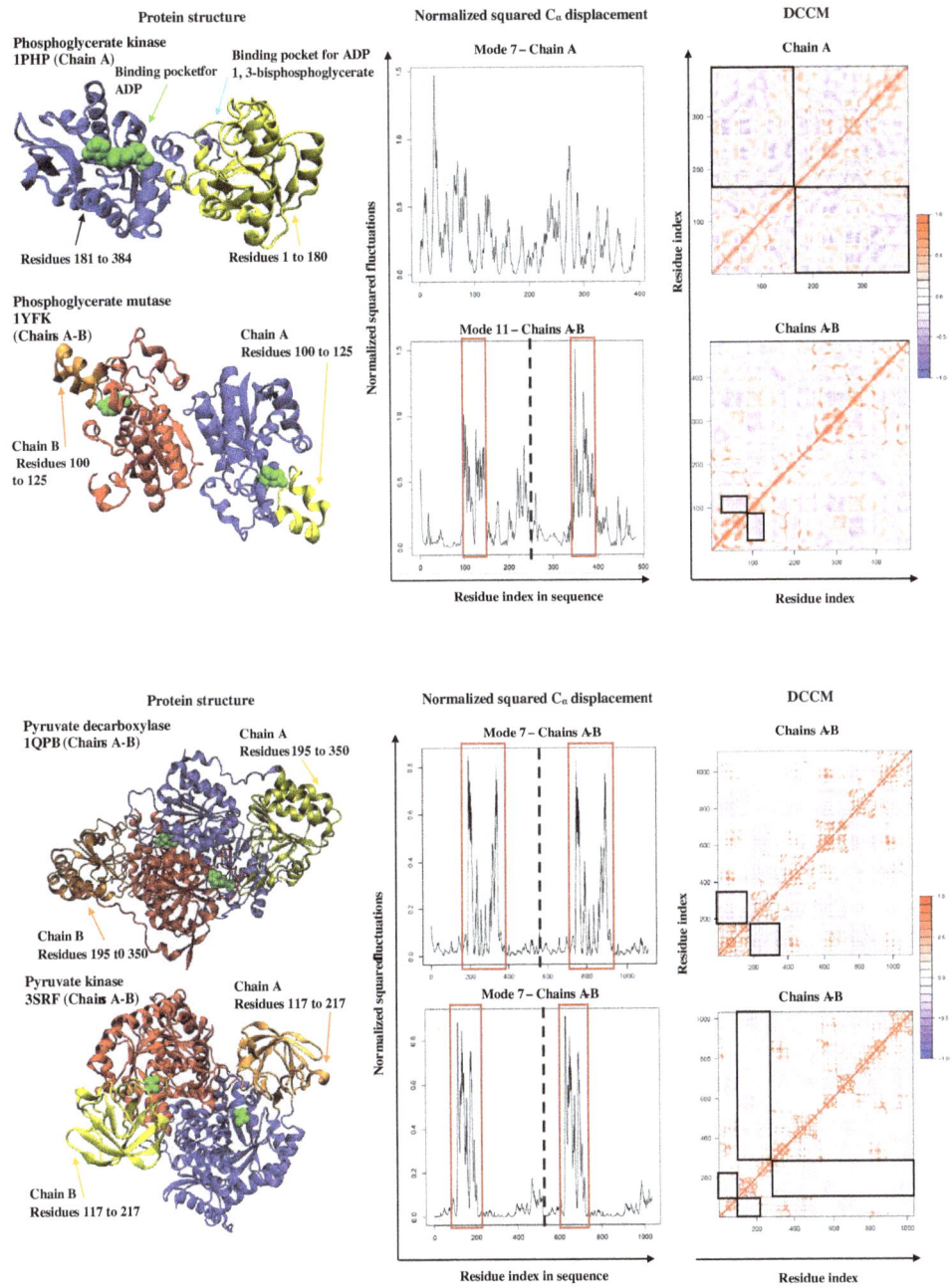

Figure 6. (*Continued*).

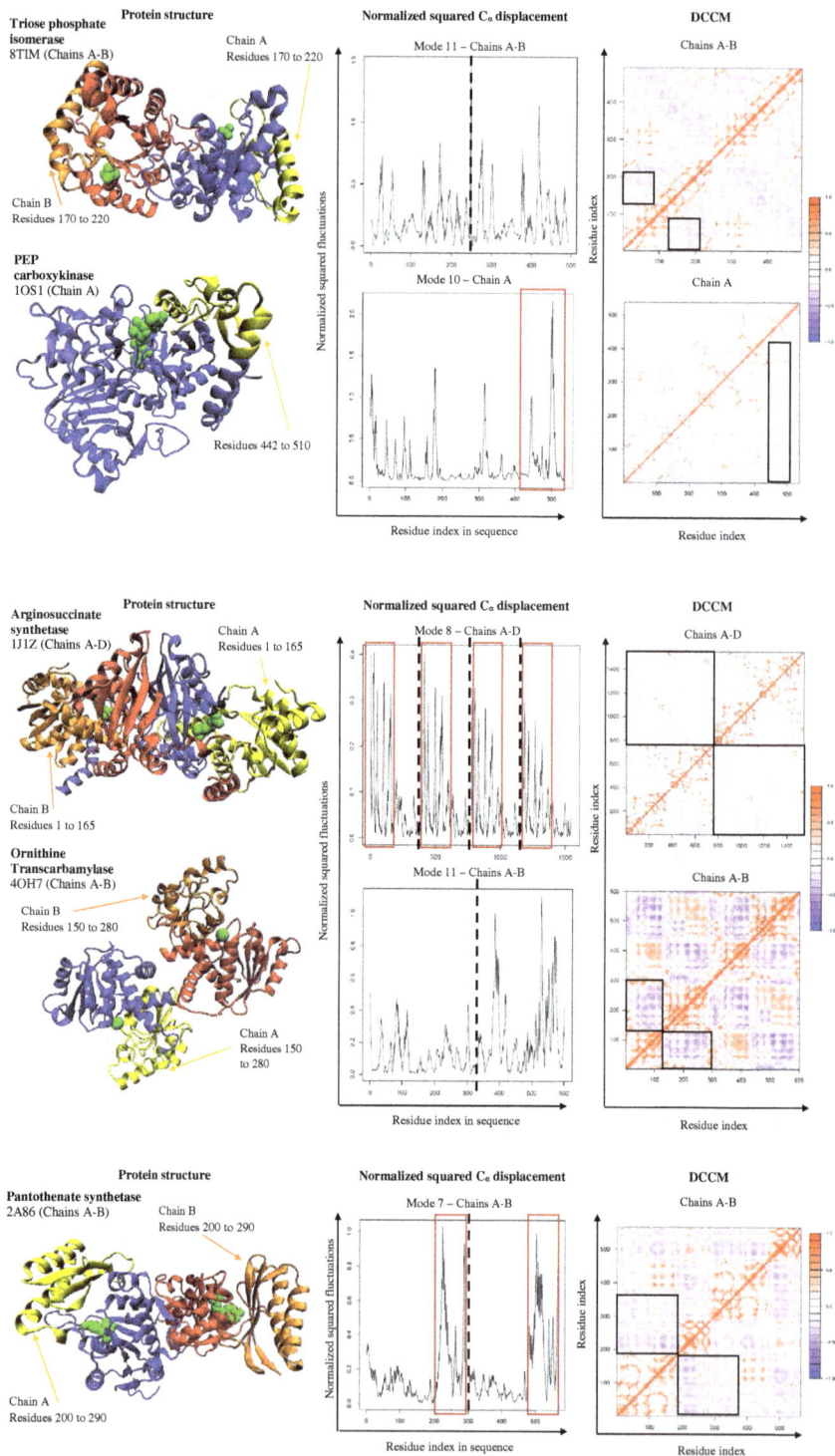

Figure 6. A list of proteins with corresponding PDB code and their structural visualization, the functionally relevant C_α atom displacement graph, and DCCM.

Notes: Chain A and chain B are shown in blue and red color, respectively. The relevant structural elements are shown in different colors. The substrate or functionally relevant active site residues are shown in green van der Waals space filling model. The dashed lines differentiate between two polypeptide chains. Red and blue boxes correlate to specific textual references.

chains were used to represent the intrinsic dynamic patterns (Table 3). However, to examine the entire dynamical nature of a protein system and the functional relevance of those intrinsic dynamics, the other chains were also taken into account. For example, MAT (Figure 1) functions as a homodimer and as shown in Figure 1, the flexible loop appears to gate the active site. The majority of the active site consists of residues 247 to 258, and the flexible loop consists of residues 101–109. Interestingly, the flexible loop shows anticorrelated motion with the active site (Figure 4). This anticorrelated motion suggests that the loop would open, which will allow the substrate to have proper access to the active site. The anticorrelated motion between active site pocket and the loop region also suggests that the two subunits move apart to make room for substrate molecules to enter active sites. These motions are seen across all species, suggesting that these functional dynamics have been conserved throughout MAT evolution.

3.6. Bhattacharyya coefficients
The average BC of the 24 protein groups remained remarkably consistent (Table 2), with the lowest score of 0.73 and the highest score of 0.89, where BC of 1.0 represents the maximum similarity in dynamics of proteins used for comparison. It appears that even for the lower sequence identities and similarities, the dynamic similarity of each protein group remains relatively high. Similar observations were made in the case of the cytochrome P450 family of enzymes, where these enzymes share a strong dynamic similarity (BC > 80%) despite low sequence identity (<25%) (Dorner et al., 2015).

   In the present study, the comparison of the dynamic similarities among proteins within a group was also made. Table 3 provides a representative heat map of the BCs for each enzyme. Heat map data describe the pairwise comparison of protein structures using the Bhattacharyya coefficients, with red shading representing identical dynamics between proteins being compared, and blue shading showing the least similar intrinsic dynamics. The heat map also groups the individual proteins into clusters, based upon the relatedness of enzyme dynamics. This provides additional information to enable the investigation of protein evolution with regard to protein dynamics. For example, the heat map for the BC of MAT is shown in Figure 5 (the sequence alignment file used for dynamic comparison is shown in Supplementary Table S2). The cluster data is shown on the x- and y-axis, grouping the dynamics of E. coli (1FUG), B. peudomallei (3IML), and M. avium (3S82) MATs together. The dynamics of 3IML are more similar to the dynamics of 1FUG, than to the dynamics of 3S82. The dynamics of 2P02 (H. sapiens), 3SO4 (E. hystolytica), and 1QM4 (R. norvegicus) form another cluster, with 1QM4 and 3SO4 being more related to each other than to 2P02.

3.7. Functional relevance of the existing intrinsic dynamics
Close analysis of the intrinsic dynamics of the metabolic enzymes studied here revealed some functionally relevant motions, which are described below. For this study, depending upon the protein system, single or multiple polypeptide chains were considered to analyze the functional dynamics. In particular, an effort has been made to consider as many polypeptide chains as possible to maintain the dynamic integrity of the given protein system and show their functional relevance. Also, for simplicity, the C_α atom displacement graph of a specific mode(s) that was functionally more relevant has been considered here. The online NMA server failed to calculate and analyze the normal modes of five enzyme families when the multimeric proteins, instead of single polypeptide chains, were used. Therefore, intrinsic dynamics and their functional relevance of 19 out of 24 enzyme systems were analyzed here. Information regarding the general enzymatic properties and catalytic roles of various enzymes studied here were taken from *Biochemistry 7th Edition* (Berg, Tymoczko, & Stryer, 2012).

   Citrate synthase (PDB code: 2H12; polypeptide chains A and B of the tetrameric protein were visualized for structural and dynamical analyses) is responsible for condensing acetyl-CoA and oxaloacetate to form citrate. To accomplish this, the enzyme undergoes an "open" to "closed" conformational change (Wiegand & Remington, 1986). First, the substrate, oxaloacetate, enters the enzyme in its "open" conformation; the coenzyme, acetyl-CoA, further interacts with nearby residues in the "closed" conformation. These residues are located too far to facilitate the reaction while in the

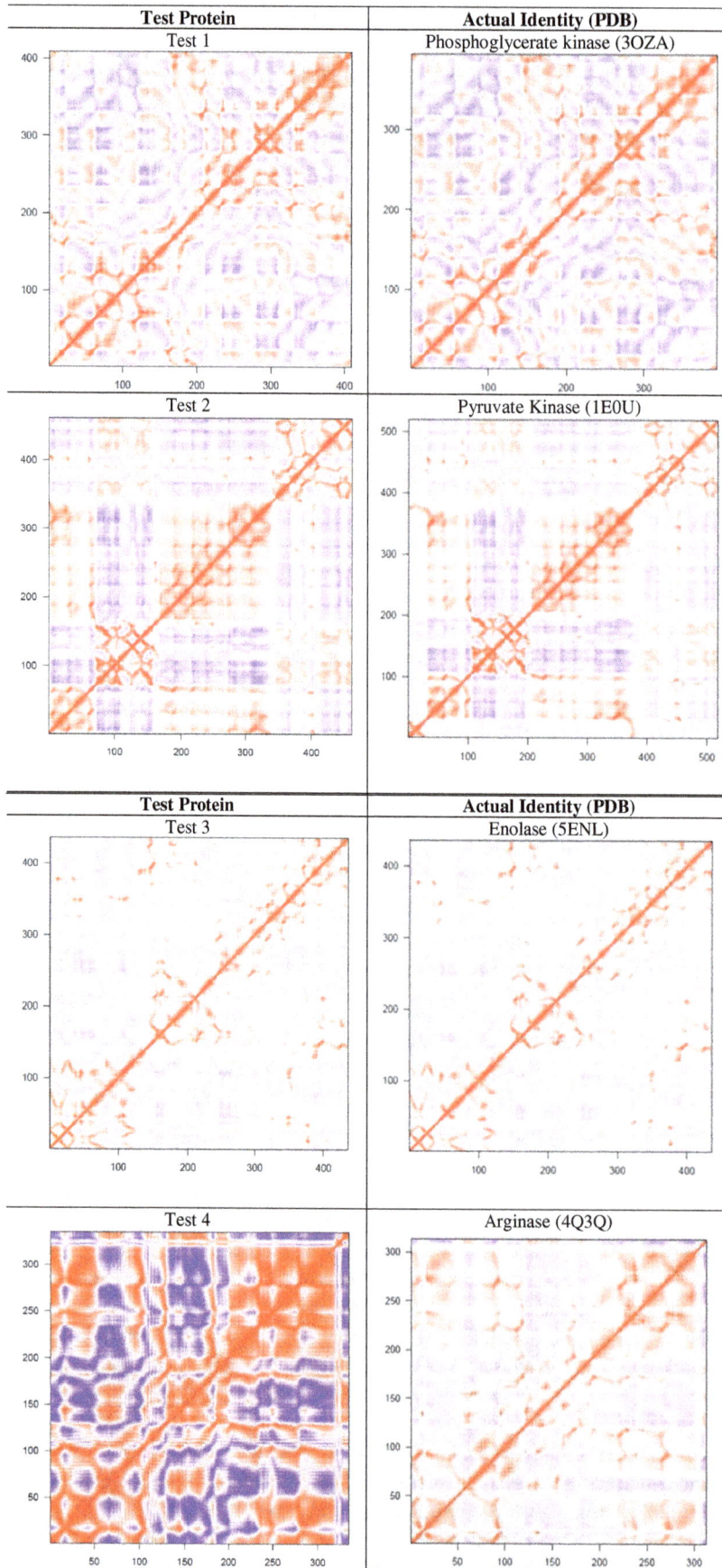

Test Protein	Actual Identity (PDB)
Test 1	Phosphoglycerate kinase (3OZA)

Test 2	Pyruvate Kinase (1E0U)

Test Protein	Actual Identity (PDB)
Test 3	Enolase (5ENL)

Test 4	Arginase (4Q3Q)

Figure 7. (*Continued*)

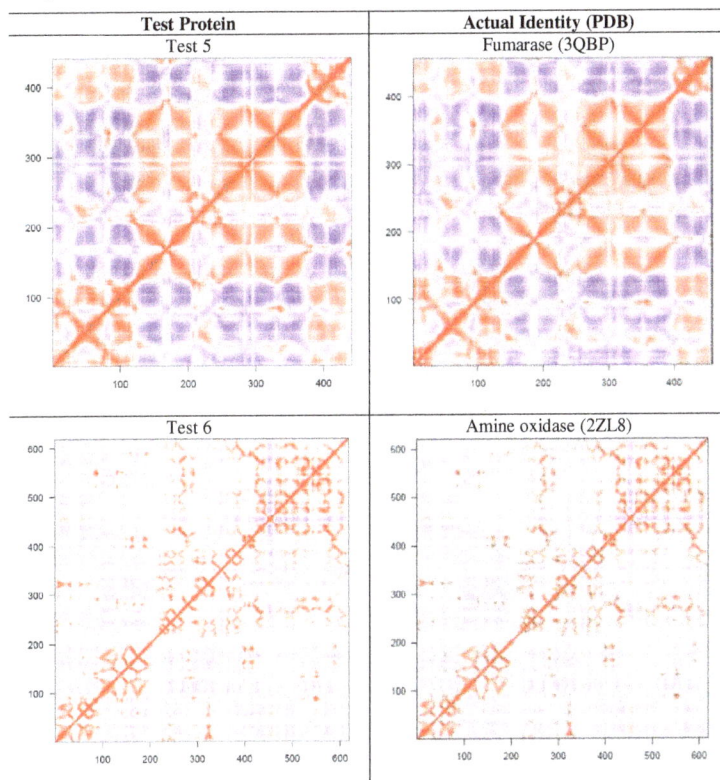

Test Protein	Actual Identity (PDB)
Test 5	Fumarase (3QBP)
Test 6	Amine oxidase (2ZL8)

Figure 7. The six test proteins' DCCMs are shown along with their matched protein representative DCCM.

Notes: Each test protein was correctly identified using the created catalog of DCCMs obtained from the single NMA calculations. The x- and y-axis represent the residue index of the protein.

"open" state. The enzyme accomplishes this by having residues 274 to 364, in conjunction with surrounding residues, exhibit an anticorrelated motion with greater flexibility (boxes in the fluctuations plot and DCCM, Figure 6) to the active site and residues 423 to 436, as it opens and closes around the active site of citrate synthase. Residues 423 to 436 on both subunits are shown to interact with the opposing subunit, suggesting that these residues assist in conformational changes of each other, as well as facilitate substrate binding and catalysis.

Fumarase (PDB code: 4ADL, a tetrameric protein; in the present study chains A and B and chains C and D are visualized separately for structural and dynamical analyses) catalyzes the reversible reaction of fumarate to malate. This reaction occurs with a conformational change after the substrate enters the active site, which promotes the enzyme to transition into its "closed" conformation (Stoddard, Dean, & Koshland, 1993). For fumarase to carry out its function, it exists as a tetramer with two active sites that interact with its substrates. However, only three of the four chains interact with each substrate (chains A, B, C and chains A, B, D interact with each substrate). For chains A and B, interactions with the substrate occur between residues 185 to 193 and 403 to 460, along with surrounding residues (red boxes in the displacement graph; Figure 6). These two groups of residues, 185 to 193 and 403 to 460, are also highly flexible and anticorrelated to each other, allowing for the greatest range of motion, so that the substrate can both easily diffuse in or out, in addition to binding tightly to the enzyme (boxes in DCCM, Figure 6). Chains C and D only interact with one of the two substrates through residues 315 to 325, which are shown in the C_{α} atom displacement graph as having the greatest movement (blue boxes in displacement graph; Figure 6).

Isocitrate dehydrogenase (PDB code: 2DHT, chains A and B was visualized for structural and dynamical analyses) is a homodimeric protein that is responsible for catalyzing the oxidative decarboxylation of isocitrate. The substrate interacts with the active site of isocitrate dehydrogenase via the three arginine residues, R115, R125 and R149 as well as five other residues within the active site (arginine residues shown in green space filled spheres, Figure 6) (Fedøy, Yang, Martinez, Leiros, & Steen, 2007). The two active sites of isocitrate dehydrogenase are on opposite sides of each other, meaning that when one active site is in the "open" state, the other is in its "closed" state conformation (Figure 6). The C_α atom displacement graph of isocitrate dehydrogenase shows general fluctuations throughout the whole enzyme, which is consistent to the overall motion. The active site for isocitrate dehydrogenase is unique in the fact that it is located between the two subunits. The mechanism for catalysis then relies on the concerted movement of both subunits. This is reflected by residues 150 to 196 on both subunits, that interact with each other, acting as a hinge site for both subunits to pivot around. Residues 1 to 100 (red box in displacement graph; Figure 6) on either chain interacts with the residues 200 to 300 (blue box in displacement graph; Figure 6) on the opposing chains, portraying a pincer-type mechanism around the active site by moving in an anti-correlated motion to each other (DCCM, Figure 6).

Malate dehydrogenase (PDB code: 4CL3; chain A of the tetrameric protein was visualized) has the role of converting malate to oxaloacetate. To do so, the enzyme is comprised of two or four chains, forming a homodimer or homotetramer complex that moves anticorrelated to each other. Malate dehydrogenase has a highly flexible loop between residues 80 and 90 (Figure 6) that opens and closes around the active site, allowing entry and exit of reactants and products as well as excludes surrounding solvents during the reaction. This flexible loop region is highly conserved within malate dehydrogenases, emphasizing its importance in catalysis (Goward & Nicholls, 1994). The overall observed anticorrelated movement within a chain appeared to be quite weak because the chain revolves in a circular pattern, making the cross-correlation map appear whiter (DCCM, Figure 6). There is a weak anticorrelated motion between residues 1 to 75 and residues 80 to 90, since they are on opposing sides due to the enzyme's helical movement (boxes in DCCM, Figure 6).

Succinyl-CoA synthetase (PDB code: 1CQI; visualized chains A and B only for structural and dynamical analyses) is a heterotetramer made up of two identical heterodimers. The role of the enzyme is to convert succinyl-CoA to succinate. This reaction is carried out through two different binding pockets, one on each distinct subunit within a heterodimer (Wolodko, Fraser, James, & Bridger 1994). Chain A contains the succinyl-CoA binding pockets, while chain B contains the ATP/GTP binding pockets. For succinate to form, ADP needs to be present in the binding pocket of chain A. The enzyme accommodates this by having chain A and chain B rotate in an anticorrelated motion to each other (DCCM, Figure 6), allowing ATP to become closer to the chain A, which is responsible for dephosphorylating ATP to ADP and the reaction to occur.

Alcohol dehydrogenase (PDB code: 6ADH; both subunits of the dimeric protein were visualized for structural and dynamical analyses) is a homodimeric protein with one active site on each identical subunit. Alcohol dehydrogenase undergoes a conformational change, not in the presence of alcohols, but rather the coenzyme NAD (Eklund et al., 1976). Within each subunit, there is a highly flexible loop between residues 240 and 250 that interacts with the active site of the enzyme, indicative of a lid-type mechanism (boxes in C_α atom displacement graph, Figure 6). The loop and surrounding residues move in an anticorrelated motion (DCCM, Figure 6). This anticorrelated motion results in a wide opening that allows substrate binding and catalysis.

Enolase (PDB code: 1IYX; both subunits of the dimeric protein were visualized for structural and dynamical analyses) is an enzyme that is part of the glycolysis pathway, converting 2-phosphoglycerate to phosphoenolpyruvate. The mechanism for this reaction is rather simple. Enolase has a very generic opening and closing motion, using flexible loops to interact with the active site (Lebioda & Stec, 1991). The first set of residues, residues 1 to 140, move in an anticorrelated motion with the other residues, residues 141 to 300 (boxes in DCCM, Figure 6). Residues 250 to 255 have the greatest flexibility (boxes in C_α atom displacement graph, Figure 6), suggesting a lid-type mechanism protecting the substrate in the active site. This simple motion is needed for fast turnover rates in the glycolysis pathway to supply the organism with energy.

Fructose 1–6 bisphosphate aldolase (PDB code: 2FJK; visualized chains A and B of the tetrameric protein) catalyzes the reversible step in glycolysis that converts fructose 1,6-bisphosphate, which is a six carbon molecule, into dihydroxyacetone phosphate and glyceraldehyde 3-phosphate, which are both three carbon molecules. Fructose 1–6 bisphosphate aldolase is a homotetramer with four identical active sites, one within each monomer. The monomers of the enzyme shift between an "open" and "closed" conformation (Hester et al., 1991). Within the tetramer, chains A and B work in concert and chains C and D mirror the motions of chains A and B. Chains A, B and chains C, D have similar helix-turn-helix residues (residues 258 to 289) where they overlap with each other and with the active sites on opposing sides. This allows the overlapping helix-turn helix residues to assist each other in facilitating the conformational change from "open" to "closed". For example, when chain A opens, its helix-turn-helix moves out of the way and allows the helix-turn-helix of chain B to close, and vice versa. The overall circular motion of the enzyme takes place because residues 135 and 150 (red boxes in C_a atom displacement graph, Figure 6) rotate with a greater range of motion than the rest of the enzyme, allowing them to act as a lid over the active site. The motion of this lid, which is constituted of residues 135 and 150, is important for substrate binding and catalysis.

Glyceraldehyde 3-phosphate dehydrogenase (PDB code: 4QX6) is an important enzyme necessary for glycolysis. This enzyme converts glyceraldehyde 3-phosphate into 1,3-bisposphoglycerate. Glyceraldehyde phosphate dehydrogenase is a homotetramer made up of 4 identical subunits, each with their own active sites. Although they are identical subunits, each chain interacts differently with each other. Interactions between chains A and B are mirrored by chains C and D, while interactions between chains A and C are mirrored by chains B and D. The openings of the active site pockets on chain A and chain B overlap, so the opening and closing of both chains occur simultaneously. Additionally each chain has a loop within residues 299 to 307 that interacts with the same residues in another chain (chains A, C and chains B, D), that further facilitates this simultaneous open and closing motion required for substrate binding (Lin et al., 1990).

Hexokinase (PDB code: 1BDG; the monomeric protein is visualized) is the first enzyme in the glycolysis pathway and also catalyzes the rate limiting reaction for glycolysis. The enzyme is responsible for converting glucose to glucose-6-phosphate, using one molecule of ATP in the process. The enzyme is a monomer with one active site where the enzyme opens and closes with a hinge like motion around the substrate. The flexibility of the enzyme allows the residues 50 to 260 (red boxes in C_a atom displacement graph, Figure 6) to move anticorrelated with the rest of the enzyme to ensure a tight "closed" conformation and the loose "open" conformation (boxes in DCCM, Figure 6). This flexibility assists in the mechanism of the enzyme to tightly bind the substrate and prevent wasteful hydrolysis of ATP (Bennett & Steitz, 1978).

Phosphoglycerate kinase (PDB code: 1PHP) is a monomeric protein that is responsible for catalyzing the transfer of phosphate group from 1,3-biphosphoglycerate to ADP in order to form 3-phosphoglycerate and ATP. The two substrates bind in two separate pockets of phosphoglycerate kinase. The ADP binds at the site formed by residues 181 to 384, whereas the second binding pocket is formed by residues 1 to 180, which binds 1,3-biphosphoglycerate. After both substrates are bound, it allows the enzyme to undergo a conformational change in an anticorrelated rotating motion that brings the two substrate binding sites close enough for the catalysis to take place, as well as to exclude water molecules to prevent unwanted ATP hydrolysis (Banks et al., 1979).

Phosphoglycerate mutase (PDB code: 1YFK; visualized chains A and B of dimer for structural and dynamical analyses) is involved in glycolysis and is responsible for catalyzing the transfer of a phosphate group from C_3 to C_2 to form 2-phosphoglycerate from 3-phosphoglycerate. Phosphoglycerate mutase is a homodimer with two identical active sites within each subunit that allows phosphate and 3-phosphoglyceric acid to bind (Winn Watson, Harkins, & Fothergill 1981). The enzyme catalyzes the transfer of a phosphate group from a specific histidine residue, to the C_2 carbon, taking the phosphate from the C_3 carbon of the 3-phosphoglycerate, through a formation of a 2, 3-bisphosphoglycerate intermediate. At the active site of phosphoglycerate mutase, is a very flexible loop (residues 100 to 125) that moves in an anticorrelated motion to the rest of the active site, acting like a lid (boxes in C_α atom displacement graph, Figure 6). The anticorrelated movement of the lid allows the entering of substrate and exiting of product from the active site.

Pyruvate decarboxylase (PDB code: 1QPB; chains A and B of tetramers were visualized for structural and dynamical analyses) catalyzes the last step in glycolysis where phosphoenolpyruvate is converted into pyruvate. Pyruvate decarboxylase is a homotetramer with identical active sites within each subunit (Lu, Dobritzsch, Baumann, Schneider, & König, 2000). For catalysis, a pair of subunits, A and B functions together as residues 1 to 194 and 351 to 560 of opposing chains anchor to each other. This allows residues 195 to 350 to have greater movement and flexibility since the rest of the enzyme is stationary. Residues 195 to 350 then move anticorrelated to the other residues allowing it to open and close around the active site of the enzyme. This suggests that residues 195 to 350 can act as a lid mechanism to the active site (boxes in the C_α atom displacements, Figure 6).

Pyruvate kinase (PDB code: 3SRF; chains A and B of the tetramers were visualized for structural and dynamical analyses) catalyzes the final step in glycolysis. The enzyme has identical active sites within each subunit of its homotetramer, where it catalyzes the conversion of phosphoenolpyruvate into pyruvate. A lid-type mechanism is employed by each subunit to allow for the catalysis to occur (Mattevi, Bolognesi, & Valentini, 1996). Residues 117 to 217 (boxes in C_α atom displacement graph, Figure 6) are highly flexible as they move from an "open" to "closed" conformation relative to the active site, as shown in the box in the C_α atom displacement graph. The DCCM (see regions in boxes) further shows that residues 117 to 217 are involved in an anticorrelated motion with respect to the rest of the enzyme to have the greatest movement during its conformational changes.

Triose phosphate isomerase (PDB code: 8TIM; both chains of the homodimer were visualized) catalyzes the reversible interconversion of different triose phosphate isomers. Triose phosphate isomerase exists as a homodimer with identical active sites within each subunit. The mechanism of catalysis is indicative of a pincer-type mechanism, with loops opening and closing over the active site (Davenport et al., 1991). Residues 170 to 220 of the enzyme open and close around the active site, and move in an anticorrelated motion to the rest of the enzyme (boxes in the DCCM, Figure 6). This anticorrelated motion alone does not capture the full catalytic mechanism, as the displacement graph shows a general flexibility exists throughout the whole enzyme to assist in catalysis.

Phosphoenolpyruvate carboxykinase (PDB code: 1OS1; only chain A was visualized) catalyzes the reaction from oxaloacetate to phosphoenolpyruvate (PEP) and CO_2. PEP carboxykinase does this by employing an induced-fit mechanism. The enzyme opens and closes around the substrates, oxaloacetate and ATP, taking on a different conformation in the presence of ATP (Dunten et al., 2002). In doing so, the enzyme is able to exclude water molecules that could potentially hydrolyze ATP. This mechanism also allows residues on the enzyme that could not originally reach the active site to interact with the substrates and favor catalysis. Residues 442 to 510 (box in displacement graph, Figure 6) and surrounding residues move anticorrelated with the active site and the rest of the enzyme, as shown in the DCCM (Figure 6).

Argininosuccinate synthetase (PDB code: 1J1Z; all four chains of the tetrameric protein were visualized for structural and dynamical analyses) catalyzes the synthesis of arginosuccinate, which is a necessary precursor for arginine and fumarate. Arginosuccinate synthetase is a homotetramer with identical active sites within each subunit. The motion of the enzyme is indicative of a lid-type mechanism with "open" and "closed" conformations, depending on substrate presence around the active site (Lemke & Howell, 2001). Residues 1 to 165 show a greater range of flexibility (boxes in C_α atom displacement graph, Figure 6) as it moves in an anticorrelated motion to the rest of the enzyme and active site (black square boxes in DCCM, Figure 6) to assist in the "open" and "closed" conformations of the enzyme.

Ornithine transcarbamylase (PDB code: 4OH7; chains A and B of the homotrimer were visualized for structural and dynamical analyses) is an enzyme necessary to catalyze the reaction between carbamoyl phosphate and ornithine to form citrulline. Visualization of the structure suggests that the two subunits have little interaction with one another. Overall, ornithine transcarbamylase is very flexible and the mechanism of catalysis is quite simplistic, where half the residues, residues 150 to 280, move in an anticorrelated motion to the rest of the enzyme (black square boxes in the DCCM, Figure 6). This indicates that the enzyme has a pincer-type mechanism around the active site, which might favor substrate binding and product release (Allewell, Shi, Morizono, & Tuchman, 1999).

Pantothenate synthetase (PDB code: 2A86; both chain A and chain B of the dimer were visualized for structural and dynamical analyses) functions as a homodimer with identical domains and active sites where it carries out the catalysis of pantothenate from condensation of pantoate and β-alanine. Pantothenate synthetase also has a flexible active site cavity to allow substrate and product entry and exit to assist catalysis (Wang & Eisenberg, 2003). The dynamics of pantothenate synthetase exhibit a circular motion where all the domains rotate back and forth around the active site, which allows residues 200 to 290 to rotate with a greater extent of flexibility (boxes in displacement graph, Figure 6), to dictate the opening and closing of the active site. This lid-type mechanism of residues 200 to 290 is further confirmed based on its anticorrelated motion with the rest of the enzyme and active site, which lies within residues 30 to 50 (black square boxes in DCCM, Figure 6).

3.8. Identification of test proteins

A set of 12 proteins (test No. 1–12) were used to examine if dynamic features of a protein could be used for its functional identification. Six of them (test No. 1–6) belong to the protein families reported in Figure 7 but are from different species, and six of them (test No. 7–12) were chosen randomly. Using the Cartesian coordinates of these 12 proteins, DCCMs from single NMA were obtained and compared with the catalog of DCCMs reported in Table 3. We were able to correctly identify six proteins (test No. 1–6) by simple comparison of DCCMs. The test proteins No. 1–6 were matched to phosphoglycerate kinase, pyruvate kinase, enolase, arginase, fumarase, and amine oxidase, respectively (Figure 7). A difference in the extent of correlation/anticorrelation can be seen in some cases such as in test protein No. 4 and arginase, while the DCCMs for test protein No. 6 and amine oxidase appear nearly identical. As expected, the DCCM of the six proteins, which is chosen randomly, did not match with any of the DCCMs reported in Table 3, suggesting that they belong to different families of proteins. As these test proteins (No. 7–12) exhibit different patterns of motions (Table S3) compared to the 24 proteins studied here, they must also possess different function. For example, we observed that aminoacyl-tRNA synthetases (Warren et al., 2014) and cytochrome P450 enzymes (Dorner et al., 2015) display very unique patterns of motion, and their functions are distinctly different from the set of proteins considered in the present study.

4. Conclusion

The NMA study of 24 important enzymes has demonstrated that each enzyme has signature dynamic patterns and the intrinsic dynamics of proteins are conserved across species, from simple organisms like *E. coli* to the complex *H. sapiens*. As expected, it was observed that for a given family of proteins, as protein sequences and structures become more similar, the protein dynamics become more related as well. We successfully created a catalog of DCCMs for the 24 protein groups. These patterns are unique to each protein and have the potential to act as fingerprints of identification for unknown proteins. This catalog of intrinsic dynamics assisted us in the correct identification of six test proteins. These test proteins were successfully identified solely by comparing their respective dynamical patterns. Similar attempt to compare DCCMs of proteins was made earlier for dynamic comparison of proteins (Munz et al., 2010). However, Munz et al. have performed MD simulations and essential dynamic analysis to develop and compare DCCM of proteins. The present approach of using coarse-grained normal mode simulation is relatively faster than MD simulation-based protein dynamic comparison and could be useful for functional identification of uncharacterized proteins. Currently, one of the main challenges in the study of the proteome is the lack of an effective and timely method of protein function identification. A majority of known protein sequences and structures have remained largely unidentified. The Protein Structure Initiative (PSI), established in 2000 by the National Institutes of General Medical Sciences, resulted in thousands of structures of proteins whose functions remain unknown. Experimental investigation of protein function is costly and time-consuming. Here we have presented a computational method for predicting protein function. Coarse-grained NMA is a fast approach and effective way to identify and characterize the function of unknown proteins. Although coarse-grained NMA has some limitations, as it provides less detailed information about the local fluctuations compared to that obtained from all-atom NMA or MD simulations (Strom et al., 2014), the method is very useful for establishing this unique dynamical fingerprint, and it only requires the PDB coordinates of a protein. Therefore, this work could be extended to the greater scope of identifying the function of uncharacterized proteins.

In the present study, DCCMs of various proteins were only visually compared; however the advancement of computational analytical methods opens the possibility of an online catalog of dynamics. This study could be extended to include more metabolic enzymes and proteins from various functional classes. For the future, we anticipate the use of NMA and the role of protein dynamics to play a larger part in protein identification and classification.

Abbreviations and symbols

BLAST	Basic local alignment search tool
BC	Bhattacharyya coefficient
DCCM	Dynamic cross-correlation matrix
ENM	Elastic network model
MAT	Methionine adenosyltransferase
NMA	Normal mode analysis
PDB	Protein Data Bank
RMSD	Root-mean-square deviation
RMSIP	Root-mean-square inner product
VMD	Visual molecular dynamics

Funding
The authors received no direct funding for this research.

Competing Interests
The authors declare no competing interest.

Author details
Sarah M. Meeuwsen[1]
E-mail: meeuwsen.sarah@gmail.com
ORCID ID: http://orcid.org/0000-0002-8178-0517
An N. Hodac[1]
E-mail: hodacan@uwec.edu
Lauren M. Adams[1]
E-mail: ADAMSLM@uwec.edu
Ryan D. McMunn[1]
E-mail: mcmunnrd@uwec.edu

Maxwell S. Anschutz[1]
E-mail: anschutz.maxwell@gmail.com
Kari J. Carothers[1]
E-mail: karijocarothers@gmail.com
Rachel E. Egdorf[1]
E-mail: r_egdorf@msn.com
Peter M. Hanneman[1]
E-mail: phanneman7@gmail.com
Jonathan P. Kitzrow[1]
E-mail: kitzrojp@gmail.com
Cynthia K. Keonigsberg[1]
E-mail: cynthiakoenigsberg@gmail.com
Oscar Lopez-Martinez[1]
E-mail: osclopez12@gmail.com
Paul A. Matthew[1]
E-mail: Paul.Mattchews@gmail.com
Ethan H. Richter[1]
Jonathan E. Schenk[1]
E-mail: Jschenk@umd.edu

Heidi L. Schmit[1]
E-mail: heidischmit2015@u.northwestern.edu
Matthew A. Scott[1]
E-mail: mattscot11@gmail.com
Eva M. Volenec[1]
Sanchita Hati[1]
E-mail: hatis@uwec.edu
[1] Chemistry Department, University of Wisconsin – Eau Claire,
Eau Claire, WI 54702, USA.

Citation information
Cite this article as: Investigation of intrinsic dynamics of
enzymes involved in metabolic pathways using coarse-
grained normal mode analysis, Sarah M. Meeuwsen, An
N. Hodac, Lauren M. Adams, Ryan D. McMunn, Maxwell
S. Anschutz, Kari J. Carothers, Rachel E. Egdorf, Peter M.
Hanneman, Jonathan P. Kitzrow, Cynthia K. Keonigsberg,
Oscar Lopez-Martinez, Paul A. Matthew, Ethan H. Richter,
Jonathan E. Schenk, Heidi L. Schmit, Matthew A. Scott, Eva
M. Volenec, Cogent Biology (2017), 3: 1291877.

Cover image
Source: Authors.

References
Agarwal, P. K., Billeter, S. R., Rajagopalan, P. R., Benkovic, S. J., &
    Hammes-Schiffer, S. (2002). Network of coupled
    promoting motions in enzyme catalysis. Proceedings of
    the National Academy of Sciences, 99, 2794–2799.
Allewell, N. M., Shi, D., Morizono, H., & Tuchman, M. (1999).
    Molecular recognition by ornithine and aspartate
    transcarbamylases. Accounts of Chemical Research, 32,
    885–894. http://dx.doi.org/10.1021/ar950262j
Altschul, S. F., Madden, T. L., Schäffer, A. A., Zhang, J., Zhang, Z.,
    Miller, W., & Lipman, D. J. (1997). Gapped BLAST and PSI-
    BLAST: A new generation of protein database search
    programs. Nucleic Acids Research, 25, 3389–3402.
    http://dx.doi.org/10.1093/nar/25.17.3389
Amadei, A., Ceruso, M. A., & Di Nola, A. (1999). On the convergence
    of the conformational coordinates basis set obtained by the
    essential dynamics analysis of proteins' molecular dynamics
    simulations. Proteins: Structure, Function, and Genetics, 36,
    419–424. http://dx.doi.org/10.1002/(ISSN)1097-0134
Bahar, I., & Rader, A. J. (2005). Coarse-grained normal mode
    analysis in structural biology. Current Opinion in
    Structural Biology, 15, 586–592. http://dx.doi.
    org/10.1016/j.sbi.2005.08.007
Banks, R. D., Blake, C. C. F., Evans, P. R., Haser, R., Rice, D. W.,
    Hardy, G., ... Phillips, A. W. (1979). Sequence, structure and
    activity of phosphoglycerate kinase: A possible hinge-
    bending enzyme. Nature, 279, 773–777.
    http://dx.doi.org/10.1038/279773a0
Bennett, Jr, W. S., & Steitz, T. A. (1978). Glucose-induced
    conformational change in yeast hexokinase. Proceedings
    of the National Academy of Sciences, 75, 4848–4852.
    http://dx.doi.org/10.1073/pnas.75.10.4848
Berg, J. M., Tymoczko, J. L., & Stryer, L. (2012). Biochemistry
    (7th ed.). New York, NY: W.H. Freeman and Company.
Berman, H. M., Westbrook, J., Feng, Z., Gilliland, G., Bhat, T. N.,
    Weissig, H., ... Bourne, P. E. (2000). The protein data bank.
    Nucleic Acids Research, 28, 235–242.
    http://dx.doi.org/10.1093/nar/28.1.235
Bhabha, G., Lee, J., Ekiert, D. C., Gam, J., Wilson, I. A., Dyson, H.
    J., ... Wright, P. E. (2011). A dynamic knockout reveals that
    conformational fluctuations influence the chemical step
    of enzyme catalysis. Science, 332, 234–238.
    http://dx.doi.org/10.1126/science.1198542
Bhattacharyya, A. (1943). On a measure of divergence
    between two statistical populations defined by their
    probability distributions. Bulletin of Calcutta Mathematical
    Society, 35, 99–109.

Brooks, B., & Karplus, M. (1983). Harmonic dynamics of
    proteins: Normal modes and fluctuations in bovine
    pancreatic trypsin inhibitor. Proceedings of the National
    Academy of Sciences, 80, 6571–6575.
    http://dx.doi.org/10.1073/pnas.80.21.6571
Carnevale, V., Pontiggia, F., & Micheletti, C. (2007). Structural
    and dynamical alignment of enzymes with partial
    structural similarity. Journal of Physics: Condensed Matter,
    19, 285206–285214.
    http://dx.doi.org/10.1088/0953-8984/19/28/285206
Chennubhotla, C., & Bahar, I. (2007). Signal propagation in
    proteins and relation to equilibrium fluctuations. PLOS
    Computational Biology, 3, 1716–1726.
Chennubhotla, C., Yang, Z., & Bahar, I. (2008). Coupling
    between global dynamics and signal transduction
    pathways: a mechanism of allostery for chaperonin
    GroEL. Molecular BioSystems, 4, 287–292.
    http://dx.doi.org/10.1039/b717819k
Chothia, C., & Lesk, A. M. (1986). The relation between the
    divergence of sequence and structure in proteins. The
    EMBO Journal, 5, 823–826.
Davenport, R. C., Bash, P. A., Seaton, B. A., Karplus, M., Petsko, G.
    A., & Ringe, D. (1991). Structure of the triosephosphate
    isomerase-phosphoglycolohydroxamate complex: an
    analog of the intermediate on the reaction pathway.
    Biochemistry, 30, 5821–5826.
    http://dx.doi.org/10.1021/bi00238a002
Devos, D., & Valencia, A. (2000). Practical limits of function
    prediction. Proteins: Structure, Function, and Genetics, 41,
    98–107. http://dx.doi.org/10.1002/(ISSN)1097-0134
Dorner, M. E., McMunn, R. D., Bartholow, T. G., Calhoon, B. E.,
    Conlon, M. R., Dulli, J. M., ... Hati, S. (2015). Comparison of
    intrinsic dynamics of cytochrome P450 proteins using
    normal mode analysis. Protein Science.
Dunten, P., Belunis, C., Crowther, R., Hollfelder, K., Kammlott, U.,
    Levin, W., ... Wertheimer, S. J. (2002). Crystal structure of
    human cytosolic phosphoenolpyruvate carboxykinase
    reveals a new GTP-binding site. Journal of Molecular Biology,
    316, 257–264. http://dx.doi.org/10.1006/jmbi.2001.5364
Eisenmesser, E. Z., Millet, O., Labeikovsky, W., Korzhnev, D. M.,
    Wolf-Watz, M., Bosco, D. A., ... Kern, D. (2005). Intrinsic
    dynamics of an enzyme underlies catalysis. Nature, 438,
    117–121. http://dx.doi.org/10.1038/nature04105
Eklund, H., Nordström, B. Zeppezauer, E., Söderlund, G., Ohlsson,
    I., Boiwe, T., ... Åkeson, Å (1976). Three-dimensional
    structure of horse liver alcohol dehydrogenase at 2.4 Å
    resolution. Journal of Molecular Biology, 102, 27–59.
Erdin, S., Lisewski, A. M., & Lichtarge, O. (2011). Protein function
    prediction: towards integration of similarity metrics.
    Current Opinion in Structural Biology, 21, 180–188.
    http://dx.doi.org/10.1016/j.sbi.2011.02.001
Fedøy, A. E., Yang, N., Martinez, A., Leiros, H. K. S., & Steen, I. H.
    (2007). Structural and functional properties of isocitrate
    dehydrogenase from the psychrophilic bacterium
    desulfotalea psychrophila reveal a cold-active enzyme
    with an unusual high thermal stability. Journal of
    Molecular Biology, 372, 130–149.
    http://dx.doi.org/10.1016/j.jmb.2007.06.040
Frank, J., & Agrawal, R. K. (2000). A ratchet-like inter-subunit
    reorganization of the ribosome during translocation.
    Nature, 406, 318–322.
    http://dx.doi.org/10.1038/35018597
Fu, Z., Hu, Y., Markham, G. D., & Takusagawa, F. (1996). Flexible
    loop in the structure of s-adenosylmethionine synthetase
    crystallized in the tetragonal modification. Journal of
    Biomolecular Structure and Dynamics, 13, 727–739.
    http://dx.doi.org/10.1080/07391102.1996.10508887
Fuglebakk, E., Echave, J., & Reuter, N. (2012). Measuring and
    comparing structural fluctuation patterns in large protein
    datasets. Bioinformatics, 28, 2431–2440.
    http://dx.doi.org/10.1093/bioinformatics/bts445

Fuglebakk, E., Tiwari, S. P., & Reuter, N. (2014). Comparing the intrinsic dynamics of multiple protein structures using elastic network models. *Biochimica et Biophysica Acta, 1850*, 911–922.

Go, N., Noguti, T., & Nishikawa, T. (1983). Dynamics of a small globular protein in terms of low-frequency vibrational modes. *Proceedings of the National Academy of Sciences, 80*, 3696–3700. http://dx.doi.org/10.1073/pnas.80.12.3696

Goward, C. R., & Nicholls, D. J. (1994). Malate dehydrogenase: A model for structure, evolution, and catalysis. *Protein Science, 3*, 1883–1888. http://dx.doi.org/10.1002/pro.v3:10

Hammes-Schiffer, S., & Benkovic, S. J. (2006). Relating protein motion to catalysis. *Annual Review of Biochemistry, 75*, 519–541. http://dx.doi.org/10.1146/annurev.biochem.75.103004.142800

Hensen, U., Meyer, T., Haas, J., Rex, R., Vriend, G., & Grubmüller, H. (2012). Exploring protein dynamics space: The dynasome as the missing link between protein structure and function. *PLoS ONE, 7*, e33931. doi:10.1371/journal.pone.0033931

Henzler-Wildman, K. A., Lei, M., Thai, V., Kerns, S. J., Karplus, M., & Kern, D. (2007). A hierarchy of timescales in protein dynamics is linked to enzyme catalysis. *Nature, 450*, 913–916. http://dx.doi.org/10.1038/nature06407

Henzler-Wildman, K., & Kern, D. (2007). Dynamic personalities of proteins. *Nature, 450*, 964–972. http://dx.doi.org/10.1038/nature06522

Hester, G., Brenner-Holzach, O., Rossi, F. A., Struck-Donatz, M., Winterhalter, K. H., Smit, J. D., & Piontek, K. (1991). The crystal structure of fructose-1,6-bisphosphate aldolase from Drosophila melanogaster at 2.5 A resolution. *FEBS Lett, 292*, 237–242.

Hinsen, K. (1998). Analysis of domain motions by approximate normal mode calculations. *Proteins: Structure, Function, and Genetics, 33*, 417–429. http://dx.doi.org/10.1002/(ISSN)1097-0134

Hinsen, K., Petrescu, A. J., Dellerue, S., Bellissent-Funel, M. C., & Kneller, G. (2000). Harmonicity in slow protein dynamics. *Chemical Physics, 261*, 25 –37. http://dx.doi.org/10.1016/S0301-0104(00)00222-6

Hinsen, K., Thomas, A., & Field, M. J. (1999). Analysis of domain motions in large proteins. *Proteins: Structure, Function, and Genetics, 34*, 369–382. http://dx.doi.org/10.1002/(ISSN)1097-0134

Hollup, S. M., Fuglebakk, E., Taylor, W. R., & Reuter, N. (2011). Exploring the factors determining the dynamics of different protein folds. *Protein Science, 20*, 197–209. http://dx.doi.org/10.1002/pro.558

Hollup, S. M., Salensminde, G., & Reuter, N. (2005). WEBnm@: A web application for normal mode analysis of proteins. *BMC Bioinformatics, 6*, 52. http://dx.doi.org/10.1186/1471-2105-6-52

Holm, L., & Rosenstrom, P. (2010). Dali server: Conservation mapping in 3D. *Nucleic Acids Research, 38*(Web Server), p. W545–W549. http://dx.doi.org/10.1093/nar/gkq366

Humphrey, W., Dalke, A., & Schulten, K. (1996). VMD: Visual molecular dynamics. *Journal of Molecular Graphics, 14*, 33–38. http://dx.doi.org/10.1016/0263-7855(96)00018-5

Ichiye, T., & Karplus, M. (1991). Collective motions in proteins: A covariance analysis of atomic fluctuations in molecular dynamics and normal mode simulations. *Proteins: Structure, Function, and Genetics, 11*, 205–217. http://dx.doi.org/10.1002/(ISSN)1097-0134

Klinman, J. P., & Kohen, A. (2013). Hydrogen Tunneling links protein dynamics to enzyme catalysis. *Annual Review of Biochemistry, 82*, 471–496. http://dx.doi.org/10.1146/annurev-biochem-051710-133623

Kosloff, M., & Kolodny, R. (2008). Sequence-similar, structure-dissimilar protein pairs in the PDB. *Proteins: Structure, Function, and Bioinformatics, 71*, 891–902. http://dx.doi.org/10.1002/prot.v71:2

Lebioda, L., & Stec, B. (1991). Mechanism of enolase: The crystal structure of enolase-magnesium-2-phosphoglycerate/phosphoenolpyruvate complex at 2.2-. ANG. resolution. *Biochemistry, 30*, 2817–2822. http://dx.doi.org/10.1021/bi00225a012

Lemke, C. T., & Howell, P. L. (2001). The 1.6 Å crystal structure of E. coli argininosuccinate synthetase suggests a conformational change during catalysis. *Structure, 9*, 1153–1164. http://dx.doi.org/10.1016/S0969-2126(01)00683-9

Lin, Y. Z., Liang, S. J., Zhou, J. M., Tsou, C. L., Wu, P., & Zhou, Z. (1990). Comparison of inactivation and conformational changes of d-glyceraldehyde-3-phosphate dehydrogenase during thermal denaturation. *Biochimica et Biophysica Acta (BBA) - Protein Structure and Molecular Enzymology, 1038*, 247–252. http://dx.doi.org/10.1016/0167-4838(90)90212-X

Lisi, G. P., & Patrick Loria, J. P. (2016). Using NMR spectroscopy to elucidate the role of molecular motions in enzyme function. *Progress in Nuclear Magnetic Resonance Spectroscopy, 92-93*, 1–17. http://dx.doi.org/10.1016/j.pnmrs.2015.11.001

Liu, X., Speckhard, D. C., Shepherd, T. R., Sun, Y. J., Hengel, S. R., Yu, L., ... Fuentes, E. J. (2016). Distinct roles for conformational dynamics in protein-ligand interactions. *Structure, 24*, 2053–2066. http://dx.doi.org/10.1016/j.str.2016.08.019

Lu, G., Dobritzsch, D., Baumann, S., Schneider, G., & König, S. (2000). The structural basis of substrate activation in yeast pyruvate decarboxylase. *European Journal of Biochemistry, 267*, 861–868. http://dx.doi.org/10.1046/j.1432-1327.2000.01070.x

Marques, O., & Sanejouand, Y. H. (1995). Hinge-bending motion in citrate synthase arising from normal mode calculations. *Proteins: Structure, Function, and Genetics, 23*, 557–560. http://dx.doi.org/10.1002/(ISSN)1097-0134

Mattevi, A., Bolognesi, M., & Valentini, G. (1996). The allosteric regulation of pyruvate kinase. *FEBS Letters, 389*, 15–19. http://dx.doi.org/10.1016/0014-5793(96)00462-0

Micheletti, C. (2013). Comparing proteins by their internal dynamics: Exploring structure–function relationships beyond static structural alignments. *Physics of Life Reviews, 10*(1), 1–26. http://dx.doi.org/10.1016/j.plrev.2012.10.009

Munz, M., Lyngsø, R., Hein, J., & Biggin, P. C. (2010). Dynamics based alignment of proteins: an alternative approach to quantify dynamic similarity. *BMC Bioinformatics, 11*, 188. http://dx.doi.org/10.1186/1471-2105-11-188

Pandini, A., Mauri, G., Bordogna, A., & Bonati, L. (2007). Detecting similarities among distant homologous proteins by comparison of domain flexibilities. *Protein Engineering Design and Selection, 20*, 285–299.

Park, J., Teichmann, S. A., Hubbard, T., & Chothia, C. (1997). Intermediate sequences increase the detection of homology between sequences. *Journal of Molecular Biology, 273*, 349–354. http://dx.doi.org/10.1006/jmbi.1997.1288

Roston, D., Kohen, A., Doron, D., & Major, D. T. (2014). Simulations of remote mutants of dihydrofolate reductase reveal the nature of a network of residues coupled to hydride transfer. *Journal of Computational Chemistry, 35*, 1411–1417. http://dx.doi.org/10.1002/jcc.v35.19

Sievers, F., Wilm, A., Dineen, D., Gibson, T. J., Karplus, K., Li, W., ... Higginsa, D. G. (2011). Fast, scalable generation of high-quality protein multiple sequence alignments using Clustal Omega. *Molecular Systems Biology, 7*, 539.

Singh, P., Francis, K., & Kohen, A. (2015). Network of remote and local protein dynamics in dihydrofolate reductase catalysis. *ACS Catalysis, 5*, 3067–3073. http://dx.doi.org/10.1021/acscatal.5b00331

Skjaerven, L., Hollup, S., & Reuter, N. (2009). Normal mode analysis for proteins. *Journal of Molecular Structure: THEOCHEM, 898*, 42–48. http://dx.doi.org/10.1016/j.theochem.2008.09.024

Stoddard, B. L., Dean, A., & Koshland, Jr, D. E. (1993). Structure of isocitrate dehydrogenase with isocitrate, nicotinamide adenine dinucleotide phosphate, and calcium at 2.5-.ANG. resolution: A pseudo-Michaelis ternary complex. *Biochemistry, 32,* 9310–9316. http://dx.doi.org/10.1021/bi00087a008

Strom, A. M., Fehling, S. C., Bhattacharyya, S., & Hati, S. (2014). Probing the global and local dynamics of aminoacyl-tRNA synthetases using all-atom and coarse-grained simulations. *Journal of Molecular Modeling, 20,* 2245. http://dx.doi.org/10.1007/s00894-014-2245-1

Tama, F., & Brooks, C. L. (2005). 3rd diversity and identity of mechanical properties of icosahedral viral capsids studied with elastic network normal mode analysis. *Journal of Molecular Biology, 345,* 299–314. http://dx.doi.org/10.1016/j.jmb.2004.10.054

Tiwari, S. P., Fuglebakk, E., Hollup, S. M., Skjærven, L., Cragnolini, T., Grindhaug, S. H., ... Reuter, N. (2014). WEBnm@ v2.0: Web server and services for comparing protein flexibility. *BMC Bioinformatics, 15,* 6597.

Tobi, D. (2013). Large-scale analysis of the dynamics of enzymes. *Proteins: Structure, Function, and Bioinformatics, 81,* 1910–1918. http://dx.doi.org/10.1002/prot.24335

Tousignant, A., & Pelletier, J. N. (2004). Protein motions promote catalysis. *Chemistry & Biology, 11,* 1037–1042. http://dx.doi.org/10.1016/j.chembiol.2004.06.007

Vöhringer-Martinez, E., & Dörner, C. (2016). Conformational substrate selection contributes to the enzymatic catalytic reaction mechanism of pin1. *The Journal of Physical Chemistry B, 120,* 12444–12453. http://dx.doi.org/10.1021/acs.jpcb.6b09187

Wang, S., & Eisenberg, D. (2003). Crystal structures of a pantothenate synthetase from M. tuberculosis and its complexes with substrates and a reaction intermediate. *Protein Science, 12,* 1097–1108. http://dx.doi.org/10.1110/(ISSN)1469-896X

Warren, N., Strom, A., Nicolet, B., Albin, K., Albrecht, J., Bausch, B., ... Gunderson, A. (2014). Comparison of the Intrinsic Dynamics of Aminoacyl-tRNA Synthetases. *The Protein Journal, 33,* 184–198. http://dx.doi.org/10.1007/s10930-014-9548-z

Weimer, K. M., Shane, B. L., Brunetto, M., Bhattacharyya, S., & Hati, S. (2009). Evolutionary basis for the coupled-domain motions in thermus thermophilus leucyl-tRNA synthetase. *Journal of Biological Chemistry, 284,* 10088–10099. http://dx.doi.org/10.1074/jbc.M807361200

Whisstock, J., & Lesk, A. (1999). Prediction of protein function from protein sequence and structure. *Quarterly Reviews of Biophysics, 36,* 307–340. http://dx.doi.org/10.1017/S0033583503003901

Wiegand, G., & Remington, S. J. (1986). Citrate synthase: Structure, control, and mechanism. *Annual Review of Biophysics and Biophysical Chemistry, 15,* 97–117. http://dx.doi.org/10.1146/annurev.bb.15.060186.000525

Winn, S. I., Watson, H. C., Harkins, R. N., & Fothergill, L. A. (1981). Structure and activity of phosphoglycerate mutase. *Philosophical Transactions of the Royal Society of London B: Biological Sciences, 293,* 121–130.

Wolodko, W. T., Fraser, M. E., James, M. N., & Bridger, W. A. (1994). The crystal structure of succinyl-CoA synthetase from Escherichia coli at 2.5-A resolution. *The Journal of Biological Chemistry, 269,* 10883–10890.

Yang, L., Song, G., Carriquiry, A., & Jernigan, R. L. (2008). Close correspondence between the motions from principal component analysis of multiple HIV-1 Protease structures and elastic network modes. *Structure, 16,* 321–330. http://dx.doi.org/10.1016/j.str.2007.12.011

Zen, A., Micheletti, C., Keskin, O., & Nussinov, R. (2010). Comparing interfacial dynamics in protein-protein complexes: an elastic network approach. *BMC Structural Biology, 10,* 26. http://dx.doi.org/10.1186/1472-6807-10-26

7

The effect of ascorbic acid on bone cancer cells *in vitro*

Gabriela Fernandes[1], Andrew W. Barone[1] and Rosemary Dziak[1]*

*Corresponding author: Rosemary Dziak, Department of Oral Biology, School of Dental Medicine, University at Buffalo, State University of New York, 3435 Main Street, Buffalo, NY 14201, USA

E-mail: rdziak@buffalo.edu

Reviewing editor: Steve Winder, University of Sheffield, UK

Additional information is available at the end of the article

Abstract: Ascorbic acid (Vitamin C) has long been known for its anti-cancer proper-ties and in the present study the effects of ascorbic acid (AsA) on osteogenic dif-ferentiation, apoptosis, and signaling pathways of the human G29 osteosarcoma cell line were studied. The expression of Runt-related transcription factor-2 (RUNX2) and osteocalcin genes were evaluated by real-time polymerase chain reaction (PCR). Osteoblastic maturation was assessed with alkaline phosphatase activity and min-eralization with alizarin red deposition, and apoptosis with a caspase-2 apoptotic assay as well as the cell viability via the cytotoxicity assay. The possible role of the MAP kinase pathway (p44/42, p38, and p-JNK signaling pathway) was also studied. Our results showed that RUNX2 and osteocalcin gene expression, mineralization, cell viability, and metabolic activity levels were increased in cells treated with low concentrations of AsA with respect to untreated cells. At higher concentrations, AsA resulted in decreases in these parameters and induced apoptosis of the G292 os-teosarcoma cells via downregulation of the MAPK pathway. The findings presented here support the ability of AsA to modulate the viability and differentiation of the G292 type of bone cancer cell with increases or decreases depending on the AsA

ABOUT THE AUTHORS

Gabriela Fernandes conducted this study as a postdoctoral fellow in the Department of Oral Biology at the University at Buffalo. She received her MS in Oral Sciences from that Institution and has also clinical training in dentistry having obtained a BDS from YCMM & RDF Dental College. Her research interests have been in tissue engineering and regulation of bone remodeling, both in normal bone as well as osteosarcomas.

Andrew W. Barone is presently pursuing a DDS at the University at Buffalo as well as conducting research in the laboratory of Rosemary Dziak. He has conducted studies on the use of a nanoscaffold in bone regeneration as well as with putative regulators of osteosarcomas.

Rosemary Dziak is a professor of Oral Biology at the University at Buffalo. Her research revolves around osteoblastic cell-mediated bone regulation and she has conducted research on the mechanism of action of many agents that have effects on pathological and physiological bone metabolism.

PUBLIC INTEREST STATEMENT

The potential role of ascorbic acid (Vitamin C) in the treatment of various forms of cancer remains controversial despite numerous studies over several decades addressing this issue in animal and tissue culture models as well as human clinical trials. The concept that there is a critical concentration of ascorbic acid needed for optimal antitumor effects has gained wide acceptance although the mechanism by which this agent regulates cell growth is still not understood. In the *in vitro* studies presented here, the effects of ascorbic acid on an osteosarcoma cell line are shown to be dose-dependent; higher concentrations decreased the cellular metabolic activity and differentiation, and increased cytotoxicity and cell death. Although research is needed to translate these *in vitro* effects to effective use of ascorbic acid in clinical treatment of osteosarcoma, the consistency of these results to those in the literature with other cancer cell lines is supportive of continued investigations.

concentration suggesting a need for further evaluation of the possible use of this vitamin in the regulation of bone cell cancer growth.

Subjects: Health Conditions; Medicine; Dentistry

Keywords: osteosarcoma; ascorbic acid; vitamin C; osteoblasts

1. Introduction

In the 1950s, vitamin C was originally hypothesized to be protective against cancer, but in the 1970s, Ewan Cameron and Linus Pauling suggested that it also had a therapeutic effect, reporting increased survival of patients with advanced cancer following high-dose IV vitamin C treatment (Pauling, 1980; Pauling et al., 1981; Pauling & Moertel, 1986). However, randomized controlled trials in which high-dose vitamin C had been administered orally in anticancer therapy had often failed. This inconsistency has been attributed to the fact that in order to elucidate a response of vitamin C in the treatment of cancer, it is necessary to maintain relatively high plasma levels, which can often be challenging with oral doses (van der Reest & Gottlieb, 2016). The rebirth of the use of Vitamin C in cancer treatment occurred with substantial contribution to research on its pharmacokinetics (Padayatty et al., 2004) as well as studies on the ability of pharmacological concentrations of the vitamin to selectively evoke death of several types of cancer cells (Chen et al., 2005). A recent systematic review of published results of 5 randomized controlled trials, 12 phase I/II trials, 6 observational studies, and 11 case reports concluded that there is no "high-quality evidence" to suggest that the use of vitamin C (ascorbic acid) supplementation either orally or IV in cancer patients enhances chemotherapeutic effects or reduces its toxicity (Jacobs, Hutton, Ng, Shorr, & Clemons, 2015). Although there is a need for additional rigorous randomized controlled trials to establish the risks and benefits of vitamin C, there is also a need to further delineate the possible mode of action of this agent in various cancer cell types for design of therapeutic approaches that might need to be unique for certain cancerous conditions.

Vitamin C is a water-soluble, essential multifunctional micronutrient that acts as an anti-oxidant and is required in its reduced form (L-ascorbic acid) for many enzymatic reactions (Golde, 2003). It is an essential supplement for osteoblastic cell differentiation from mesenchymal stem cell precursors (Langenbach & Handschel, 2013). At low doses, it can increase cellular proliferation, collagen production, and hydroxylation of proline and lysine residues in collagen (Schwarz, Kleinman, & Owens, 1987). However, in higher concentrations, it can be cytotoxic (Duarte & Lunec, 2005) and inhibit prostaglandins of the two series (arachidonic acid derived), which have been correlated with inflammation and increased cell proliferation (ElAttar & Lin, 1992). It has also been suggested that vitamin C may promote oxidative metabolism by inhibiting the utilization of pyruvate for anaerobic glycolysis (Lohmann, 1987). This can be attributed to its pro anti-oxidant effect. Previous studies have demonstrated that the anti-proliferative activity of ascorbic acid (AsA) is due to the inhibition of expression of genes involved in cell division progression (Hitomi & Tsukagoshi, 1996; Ohno, Ohno, Suzuki, Soma, & Inoue, 2009). One recent study, (Valenti et al., 2014) has assessed the effects of various concentrations of AsA in the anti-proliferation and differentiation of MG-63 cells and expanded on earlier studies in which AsA showed a growth repressive effect depending on its concentration in this same cell line (Takamizawa et al., 2004). Although these cells are often used as a prototype of normal human osteoblastic cells, MG-63 cells are an osteosarcoma lineage with an abnormal gene expression profile, altered extracellular bone matrix synthesis, and atypical bone formation (Benayahu, Shur, Marom, Meller, & Issakov, 2002).

JNK and p38 are key mediators of stress and inflammation responses, while the ERKs cascade is mostly induced by growth factors (Hu, Feng, & Cheng, 2001). The JNK stress pathway participates in many different intracellular processes, including cell growth, differentiation, transformation, and apoptosis (Niu et al., 2015). Consequently, some studies have proposed the inhibition of these pathways as targets in cancer therapy (Fliedner et al., 2014). Human osteosarcoma cell lines that have shown resistance to conventional chemotherapy, such as cisplatin and methotrexate, have been shown to be very sensitive to agents that are associated with activation of stress-response JNK signaling pathway and JNK downstream target, AP-1 transcriptional factor. In particular, the MG-63 cell line as well as the human osteosarcoma cell line, G292, have been shown to exhibit inhibition of cell viability and induction of apoptosis in respond to BBMD3, a derivative of natural product berbamine, that is an activator of JNK/AP-1 signaling (Yang et al., 2013). In view of the similarities in the response of these osteosarcoma cells to BBMD3, we proposed that G292 cells would respond to AsA with some anti-tumor responses as previously reported for MG-63 cells (Takamizawa et al., 2004; Valenti et al., 2014).

In this paper, we report, for the first time the anti-metabolic and differentiation effects of AsA on the human osteosarcoma G292 cell line. Furthermore, we studied possible pathways involved with occurrence of apoptosis of the cells post treatment with AsA.

2. Materials and methods

2.1. Cell culture conditions

The human osteosarcoma-derived cell line G292, purchased from the American Type Culture Collection (Manassas, VA) was grown in 75-cm^2 plastic culture flasks (Falcon, Oxnard, CA) containing minimum essential medium-alpha (MEM-α; Gibco Life Technologies, Grand Island, NY) supplemented with 10% fetal bovine serum (Gibco Life Technologies) and a 1% anti-mycotic solution (Gibco Life Technologies). Cultures were maintained at 37°C in a humidified incubator with 5% CO$_2$. The medium was routinely changed every 2–3 days, and cells were passaged approximately every 72–96 h.

2.2. Cell metabolic activity, cell viability, and apoptotic activity

Cell metabolic activity was evaluated by a colorimetric assay based on the reduction of the tetrazolium salt sodium 3I-[1-phenylamino-carbonyl-3,4-tetrazolium]-bis (4-methoxy-6-nitro) benzene sulfonic acid hydrate) (MTT) by mitochondrial dehydrogenase of viable cells to a formazan dye (Promega, Madison, WI, USA) as described previously (Fernandes, Barone, & Dziak, 2016). The MTT test was performed after 48 h of L-ascorbic acid (AsA) (Sigma-Aldrich, St. Louis, MO) exposure. Briefly, the cells were seeded in complete medium (MEM-α, 10% FBS, and 1% anti-mycotic) and after 80% confluence; the cells were treated with the varying concentrations (0, 62.5, 125, 250, 500, 1,000 µM) AsA for the indicated time. The cells were then incubated with 200-µl clear (without phenol red) MEM-α and 20-µl MTT assay reagent for an additional 3 h and with 200-µl DMSO at the end of the incubation period. The supernatants were then transferred to a new 96-well plate for recording the absorbance at 490 nm using a 96-well plate reader.

The cytotoxicity of AsA and a putative apoptosis effect on the G292 cells were assayed using a CellTox Green™ assay (Promega, Madison, WI) and an ApoLive-Glo™ Multiplex Assay kit (Promega, Madison, WI), respectively. Cytotoxicity was assayed in a 96-well plate format with 5,000 cells/well after 24-h incubation with the inhibitors/vehicle with the CellTox Green™ assay (Promega, Madison, WI), following the Express, No-step Addition at Seeding Method described by the manufacturer. Fluorescence (Ex510/Em532) was measured with a fluorescent plate reader (Biotek, USA).

For the measurements of viability and apoptosis using the Apo-Live-Glo™ Multiplex kit G292 cells of approximately 500/well were seeded in a flat 96-well micro-plate as triplicates with the control and experimental treatment groups of AsA. After a total of 7 days exposure to AsA, (with medium changed every 48 h in both control and AsA groups) old medium was removed from the wells and 100 µl of fresh medium was added. Based on the color of the phenol red indicator present in the medium, there was no indication of any significant changes in pH in any of the cultures under the experimental conditions used in these incubations. Twenty microliters of viability/cytotoxicity reagent containing both GF-AFC and bis-AAF-R110 substrates were added to each well, and briefly mixed by orbital shaking at 300–500 rpm for 30 s and then incubated at 37°C for 30–180 min. Fluorescence (viability) was measured at Ex400/Em505 using a fluorescent plate reader (Biotek, USA). Immediately after the readings were completed, 100 µl of Caspase-Glo 3/7 reagent was added to each well, and briefly mixed by orbital shaking at 300–500 rpm for 30 s and then incubated at room temperature for 30–180 min. Luminescence, which is proportional to the amount of caspase activity present in the samples, was measured using a Perkin Elmer Victor^3$_{TM}$V Wallac plate reader using the luminescence (1.0 s) protocol.

2.3. ALP activity and Alizarin red staining

Alkaline Phosphatase (ALP) activity was assessed with an ALP kit purchased from Sigma-Aldrich, following the manufacturer's instructions as described previously (Fernandes et al., 2016). Briefly, the cells were seeded in the complete medium (MEM-α, 10% FBS, and 1% anti-mycotic) and after 80% confluence, the cells were treated with varying concentrations of ascorbic acid for the indicated time before the ALP activity was measured. ALP activity was normalized according to protein content of the cell sample lysate. The protein concentration of the cell lysate was measured with a bicinchoninic acid (BCA) Protein Assay Kit (Sigma-Aldrich) and the ALP activity was expressed as absorbance at 405 nm/mg protein.

To measure the level of calcium mineral deposition, alizarin red staining (AR-S) was performed on the cells that were treated with the various concentrations of ascorbic acid. After 3 weeks in culture, the cells were fixed with 70% ethanol, rinsed five times with deionized water, treated with 40 mM alizarin red solution for 10 min at pH 4.2, and then washed for 15 min with PBS. Cetylpyridinium chloride (CPC) extraction was used for destaining. AR-S was removed from the cell samples by the addition of CPC (10% w/v, pH 7.0) followed by incubation at room temperature with gentle shaking for 1 h. The absorbances of the CPC extractions were measured at 550 nm (Barres, Mota Anna, Greenberg, Almojaly, & Dziak, 2015).

2.4. Total RNA extraction and reverse transcription

After inducing the G292 cells with the varying concentrations of ascorbic acid treatment, the cells were scraped, and pellets were collected by centrifugation at 1,000× g for 10 min at 4°C for RNA extraction. Total RNA was extracted from each pellet using the RNeasy minikit (Qiagen, USA) with DNAse I treatment. First-strand cDNA was generated using the High-Capacity cDNA Archive Kit, with random hexamers, (Applied Biosystems PE, Foster City, CA, USA) according to the manufacturer's protocol. The cDNA product was quantified using a NanoDrop 2000/2000c Spectrophotometer (Thermo Fisher Scientific, Wilmington, DE, USA) and then aliquoted in equal volumes and stored at −80°C.

2.5. Real-time PCR

Polymerase chain reaction (PCR) was performed in a total volume of 20 μl containing 1 × Premix Ex Taq™ (2×), 1 × Rox Reference Dye (50×) and 20 ng of cDNA. Probe sets for each gene (RUNX2 5′-F: AAGTGCGGTGCAAACTTTCT-3′ R: 5′-TCTCGGTGGCTGCTAGTGA-3′);(OPN osteocalcin—F GCAGAGTCCAGCAAAGGT; R CAGCCATTGATACAGGTAGC) were obtained from Assay-on-Demand Gene Expression Products (Applied Biosystems). Real-time PCR reactions were carried out in a two-tube system and in multiplex. The amplification conditions included 30 s at 95°C (initial denaturation), followed by 50 cycles at 95°C for 5 s (denaturation) and at 60°C for 31 s (annealing/extension). Thermocycling and signal detection were performed with ABI Prism 7000 Sequence Detector (Applied Biosystems). Signals were detected according to the manufacturer's instructions. ΔΔcT values were then calculated with respect to the control. To normalize mRNA expression for sample-to-sample differences in RNA input, quality, and reverse transcriptase efficiency, the housekeeping gene GAPDH was amplified. Endogenous control gene was abundant and remained constant proportionally to total RNA among the samples.

2.6. Western blot

G292 cells were seeded at equal concentrations and incubated with 1,000 μM of ascorbic acid for 4 days. Control group cells were not treated. Cells were washed with cold PBS and lysed in RIPA buffer with protein inhibitor cocktail (Thermo Fisher Scientific, IL, USA). The protein concentration of the cell lysate was measured with a Bicinchoninic Acid (BCA) Protein Assay Kit (Sigma-Aldrich).

Samples were run on a 15% SDS-PAGE gel and the proteins were transferred to nylon membranes. The membranes were blocked with 2% non-fat dry milk in TBS overnight at 4°C and then incubated with rabbit antibody to phospho-p44/42 MAPK (Erk1/2) (Thr202/Tyr204), 1:5000 (Cell Signaling Technology, MA, USA), rabbit antibody to p-JNK, 1:2000 (Cell Signaling Technology, MA, USA), and rabbit antibody to p-38, 1:2000 (Cell Signaling Technology, MA, USA) and housekeeping GAPDH.

2.7. Statistical analysis
Statistical analyses were performed using SPSS-17.0 software. Where indicated, experimental data were reported as mean ± standard deviation of triplicate independent samples. Data were analyzed using Student's t-test and one-way analysis of variance, and Tukey's HSD test was applied as a post hoc test if statistical significance was determined. A value of $p \leq 0.05$ was considered statistically significant.

3. Results

3.1. Effect of ascorbic acid on G29 cell metabolic, cell viability, and apoptotic activity
As shown in Figure 1(A), incubation of the G292 cells with 62.5 µM AsA significantly increased the metabolic activity of the cells compared to controls (0 µM AsA) but with higher concentrations an

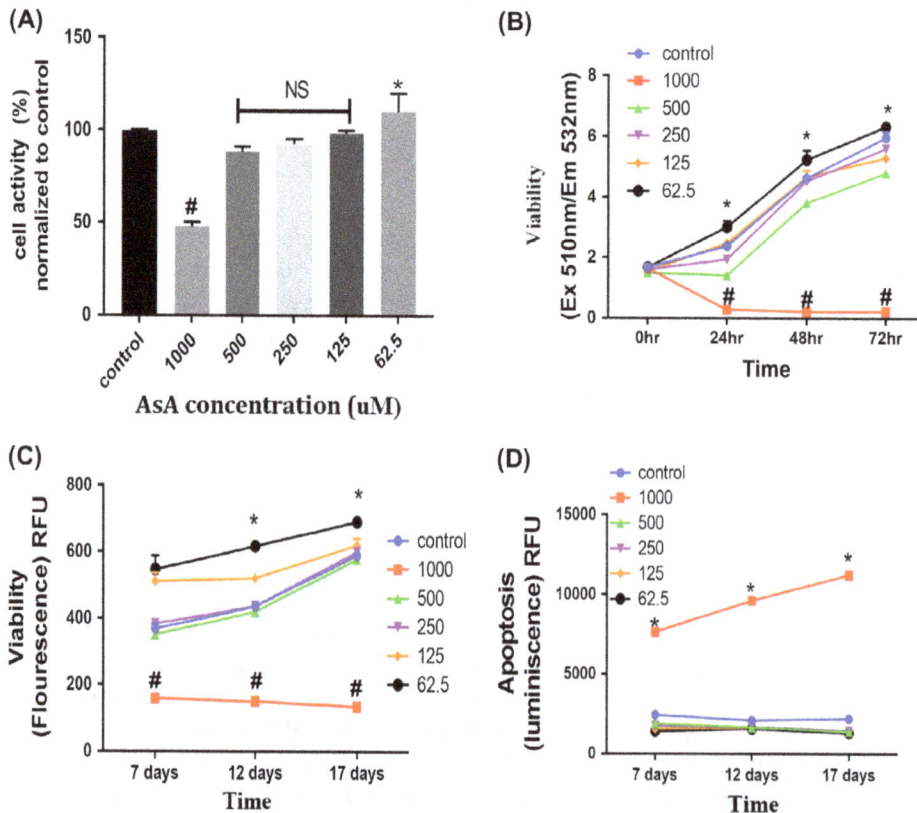

Figure 1. (A) MTT assay: cell metabolic activity was assessed with various concentrations (62.5, 125, 250, 500, 1,000 µM) of ascorbic acid (AsA) added to the G292 cells; absorbance measurements were normalized to the controls (0 µM). 1,000 µM demonstrated significantly greater inhibition of cell activity (# = $p < 0.05$), whereas, 62.5 µM AsA demonstrated a significant increase in cell activity compared to controls (* = $p < 0.05$). (B) Cytotoxicity assay: cells treated with various concentrations of AsA were evaluated via the CellTox Green™ assay for viability at 0, 24, 48, and 72 h. 1,000 µM AsA resulted in significantly lower cell viability (# = $p < 0.05$), while 62.6 µM AsA led to significantly higher cell viability as compared to other groups (* = $p < 0.05$). (C) Cell viability and (D) Apoptosis (assessed with the Apo-Live-Glo™ Multiplex kit): Incubation with 1,000 µM AsA resulted in significantly lower cell viability and higher apoptosis, whereas 62.5 µM AsA led to significantly higher cell viability in comparison to other groups (# = $p < 0.05$; * = $p < 0.05$).

inverse relationship between AsA concentration and activity was observed with 1,000 µM AsA result-ing in a significant decrease compared to the other groups.

We measured cytotoxicity with a fluorescence plate format assay that uses a cyanine dye imper-meant to live cells and stains DNA in dead cells. Using this assay, it was observed that 62. 5 µM AsA significantly increased the live G292 cells activity compared to controls, while higher concentrations of AsA reduced this parameter and the cells incubated with 1,000 µM AsA showed significantly less cellular activity compared to the other groups. These effects were observed at 24, 48, and 72 h of incubation with AsA (Figure 1(B)).

The graphs in (Figure 1(C) and (D)) depict the results of an ApoLive-Glo™ multiplex assay showing the cell viability of G292 cells treated with ascorbic acid. The viability data (Figure 1(C)) supports the results obtained with the other assays assessing cellular activity with 62.5 µM AsA significantly in-creasing this parameter compared to controls and the higher dose (1,000 µM AsA) decreasing viabil-ity. These effects were observed with incubation periods from 7 to 17 days with medium changes with fresh AsA every 3 days. The results of the apoptosis phase of the assay revealed that incubation with 1,000 µM AsA over this period of time significantly increases apoptosis of the G292 cells (Figure 1(D)).

3.2. Effects of ascorbic acid on G292 cell differentiation and mineralization

To assess the ability of AsA to induce G292 cell differentiation, ALP activity and the presence of min-eralization, detected by alizarin red, were evaluated. After 8 days of incubation, with medium chang-es every 3 days, 62.5 µM AsA significantly increased ALP activity compared to controls (0 µM AsA), while 1,000 µM AsA significantly decreased this marker of differentiation and the other lower AsA concentrations tested (125, 250, 500 µM) had no significant effects (Figure 2(A)). After 10 days of treatment incubation with AsA, as shown in Figure 2(B), we observed a significant increase in miner-alization with 62.5 µM AsA, and a significant decrease with 1,000 µM AsA and no significant effects in cells incubated with the other tested AsA concentrations.

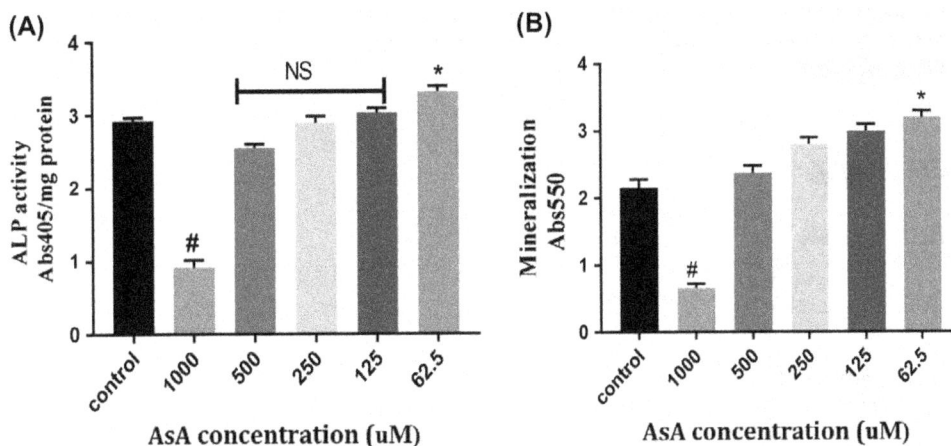

Figure 2. (A) Alkaline phosphatase (ALP) activity: Various concentrations (0, 62.5, 125, 250, 500, 1,000 µM) AsA were tested for their effects on ALP activity after 8 days of incubation. Incubation with 62.5 µM AsA resulted in significantly higher ALP activity (* = $p < 0.05$), whereas doses of 1,000 µM AsA led to lower ALP activity (# = $p < 0.05$). (B) Alizarin Red (ARS) Mineralization assay: Quantitative results from ARS staining demonstrated significantly higher mineralization activity with 62.5 µM (* = $p < 0.05$), whereas 1,000 µM AsA demonstrated significantly lower mineralization activity (# = $p < 0.05$) after 10 days of incubation.

To further evaluate the role of AsA on osteogenic differentiation, we analyzed the effects of AsA at concentrations ranging from 0 to 1,000 µM on the transcription of Runt-related transcription factor-2 (RUNX2) and on the osteosarcoma-related gene osteocalcin. AsA concentrations of 62.5 and 125 µM induced a significant increase of RUNX2 mRNA expression with respect to controls after 72 h of culture with 62.5 µM producing the highest increase in this marker of differentiation (Figure 3(A)). Incubations with concentrations of 250 and 500 µM AsA resulted in no significant effect; however, with 1,000 µM AsA, RUNX2 mRNA expression was significantly decreased ($p < 0.05$) compared to controls (Figure 3(A)). Similar results on the expression of osteocalcin mRNA in these cells were obtained after 72-h incubations with concentrations of AsA from 62.5 to 1,000 µM (Figure 3(B)).

3.3. Role of MAPK pathway on the effects of ascorbic acid on G292 osteosarcoma cells

Western blot analysis revealed that the protein levels of phospho-ERK1/2 MAPK, phospho-pJNK, p-38, and phospho-p38 in the AsA-treated group (1,000 µM) were lower than those in control groups in G292 cells (Figure 4).

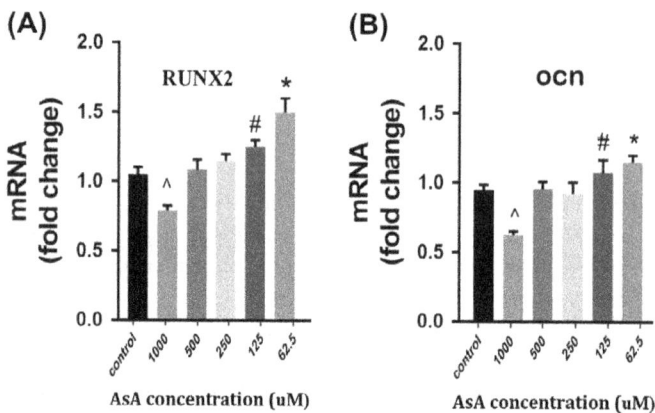

Figure 3. Real-time PCR [qPCR] (A and B): The effect of various concentrations (0, 62.5, 125, 250, 500, 1,000 µM) of AsA and the mRNA fold change of RUNX2 and osteocalcin (ocn) was evaluated in the G292 cells after 72-h incubation. 1,000 µM AsA-treated cells demonstrated significantly lower fold change (^ = $p < 0.05$), Both 125 and 62.5 µM AsA treated cells demonstrated significantly higher fold changes in RUNX2 and ocn in comparison to other groups (#, * = $p < 0.05$) with 62.5 AsA producing the greatest increases.

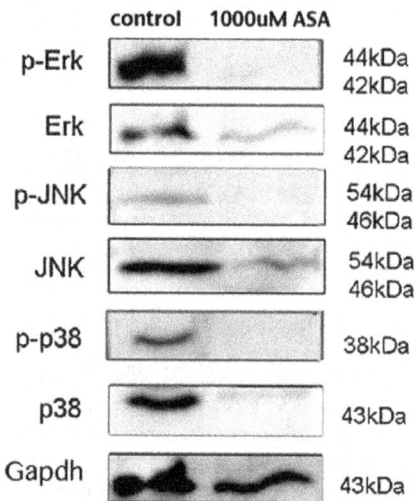

Figure 4. Western blot analysis of phospho-ERK1/2 MAPK, phospho-pJNK, and phospho-p38 shows the expression of these proteins in the AsA-treated group (1,000 µM) was lower than those in controls in the G292 cells.

4. Discussion

Several studies in the literature, recently reviewed by van der Reest and Gottlieb (2016) have suggested the role of L-ascorbic acid, also referred to as Vitamin C, as an anti-tumor agent. In this report, we provide, for the first time, evidence demonstrating that ascorbic acid has an inhibitory and apoptotic effect on G292 osteosarcoma cells. We found that high-dose AsA can inhibit the growth of G292 osteosarcoma significantly, whereas low-dose AsA significantly increases differentiation and cell metabolic activity. Furthermore, higher doses AsA decreased osteogenic differentiation gene expression in a dose-dependent manner and inhibited G292 activity via the MAPK pathway.

Previous studies have demonstrated that AsA in low doses could stimulate differentiation and proliferation of MG-63 osteosarcoma cells, whereas in high doses could induce apoptosis (Valenti et al., 2014). In our study, we extended this work to study the effect of AsA on G292 osteosarcoma cell line since the G292 cell line is a more stable cell line and less aggressive (proliferates slowly) as compared to the MG-63 cell line (Lucero et al., 2013). In the G292 cells studied here, it was observed that the cell metabolic activity, ALP activity and mineralization increased with the lower dose AsA and decreased with high doses of AsA. Furthermore, apoptosis also was shown to increase in G292 cells treated with higher doses of AsA. Apoptosis was measured in these studies with the ApoLive-Glo™ Multiplex Assay that measures both the number of viable cells as a marker of cytotoxicity and caspase activation as a marker of apoptosis to determine the mechanism of cell death (Telford, Komoriya, & Packard, 2002). The first part of the assay measures the activity of a protease marker of cell viability. The live cell protease activity is restricted to intact viable cells and is measured using a fluorogenic, cell-permeant, peptide substrate (glycyl-phenylalanyl-amino fluorocoumarin; GF-AFC). The substrate enters intact cells, where it is cleaved by the live-cell protease activity to generate a fluorescent signal proportional to the number of living cells that then becomes inactive upon loss of cell membrane integrity and leakage into the surrounding culture medium. The second part of the assay uses the Caspase-Glo® Assay technology to detect caspase activation, a key biomarker of apoptosis. The Caspase-Glo® Assay provides a luminogenic caspase-3/7 substrate, which contains the tetrapeptide sequence DEVD, in a reagent optimized for caspase activity, luciferase activity, and cell lysis. It should be noted here that the MTT assay also employed in our studies correlated well with the viability part of the ApoLive-Glo™ Multiplex assay, although the latter gave more subtle results. This assay essentially assessed the cell membrane integrity and is therefore more sensitive than the MTT assay that measures mitochondrial enzymes as an indication of cell metabolic activity (Fernandes et al., 2016).

ALP activity is a marker of osteoblastic differentiation. In the present study, low doses of AsA increased ALP activity, whereas high doses decreased this activity, indicating that high dose of AsA can inhibit osteoblastic differentiation which is in agreement with the study of Valenti et al. (2014), where they reported that low doses of AsA could increase osteoblast differentiation and high doses could induce decreases in this parameter. Moreover, in our studies here, the mineralization potential of G292 cells increased with low doses of AsA, but not at high doses of AsA with a decrease at 1,000 µM. Furthermore, RUNX2 and osteocalcin genes that are associated with osteogenic differentiation were elevated in G292 cells treated with low doses of AsA (62.5–125 µM), however, there was a significant decrease in these levels at a concentration of 1,000 µM AsA suggesting that AsA, at high doses, can inhibit terminal maturation of osteoblasts. Moreover, high doses of AsA treated osteosarcoma demonstrated a rounded morphology of the cells, which has recently been known to be associated with cells with damaged DNA undergoing mitosis which could provide an association between the process of cell cycle arrest and cell death as reported by Kubara et al. (2012).

Our data suggest that cell death induced by AsA in human osteosarcoma cells is an apoptotic process via the downregulation of the MAPK signaling pathways, consistent with reports from previous studies on cell death and survival in osteosarcomas (Niu et al., 2015).

The MAPK pathway plays a key role in extracellular signal transduction to initiate cellular responses (Burotto, Chiou, Lee, & Kohn, 2014). MAPK pathways can relay, amplify and integrate signals from a diverse range of stimuli and elicit appropriate physiological responses, including cellular proliferation, differentiation, development, inflammatory responses, and apoptosis (Na, Kim, & Park, 2012). Studies have demonstrated that the MAPK signaling pathways activate and phosphorylate the osteoblast-specific transcription factor Cbfa1 (RUNX2) in MC3T3-E1 cells and play an important role in the regulation of osteoblast-specific gene expression (Xiao et al., 2000). Other studies suggest an important role for the extracellular signal-regulated kinase (ERK)–MAPK pathway in the regulation of osteoblast differentiation and fetal bone development (Ge, Xiao, Jiang, & Franceschi, 2007). Additionally, the role of p-38 and JNK signaling pathway contributes to cell proliferation and apoptosis. JNK appears to be an intrinsic component of the mitochondrial-dependent death pathway during stress-induced apoptosis. Suggestion of JNK involvement in apoptosis came from the observation that $Jnk1^{-/-} jnk2^{-/-}$ mice were resistant to apoptosis induced by UV irradiation, anisomycin, and DNA-alkylating agent methyl methanesulfate (Baltriukiene, Kalvelyte, & Bukelskiene, 2007). The downregulation of p-44/42, p-38, and p-JNK by 1,000 µM AsA observed in our studies is consistent with a role of these components in actions of this agent on the G292 osteosarcoma cells.

There are emerging data that the pharmaceutically active form of AsA is its oxidized form, dehydroascorbate (DHA). Preclinical studies with a variety of cancer cells showed that pharmacological doses of ascorbate can act as a proxidant and produce hydrogen peroxide which results in toxic effects on cancerous cells without producing adverse effects on normal cell viability (Chen et al., 2005, 2008). In a study with colorectal cancer cells, it was reported that tumor cells with high GLUT1 glucose transporter expression along with KRAS or BRAF oncogene-induced glycolytic addiction are selectively susceptible to cytotoxic effects induced by the AsA (Yun et al., 2015). Likewise, it had been shown that increased endogenous levels of hydrogen peroxide with high concentrations of AsA decreased the tumor growth of pancreatic tumor cell lines (Espey et al., 2011). The potential role of peroxide-induced oxidative stress in the mechanism of AsA action in G292 cells was not addressed in these previous studies, or in this present study. However, this role should be pursued in future studies, particularly in a comparative analyses with other osteosarcoma cell lines in order to achieve a more thorough understanding of critical factors involved in cytotoxic effects in this type of tumor.

In conclusion, ascorbic acid has dose-related effects on G292 cells with high doses resulting in decreases in parameters of osteoblastic differentiation and maturation, and increases in apoptosis. The dose of AsA (1,000 µM) at which antiproliferative, antidifferentiation, and apoptotic effects on G292 cells were consistently observed here falls within the range of doses (600–4,000 µM) that have had such effects on other cell types *in vitro* (Belin et al., 2009; Chen et al., 2005, 2008; Valenti et al., 2014). Although more clinical studies are necessary to establish the *in vivo* efficacy of ASA as an anti-osteosarcoma therapeutic agent, pharmacokinetics conducted with this agent suggest that the doses effective here could be achieved with IV therapy (Padayatty et al., 2004). Therefore, these present studies provide a basis for further investigation on the use of AsA at high doses as an adjuvant to standard chemotherapy for some forms of osteosarcoma.

Acknowledgments
The authors wish to thank Leandra Velasquez of the Department of Oral Biology at University at Buffalo for her technical assistance with the western blot studies described here as well as Dr Michelle Visser, Assistant Professor of Oral Biology at University of Buffalo, for use of some equipment and reagents in her laboratory as well as her advice on technical aspects of the western blot analysis employed here.

Funding
The authors received no direct funding for this research. Funding was from general departmental support.

Competing Interests
The authors declare no competing interest.

Author details
Gabriela Fernandes[1]
E-mail: gfernandes@buffalo.edu
Andrew W. Barone[1]
E-mail: awbarone@buffalo.edu
Rosemary Dziak[1]
E-mail: rdziak@buffalo.edu

[1] Department of Oral Biology, School of Dental Medicine, University at Buffalo, State University of New York, 3435 Main Street, Buffalo, NY 14201, USA.

References

Baltriukiene, D., Kalvelyte, A., & Bukelskiene, V. (2007). Induction of apoptosis and activation of JNK and p38 MAPK pathways in deoxynivalenol-treated cell lines. *Alternatives to Laboratory Animals: ATLA, 35*, 53–59. Retrieved from www.ncbi.nlm.nih.gov/pubmed/17411352

Barres, L., Mota Anna, D. S., Greenberg, M., Almojaly, S., & Dziak, R. (2015). Effects of alendronate on human alveolar osteoblastic cells: Interactions with platelet-derived growth factor. *International Journal of Oral Health Dentistry, 1*, 2. doi:10.16966/2378-7090.108

Belin, S., Kaya, F., Duisit, G., Giacometti, S., Ciccolini, J., & Fontés, M. (2009). Antiproliferative effect of ascorbic acid is associated with the inhibition of genes necessary to cell cycle progression. *PLoS One, 4*, e4409. doi:10.1371/journal.pone.0004409

Benayahu, D., Shur, I., Marom, R., Meller, I., & Issakov, J. (2002). Cellular and molecular properties associated with osteosarcoma cells. *Journal of Cellular Biochemistry, 84*, 108–114. doi:10.1002/jcb.1270

Burotto, M., Chiou, V. L., Lee, J. M., & Kohn, E. C. (2014). The MAPK pathway across different malignancies: A new perspective. *Cancer, 120*, 3446–3456. doi:10.1002/cncr.28864

Chen, Q., Espey, M. G., Krishna, M. C., Mitchell, J. B., Corpe, C. P., Buettner, G. R., ... Levine, M. (2005). Pharmacologic ascorbic acid concentrations selectively kill cancer cells: Action as a pro-drug to deliver hydrogen peroxide to tissues. *Proceedings of the National Academy of Sciences of the United States of America, 102*, 13604–13609. doi:10.1073/pnas.0506390102

Chen, Q., Espey, M. G., Sun, A. Y., Pooput, C., Kirk, K. L., Krishna, M. C., ... Levine, M. (2008). Pharmacologic doses of ascorbate act as a prooxidant and decrease growth of aggressive tumor xenografts in mice. *Proceedings of the National Academy of Sciences, 105*, 11105–11109. doi:10.1073/pnas.0506390102

Duarte, T. L., & Lunec, J. (2005). Review: When is an antioxidant not an antioxidant? A review of novel actions and reactions of vitamin C. *Free Radical Research, 39*, 671–686. doi:10.1080/10715760500104025

ElAttar, T. M. A., & Lin, H. S. (1992). Effect of vitamin C and vitamin E on prostaglandin synthesis by fibroblasts and squamous carcinoma cells. *Prostaglandins, Leukotrienes and Essential Fatty Acids, 47*, 253–257. doi:10.1016/0952-3278(92)90194-N

Espey, M. G., Chen, P., Chalmers, B., Drisko, J., Sun, A. Y., Levine, M., & Chen, Q. (2011). Pharmacologic ascorbate synergizes with gemcitabine in preclinical models of pancreatic cancer. *Free Radical Biology and Medicine, 50*, 1610–1619. doi:10.1016/j.freeradbiomed.2011.03.007

Fernandes, G., Barone, A., & Dziak, R. (2016). Effects of verapamil on bone cancer cells. *Journal of Cell-Biology-&-Cell-Metabolism, 3*, 13. Retrieved from http://www.heraldopenaccess.us/fulltext/Cell-Biology-&-Cell-Metabolism/Effects-of-Verapamil-on-Bone-Cancer-Cells-In-Vitro.pdf

Fliedner, S. M., Engel, T., Lendvai, N. K., Shankavaram, U., Nölting, S., Wesley, R., ... Lehnert, H. (2014). Anti-cancer potential of mapk pathway inhibition in paragangliomas—Effect of different statins on mouse pheochromocytoma cells. *PloS One, 9*, e97712. doi:10.1371/journal.pone.0097712

Ge, C., Xiao, G., Jiang, D., & Franceschi, R. T. (2007). Critical role of the extracellular signal–regulated kinase–MAPK pathway in osteoblast differentiation and skeletal development. *The Journal of Cell Biology, 176*, 709–718. doi:10.1083/jcb.200610046

Golde, D. W. (2003). Vitamin C in cancer. *Integrative Cancer Therapies, 2*, 158–159. doi:10.1177/1534735403002002009

Hitomi, K., & Tsukagoshi, N. (1996). Role of ascorbic acid in modulation of gene expression. In *Subcellular biochemistry* (pp. 41–56). Springer US. doi:10.1007/978-1-4613-0325-1_3

Hu, J. Z., Feng, D. Y., & Cheng, R. X. (2001). Hunan yi ke da xue xue bao= Hunan yike daxue xuebao= [Expressions of p-MAPK, cyclin D1, p53 protein and their relationship in osteosarcoma]. *Bulletin of Hunan Medical University, 26*, 325–327.

Jacobs, C., Hutton, B., Ng, T., Shorr, R., & Clemons, M. (2015). Is there a role for oral or intravenous ascorbate (vitamin C) in treating patients with cancer? A systematic review. *The Oncologist, 20*, 210–223. doi:10.1634/theoncologist.2014-0381

Kubara, P. M., Kernéis-Golsteyn, S., Studény, A., Lanser, B. B., Meijer, L., & Golsteyn, R. M. (2012). Human cells enter mitosis with damaged DNA after treatment with pharmacological concentrations of genotoxic agents. *Biochemical Journal, 446*, 373–381. doi:10.1042/BJ20120385

Langenbach, F., & Handschel, J. (2013). Effects of dexamethasone, ascorbic acid and β-glycerophosphate on the osteogenic differentiation of stem cells *in vitro*. *Stem Cell Research & Therapy, 4*, 1. doi:10.1186/scrt328

Lohmann, W. (1987). Ascorbic acid and cancer. *Annals of the New York Academy of Sciences, 498*, 402–417. doi:10.1111/j.1749-6632.1987.tb23777.x

Lucero, C. M., Vega, O. A., Osorio, M. M., Tapia, J. C., Antonelli, M., Stein, G. S., ... Galindo, M. A. (2013). The cancer-related transcription factor Runx2 modulates cell proliferation in human osteosarcoma cell lines. *Journal of Cellular Physiology, 228*, 714–723. doi:10.1002/jcp.24218

Na, K. Y., Kim, Y. W., & Park, Y. K. (2012). Mitogen-activated protein kinase pathway in osteosarcoma. *Pathology-Journal of the RCPA, 44*, 540–546. doi:10.1097/PAT.0b013e32835803bc

Niu, N. K., Wang, Z. L., Pan, S. T., Ding, H. Q., Au, G. H., He, Z. X., ... Yang, T. (2015). Pro-apoptotic and pro-autophagic effects of the Aurora kinase A inhibitor alisertib (MLN8237) on human osteosarcoma U-2 OS and MG-63 cells through the activation of mitochondria-mediated pathway and inhibition of p38 MAPK/PI3 K/Akt/mTOR signaling pathway. *Drug Design, Development and Therapy, 9*, 1555. doi:10.2147/DDDT.S74197

Ohno, S., Ohno, Y., Suzuki, N., Soma, G. I., & Inoue, M. (2009). High-dose vitamin C (ascorbic acid) therapy in the treatment of patients with advanced cancer. *Anticancer Research, 29*, 809–815. Retrieved from http://ar.iiarjournals.org/content/29/3/809.long

Padayatty, S. J., Sun, H., Wang, Y., Riordan, H. D., Hewitt, S. M., Katz, A., ... Levine, M. (2004). Vitamin C pharmacokinetics: Implications for oral and intravenous use. *Annals of Internal Medicine, 140*, 533–537. doi:10.7326/0003-4819-140-7-200404060-00010

Pauling, L. (1980). Vitamin C therapy of advanced cancer. *The New England Journal of Medicine, 302*, 694–695. doi:10.1056/NEJM198003203021219

Pauling, L., Anderson, R., Banic, S., Basu, T. K., Kallistratos, G., Murata, A., ... Siegel, B. V. (1981). Workshop on vitamin C in immunology and cancer. International journal for vitamin and nutrition research. *Supplement= Internationale Zeitschrift fur Vitamin-und Ernahrungsforschung, 23*, 209–219. Retrieved from http://europepmc.org/abstract/med/6180999

Pauling, L., & Moertel, C. (1986). A proposition: Megadoses of vitamin C are valuable in the treatment of cancer.

Nutrition Reviews, 44, 28–29.
doi:10.1111/j.1753-4887.1986.tb07553.x

Schwarz, R. I., Kleinman, P., & Owens, N. (1987). Ascorbate can act as an inducer of the collagen pathway because most steps are tightly coupled. *Annals of the New York Academy of Sciences, 498*, 172–185. doi:10.1111/j.1749-6632.1987. tb23760.x

Takamizawa, S., Maehata, Y., Imai, K., Senoo, H., Sato, S., & Hata, R. I. (2004). Effects of ascorbic acid and ascorbic acid 2-phosphate, a long-acting vitamin C derivative, on the proliferation and differentiation of human osteoblast-like cells. *Cell Biology International, 28*, 255–265. doi:10.1016/j.cellbi.2004.01.010

Telford, W. G., Komoriya, A., & Packard, B. Z. (2002). Detection of localized caspase activity in early apoptotic cells by laser scanning cytometry. *Cytometry, 47*, 81–88. doi:10.1002/cyto.10052

Valenti, M. T., Zanatta, M., Donatelli, L., Viviano, G., Cavallini, C., Scupoli, M. T., & Dalle Carbonare, L. U. C. A. (2014). Ascorbic acid induces either differentiation or apoptosis in MG-63 osteosarcoma lineage. *Anticancer Research, 34*,

1617–1627. Retrieved from http://ar.iiarjournals.org/content/34/4/1617.long

van der Reest, J., & Gottlieb, E. (2016). Anti-cancer effects of vitamin C revisited. *Cell Research, 26*, 269–270. doi:10.1038/cr.2016.7

Xiao, G., Jiang, D., Thomas, P., Benson, M. D., Guan, K., Karsenty, G., & Franceschi, R. T. (2000). MAPK pathways activate and phosphorylate the osteoblast-specific transcription factor, Cbfa1. *Journal of Biological Chemistry, 275*, 4453–4459. doi:10.1074/jbc.275.6.4453

Yang, F., Nam, S., Zhao, R., Tian, Y., Liu, L., Horne, D. A., & Jove, R. (2013). A novel synthetic derivative of the natural product berbamine inhibits cell viability and induces apoptosis of human osteosarcoma cells, associated with activation of JNK/AP-1 signaling. *Cancer Biology & Therapy, 14*, 1024–1031. doi:10.4161/cbt.26045

Yun, J., Mullarky, E., Lu, C., Bosch, K. N., Kavalier, A., Rivera, K., … Muley, A. (2015, December). Vitamin C selectively kills KRAS and BRAF mutant colorectal cancer cells by targeting GAPDH. *Science, 350*, 1391–1396. doi:10.1126/science.aaa5004.

8

Effects of pre-exposure vaccination and quarantine in the fight against ebola

C.P. Bhunu[1]*, M. Masocha[2] and C.W. Mahera[3]

*Corresponding author: C.P. Bhunu, Department of Mathematics, University of Zimbabwe, Box MP 167, Mount Pleasant, Harare, Zimbabwe

E-mails: cpbhunu@gmail.com, cp-b@hotmail.co.uk

Reviewing editor: Wolfgang Weninger, Centenary Institute for Cancer Medicine and Cell Biology, Australia

Additional information is available at the end of the article

Abstract: Swift quarantine offers some hope in the fight against Ebola but its implementation faces some resistance in many settings. Hence, it is critical to explore whether introducing pre-exposure vaccination in an area where quarantine for the exposed and infected is already practiced would benefit the community with regard to controlling Ebola virus disease and vice versa. We present a mathematical model that explores the potential role of pre-exposure vaccination and quarantine in the fight against Ebola. Threshold parameter of the model is computed and rigorously analysed. Sensitivity analysis is carried out in an effort to understand the effects of constituent parameters on the threshold parameter. The results indicate that pre-exposure and quarantine are able to reduce the disease threshold parameter suggesting they offer hope of Ebola virus disease control.

Subjects: Bioscience; Health and Social Care; Mathematics & Statistics

Keywords: mathematical model; pre-exposure vaccination and quarantine

1. Introduction

The current Ebola virus disease outbreak in west Africa threatens the global human health hence it urgently needs to be controlled. Since its discovery in 1976 in Zaire—now the Democratic Republic of Congo (CDC, 2014)—more than 25 outbreaks of Ebola virus disease have been recorded (Camacho et al., 2014). From 11 January 2014, there have been 21,296 probable, confirmed and suspected Ebola cases, and 8,429 deaths, (WHO, 2014a) with a case fatality ratio of about 70%. So far, quarantine is one of the intervention strategies that is widely employed to bring the epidemic under control.

ABOUT THE AUTHOR

C.P. Bhunu (BSc Hons, MSc, DPhil) is a professor in the Department of Mathematics, University of Zimbabwe. He currently serves as an external examiner for Chinhoyi University of Technology, Harare Institute of Technology and Zimbabwe Open University. He was a visiting African scientist at Cambridge Infectious Disease Consortium (2010) and a visiting professor at the University of Venda (2012), respectively. He is a life member of the Clare Hall College, University of Cambridge. He also serves as an editor and reviewer of several international journals in Applied Mathematics. His research interests lie in the field of mathematical modelling of issues affecting mankind ranging from social issues to biological issues as well as the theoretical analysis of the mathematical models that arise in all these applications.

PUBLIC INTEREST STATEMENT

Currently there is no vaccine against Ebola epidemics, in this manuscript the potential benefits of a yet to be developed; pre-exposure vaccine against Ebola is explored using a mathematical model. The manuscript will also explore other possible intervention strategies (quarantine and quick and safe burial of the infected dead).

While quarantine has proved successful in some settings it remains controversial in others. This then calls for the need to explore other intervention strategies.

Currently, there is no vaccine for Ebola virus licenced for human use but some advances made and efficacy studies in non-human primates on several platforms have been encouraging (CIDRAP, 2015; Feldmann et al., 2007; Sullivan, Sanchez, Rollin, & Yang, Nabel, 2000; Sullivan et al., 2003). It is worth reporting that very little progress has been made in developing treatment interventions for Ebola virus infections (Bray & Paragas, 2002; Feldmann, Jones, Schnittler, & Geisbert, 2005; Geisbert & Hensley, 2004). Currently, there are a number of vaccines at different stages of development. These include, but are not limited to: (a) cAd3-ZEBOV (also known as the NIAID/GSK Ebola vaccine or cAd3-EBO Z), a GlaxoSmithKline (GSK) vaccine which began in 2011 with clinical trials currently under way in USA, Europe and Africa (CIDRAP, 2015). (b) Recombinant Vesicular Stomatitis Virus-Zaire Ebola Virus (rVSV-ZEBOV), a live-virus replication competent vaccine and some experts have raised concern about viral shedding which could pose a threat to livestock and possibly humans (CIDRAP, 2015). However, animal studies have not demonstrated VSV shedding post-vaccination (Geisbert & Fieldmann, 2004) and rVSV vaccines on non-human primates have demonstrated efficacy against infection, both pre- and post-exposure (Geisbert & Fieldmann, 2004). Phase 1 clinical trials of this vaccine were stopped when 20% of the participants had experienced joint symptoms (WHO, 2014b). (c) Ad26.ZEBOV/MVA-BN-filo, which is currently under phase 1 clinical trials in UK (CIDRAP, 2015).

Following up from previous mathematical studies on Ebola (Camacho et al., 2014; Fasina et al., 2014; Ndanguza, Tchuenche, & Haario, 2013 and other references cited there in) we explore the role of dual intervention strategies, that is pre-exposure vaccination and quarantine, in the control of Ebola virus disease. While we acknowledge the fact that to date no vaccine has been licenced for human use, efforts are underway to develop novel pre-exposure drugs. This justifies the need to develop a mathematical model that gives insight into the potential role of pre-exposure vaccination in the fight against Ebola. Here we present such a model that allows us to test the impact of pre-exposure vaccination in the presence of quarantine or quarantine in the presence pre-exposure vaccination. To our knowledge, this is the first time these intervention strategies have been explored simultaneously.

2. Model description

The population is divided into the following classes: unvaccinated susceptibles $S_u(t)$, vaccinated susceptibles $S_v(t)$, exposed and not yet detected individuals $E_1(t)$, exposed and quarantined individuals $E_2(t)$, infectious not yet quarantined $I_1(t)$, quarantined infectives $I_2(t)$, recovered individuals $R(t)$, dead and not yet buried $D(t)$. The total human population is given by

$$N(t) = S_u(t) + S_v(t) + E_1(t) + E_2(t) + I_1(t) + I_2(t) + R(t)(t).$$

Unvaccinated and vaccinated susceptibles are infected with Ebola virus at rates $\lambda(t)$ and $\sigma\lambda(t)$, ($\sigma \in (0,1)$ accounting for reduction in susceptibility to infection for the vaccinated), respectively, to enter the exposed class $E_1(t)$. Here,

$$\lambda(t) = \frac{\beta_I(\theta E_1(t) + I_1(t)) + \beta_d D(t)}{N(t)}, \tag{1}$$

where β_I is the community contact rate, β_d is the funeral contact rate, $\theta \in (0,1)$ accounts for the reduction in infectivity for those not yet displaying the clinical signs of the disease. Unvaccinated susceptibles are vaccinated at a rate ρ_v and for the vaccinated the vaccine wanes at a rate ρ_w. Individuals in $E_1(t)$-class are detected and put on quarantine at a rate τ to enter the $E_2(t)$-class. Individuals in the $E_1(t)$ and $E_2(t)$ classes develop clinical Ebola symptoms at a rate γ to enter the $I_1(t)$ and $I_2(t)$-classes, respectively. Infectives not yet quarantined $I_1(t)$ are quarantined at a rate ϕ to enter the $I_2(t)$-class. Those $I_1(t)$ and $I_2(t)$-classes die at a rate v due to the disease. However, for Those who die in class $I_1(t)$ and $I_2(t)$ they enter the class of the dead $D(t)$ and are buried at a rate μ_b.

It is important to note that some infectives ($I_1(t)$ and $I_2(t)$) recover at a rate α to enter the recovered class $R(t)$. Unless otherwise stated, values used in the analysis and simulations are given in Table 1.

The following system of equations describe the model.

$$
\begin{aligned}
S'_u(t) &= -\lambda S_u - \rho_v S_u + \rho_w S_v, \\
S'_v(t) &= -\sigma \lambda S_v - \rho_w S_v + \rho_v S_u, \\
E'_1(t) &= \lambda S_u + \sigma \lambda S_v - (\tau + \gamma)E_1, \\
E'_2(t) &= \tau E_1 - \gamma E_2, \\
I'_1(t) &= \gamma E_1 - (\alpha + v + \phi)I_1, \\
I'_2(t) &= \gamma E_2 + \phi I_1 - (\alpha + v)I_2, \\
R'(t) &= \alpha(I_1 + I_2), \\
D'(t) &= v(I_1 + I_2) - \mu_b D.
\end{aligned}
\tag{2}
$$

The threshold parameter of model system (Equation (2)) that governs the spread of Ebola is given by

$$
\begin{aligned}
\mathcal{R}_{E_v} &= \mathcal{R}_{E_D} + \mathcal{R}_{E_L} \\
&= \frac{\beta_d \gamma (\sigma \rho_v + \rho_w)}{(\gamma + \tau)(\alpha + v + \phi)(\rho_v + \rho_w)\mu_b} + \frac{\beta_l(\theta(\alpha + \phi + v) + \gamma)(\sigma \rho_v + \rho_w)}{(\gamma + \tau)(\alpha + v + \phi)(\rho_v + \rho_w)}, \\
\mathcal{R}_{E_D} &= \frac{\beta_d \gamma (\sigma \rho_v + \rho_w)}{(\gamma + \tau)(\alpha + v + \phi)(\rho_v + \rho_w)\mu_b}, \quad \mathcal{R}_{E_L} = \frac{\beta_l(\theta(\alpha + \phi + v) + \gamma)(\sigma \rho_v + \rho_w)}{(\gamma + \tau)(\alpha + v + \phi)(\rho_v + \rho_w)}
\end{aligned}
\tag{3}
$$

where which is a threshhold parameter which determines whether or not the disease (Ebola) will invade the host population the presence of pre-exposure vaccination for the uninfected and quarantine for the infected. If $\mathcal{R}_{E_v}$ is less than unity then Ebola will be under control and if it is not then there will be an outbreak of it. $\mathcal{R}_{E_D}$ and $\mathcal{R}_{E_L}$ represent the contribution of infected corpses and the infected humans to Ebola epidemics.

Table 1. Default (baseline) model parameters used in the analysis and simulations

Definition	Symbol	Value (Range)	Source
Community contact rate	β_l	0.10 (0.01–0.2)	Camacho et al. (2014)
Funeral contact rate	β_d	0.78 (0.08–2.0)	Camacho et al. (2014)
Modification parameter	σ	0.25 (0.01–1.0)	Assumed
Modification parameter	θ	0.025 (0.0–1.0)	Assumed
Vaccination rate	ρ_v	0.05 (0.0–1.0)	Bhunu (2015)
Vaccine waning rate	ρ_w	0.025 (0.0–1.0)	Bhunu (2015)
Susceptibles quarantine rate	τ	0.01 (0.0–1.0)	Assumed
Infectives quarantine rate	ϕ	0.01(0.0–1.0)	Assumed
Mean time from death to burial (days)	$\frac{1}{\mu_b}$	0.99 (0.80–1.18)	Camacho et al. (2014)
Mean time from infection to recovery (days)	$\frac{1}{\alpha}$	10.00 (9.80–10.19)	Camacho et al. (2014)
Disease induced death rate	v	0.88 (0.80–0.94)	Camacho et al. (2014)
Incubation period (days)	$\frac{1}{\gamma}$	5.99 (5.80–6.18)	Camacho et al. (2014)

Note: Vaccination and wanning rates taken from Bhunu (2015) are for illustration only as they are for vector borne infections.

2.1. Analysis of the threshold parameter, $\mathcal{R}_{E_v}$

In the absence of any intervention strategy we have

$$\lim_{(\rho_v,\tau,\phi)\to(0,0,0)} \mathcal{R}_{E_v} = \mathcal{R}_0 = \frac{\beta_d}{(\alpha+v)\mu_b} + \frac{\beta_l(\gamma+(\alpha+v)\theta)}{(\alpha+v)\gamma}. \tag{4}$$

2.1.1. Effects of pre-exposure vaccination

If pre-exposure vaccination is the only intervention then

$$\lim_{(\tau,\phi)\to(0,0)} \mathcal{R}_{E_v} = \mathcal{R}_V$$

$$= \frac{\beta_d(\sigma\rho_v+\rho_w)}{(\alpha+v)(\rho_v+\rho_w)\mu_b} + \frac{\beta_l(\gamma+(\alpha+v)\theta)(\sigma\rho_v+\rho_w)}{(\alpha+v)(\rho_v+\rho_w)\gamma}. \tag{5}$$

Subtracting Equation (5) from Equation (4) we obtain

$$\Delta_V = \mathcal{R}_0 - \mathcal{R}_V$$

$$= \frac{\rho_v(1-\sigma)}{(\alpha+v)(\rho_v+\rho_w)}\left(\frac{\beta_d}{\mu_b} + \frac{\beta_l(\gamma+(\alpha+v)\theta)}{\gamma}\right) > 0. \tag{6}$$

Thus, the limiting value of $\mathcal{R}_{E_v}$ when there is no intervention is greater than the limiting value of $\mathcal{R}_{E_v}$ when pre-exposure vaccination is the only intervention ($\Delta_V > 0$). Thus, pre-exposure vaccination reduces $\mathcal{R}_V$. A reduction in $\mathcal{R}_V$ translates to a decrease in disease prevalence. This result points to need to expedite research on the development of pre-exposure vaccines in order for the world to defeat the spread of the deadly Ebola virus.

2.1.2. Effects of quarantine

If quarantine is the only intervention then

$$\lim_{\rho_v\to 0} \mathcal{R}_{E_v} = \mathcal{R}_Q = \frac{\beta_d\gamma}{(\gamma+\tau)(\alpha+v+\phi)\mu_b} + \frac{\beta_l(\gamma+(\alpha+\phi+v)\theta)}{(\alpha+\phi+v)(\gamma+\tau)}. \tag{7}$$

Subtracting Equation (7) from Equation (4) we obtain

$$\Delta_Q = \mathcal{R}_0 - \mathcal{R}_Q$$

$$= \frac{\beta_d(\tau(\alpha+v+\phi)+\gamma\phi)}{v_b(\gamma+\tau)(\alpha+v+\phi)(\alpha+v)} + \frac{\beta_l(\tau(\alpha+v+\phi)(\gamma+(\alpha+v)\theta)+\gamma^2\phi)}{\gamma(\gamma+\tau)(\alpha+v+\phi)(\alpha+v)} > 0. \tag{8}$$

Thus, the limiting value of $\mathcal{R}_{E_v}$ when there is no intervention is greater than the limiting value of $\mathcal{R}_{E_v}$ when quarantine for the exposed and infected individuals is the only intervention ($\Delta_Q > 0$). Thus, quarantine reduces $\mathcal{R}_Q$. A reduction in $\mathcal{R}_Q$ translates to a decrease in disease prevalence. From this result we can conclude that even as the only intervention strategy, quarantining the exposed and infected individuals has the potential to control this deadly infection. However, it's implementation faces some resistance challenges. There is need to understand if introducing pre-exposure vaccination in an area where quarantine for the exposed and infected is already practiced would benefit the community with regard to controlling Ebola virus disease. Comparing $\mathcal{R}_Q$ and $\mathcal{R}_{E_v}$ we have

$$\Delta_{Q_v} = \mathcal{R}_Q - \mathcal{R}_{E_v}$$

$$= \frac{\rho_v(1-\sigma)(\beta_d\gamma + \mu_b\beta_l(\theta(\alpha+v+\phi)+\gamma))}{\mu_b(\gamma+\tau)(\rho_v+\rho_w)(\alpha+v+\phi)} > 0. \tag{9}$$

Thus Equation (9) is positive ($\Delta_{Q_v} > 0$) suggesting that introducing pre-exposure vaccination in an area where quarantine for the exposed and infected individuals is already present will be beneficial to the community. Furthermore, we explore the implications of introducing quarantine for exposed and the infected individuals in an area where pre-exposure vaccination is practiced. Comparing $\mathcal{R}_V$ and $\mathcal{R}_{E_v}$ we have

$$\Delta_{V_Q} = \mathcal{R}_V - \mathcal{R}_{E_V}$$
$$= \frac{\rho_v(1 - \sigma)(\beta_d\gamma + \mu_b\beta_l(\theta(\alpha + v + \phi) + \gamma))}{\mu_b(\gamma + \tau)(\rho_v + \rho_w)(\alpha + v + \phi)} > 0. \tag{10}$$

Thus, the limiting value of $\mathcal{R}_{E_V}$ when pre-exposure vaccination is the only intervention is greater than $\mathcal{R}_{E_V}$ ($\Delta_{V_Q} > 0$). Thus, quarantine reduces $\mathcal{R}_{E_V}$. This result suggests that introducing quarantine in an environment in which pre-exposure vaccination is in place will assist in reducing Ebola transmission in the community. Finally, we explore the implications of introducing pre-exposure vaccination in an environment where quarantine for the exposed and infected is in place. Comparing Comparing $\mathcal{R}_Q$ and $\mathcal{R}_{E_V}$ we have

$$\Delta_{Q_V} = \mathcal{R}_Q - \mathcal{R}_{E_V}$$
$$= \frac{\beta_d(\sigma\rho_v + \rho_w)(\gamma\phi + (\alpha + v + \phi)\tau)}{\mu_b(\alpha + \tau)(\rho_v + \rho_w)(\alpha + v + \phi)(\alpha + v)} \tag{11}$$
$$+ \frac{\beta_l(\sigma\rho_v + \rho_w)(\gamma^2\phi + (\alpha + v + \phi)(\gamma + (\alpha + v)\theta)\tau)}{\gamma(\alpha + \tau)(\rho_v + \rho_w)(\alpha + v + \phi)(\alpha + v)} > 0.$$

The fact that the limiting value of $\mathcal{R}_{E_V}$ when quarantine is the only intervention is greater than $\mathcal{R}_{E_V}$ ($\Delta_{Q_V} > 0$), suggests that introducing pre-exposure in an environment in which quarantine was already would be beneficial to the community.

3. Sensitivity analysis

There are many types of sensitivity analyses that may be used to assess uncertainty associated with a cost–benefit study which includes partial derivatives, a sensitivity index, a relative deviation ratio, partial rank correlation coefficients, rank regression coefficients and the squared-ranks test among others. However, in this manuscript, we are to use Latin Hypercube Sampling and Partial Rank Correlation Coefficients (PRCCs) with 1,000 simulations per run to account for the effects of variations in $\mathcal{R}_{E_V}$ to its constituent parameters.

PRCCs illustrate the degree of the effect that each parameter has on the outcome. Figure 1 illustrates the PRCCs using $\mathcal{R}_{E_V}$ as an output variable. The parameters with the greatest effect on the outcome are burial and quarantine for of the exposed suggesting that their increase will have a positive impact in controlling the spread of Ebola. This result calls for quick and safe burial of all corpses suspected to have been killed by Ebola virus. Furthermore, pre-exposure vaccination is also shown to have an effect on the outcome, its increase results in a decrease in the threshold parameter, $\mathcal{R}_{E_V}$. A decrease in $\mathcal{R}_{E_V}$ suggests a decrease in disease prevalence.

Figure 2 illustrates the effect that varying six sample parameters will have on $\mathcal{R}_{E_V}$. Results from Figure 2 show that increase in quarantining the exposed ($\geq 60\%$) and quick and safe burial of corpses who died due to Ebola ($\geq 30\%$) reduces $\mathcal{R}_{E_V}$ to levels below unity. A reduction in $\mathcal{R}_{E_V}$ to levels

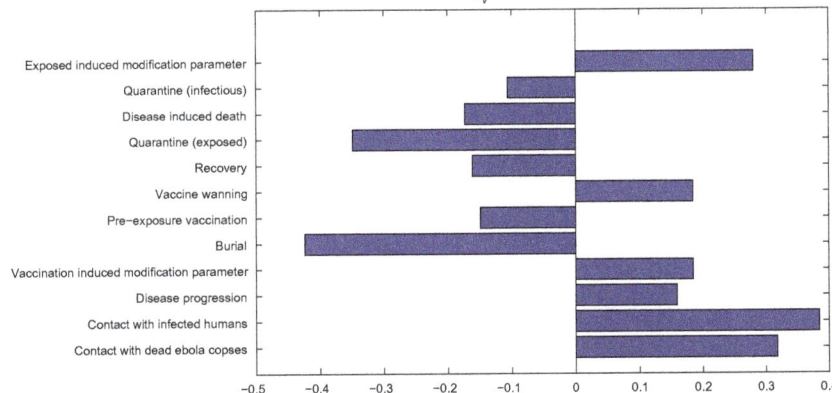

Figure 1. Partial rank correlation coefficients showing the effect of parameter variations on $\mathcal{R}_{E_V}$ using ranges in the table.

Notes: Parameters with positive PRCCs will increase $\mathcal{R}_{E_V}$ when they are increased, whereas parameters with negative PRCCs will decrease $\mathcal{R}_{E_V}$ when they are increased.

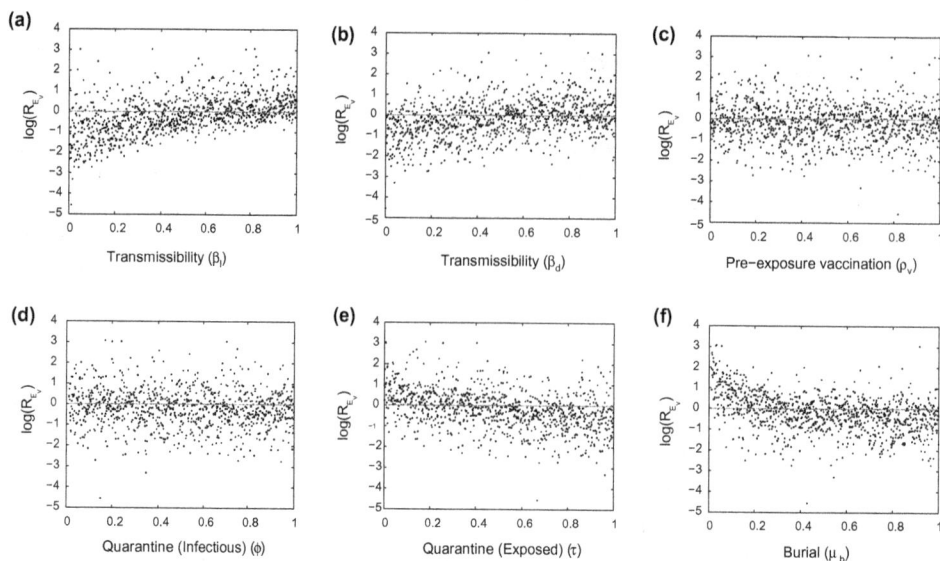

Figure 2. Monte Carlo simulations of 1,000 sample values for six illustrative parameters $(\beta_i, \beta_d, \rho_v, \phi, \tau$ **and** $\mu_b)$ **chosen via Latin hypercube sampling.**

below unity is a sufficient condition for disease control. Results from Figure 2(a) and (b) show that Ebola transmission is mostly driven by live infectious cases as opposed to infected corpses.

4. Discussion

A mathematical model is developed to explore the potential impact of the "new" pre-exposure vaccine and quarantine of the infected and the exposed in controlling the spread of Ebola virus. Analysis of the threshold parameter $(\mathcal{R}_{E_V})$ suggests that pre-exposure vaccination or quarantine for the exposed/infected can single handedly afford to control Ebola virus transmission under control. Furthermore, analysis of the threshold parameter $(\mathcal{R}_{E_V})$ shows that quarantine (pre-exposure vaccination) coupled with pre-exposure vaccination (quarantine) does better than either of the single strategies. Sensitivity analysis (PRCCs and Monte Carlo simulations) show that the threshold parameter $(\mathcal{R}_{E_V})$ is a decreasing function of quarantine (for the exposed/infectious), pre-exposure vaccination and quick and safe burial of the infected corpses. This is in agreement with the analytic results obtained, that is sensitivity analysis and analytic results go hand in hand. Against this background, there is need for researchers to strengthen their research towards producing a novel pre-exposure vaccine which when used together with other conventional intervention strategies like will be able to keep Ebola virus infections under control. However, in the current set-up there is a need to upscale quarantine as well as quick and safe burial of the corpses suspected to have died due to Ebola virus infection. The model presented in this manuscript gives insight into the dynamics of Ebola in the presence of quarantine and vaccination. The results assist decision-makers to make informed decisions regarding Ebola control. This study can be further improved by incorporating the aspects associated with the costs and benefits of exploring new pre-exposure vaccine and quarantine in the fight against Ebola.

Acknowledgements
The author thanks the handling editor and reviewers for their insighful comments which improved the manuscript.

Funding
The authors received no direct funding for this research.

Competing Interests
The authors declare no competing interests.

Author details
C.P. Bhunu[1]
E-mails: cpbhunu@gmail.com, cp-b@hotmail.co.uk
M. Masocha[2]
E-mail: masocha@itc.nl

C.W. Mahera[3]
E-mail: maheraii@yahoo.com
[1] Department of Mathematics, University of Zimbabwe, Box MP 167, Mount Pleasant, Harare, Zimbabwe.
[2] Department of Geography and Environmental Science, University of Zimbabwe, Box MP 167, Mount Pleasant, Harare, Zimbabwe.
[3] African Institute for Mathematical Sciences (AIMS), Box 176, Bagamoyo, Tanzania.

Citation information
Cite this article as: Effects of pre-exposure vaccination and quarantine in the fight against ebola, C.P. Bhunu, M. Masocha & C.W. Mahera, *Cogent Biology* (2016), 2: 1199176.

Cover image
Source: Authors.

References

Bhunu, C. P. (2015). Assessing the potential of pre-exposure vaccination and chemoprophylaxis in the control of lymphatic filariasis. *Applied Mathematics and Computation, 250,* 571–579.

Bray, M., & Paragas, J. (2002). Experimental therapy of lovirus infections. *Antiviral Research, 54,* 1–17.

Camacho, A., Kucharski, A. J., Funk, S., Breman, J., Piot, P., Edmunds, W. J., (2014). Potential for large outbreaks of Ebola virus disease. *Epidemics, 9,* 70–78.

CIDRAP (Centre for Infectious Disease Research and Policy). (2015). *Recommendations for accelerating the development of Ebola vaccines: Report and analysis.*

CDC. (2014). *Outbreaks chronology: Ebola Hemorrhagic fever.* Retrieved from http://www.cdc.gov/vhf/ebola/resourses/outbreak-table.html

Fasina, F. O., Shittu, A., & Lazarus, D., Tomori, O., Simonsen, L., Viboud, C., Chowell, G. (2014). Transmission dynamics and control of Ebola virus disease disease in Nigeria. *Rapid Communications.*

Feldmann, H., Jones, S. M., Schnittler, H. J., & Geisbert, T. (2005). Therapy and prophylaxis of Ebola virus infections. *Current Opinion in Investigational Drugs, 6,* 823–830.

Feldmann, H., Jones, S. M., Daddario-DiCaprio, K. M., Geisbert, J. B., Ströher, U., Grolla, A., ... Geisbert, T. W. (2007). Effective post-exposure treatment of Ebola infection. *PLOS Pathogens, 3*(1), e2. doi:10.1371/journal.ppat.0030002

Geisbert, T. W., & Hensley, L. E. (2004). Ebola virus: New insights into disease aetiopathology and possible therapeutic interventions. *Expert Reviews in Molecular Medicine, 6,* 1–24.

Geisbert, T. W., & Fieldmann, H. (2004). Recombinant vesicular stomatis virus-based vaccines against Ebola and Marburg virus infections. *Journal of Infectious Diseases, 204,* S1075–S1081.

Ndanguza, D., Tchuenche, J. M., & Haario, H. (2013). Statistical data analysis of the 1995 Ebola out-break in the Democratic Republic of Congo. *Afrika Matematika, 24,* 55–68.

Sullivan, N. J., Geisbert, T. W., Geisbert, J. B., Xu, L., Yang, Z. Y., Roederer, M.,... Nabel, G. J. (2003). Accelerated vaccination for Ebola virus haemorrhagic fever in non-human primates. *Nature, 424,* 681–684.

Sullivan, N. J., Sanchez, A., Rollin, P. E., Yang, Z. Y., & Nabel, G. J. (2000). Development of a preventive vaccine for Ebola virus infection in primates. *Nature, 408,* 605–609.

WHO. (2014a). *Disease outbreak news.* Geneva: Author.

WHO. (2014b). *Third teleconference on vaccine clinical trials design for Guinea, Liberia and Sierra Leon.* Geneva: Author.

9

Melaleuca (*Melaleuca alternifolia*) essential oil demonstrates tissue-remodeling and metabolism-modulating activities in human skin cells

Xuesheng Han[1]* and Tory L. Parker[1]

*Corresponding author: Xuesheng Han, dōTERRA International, LLC, 389 S. 1300 W., Pleasant Grove, UT 84062, USA
E-mails: lhan@doterra.com, lawry.han@gmail.com

Reviewing editor: Dominic Ng, University of Queensland, Australia
Additional information is available at the end of the article

Abstract: Melaleuca (*Melaleuca alternifolia*) essential oil (MEO), commonly known as tea tree oil, is popularly used in skincare products. In the current study, we investigated the biological activity of a commercially available MEO (with terpinen-4-ol as the major active component) in pre-inflamed human dermal fibroblasts, which were designed to simulate chronic inflammation. We analyzed the levels of seventeen biomarkers that are important in inflammation and tissue remodeling. Additionally, we studied the effect of MEO on genome-wide gene expression. MEO showed a robust antiproliferative activity against the cells. It also increased the levels of monocyte chemoattractant protein 1, an inflammatory chemokine, and several tissue remodeling molecules such as epidermal growth factor receptor, matrix metalloproteinase 1, and tissue inhibitor of metalloproteinase-1 and -2. It was also noted that MEO diversely modulated global gene expression. Furthermore, Ingenuity Pathway Analysis showed that MEO affects many important signaling pathways that are closely related to metabolism, which suggests its potential modulation of metabolism. The results provide an important evidence of the biological activity of MEO in human dermal fibroblasts. They also suggest that MEO plays useful roles in tissue remodeling and metabolism; however, further research is needed to explore the mechanisms underlying these actions.

ABOUT THE AUTHORS

Dr Han's group primarily studies the health benefits of essential oils. We are specifically interested in the efficacy and safety of essential oils and their active components. Our studies of essential oils in both *in vitro* and clinical settings utilize a variety of experimental approaches, including analytical, biological, biochemical, and biomedical methodologies. We work closely with research institutes, hospitals, and clinics to move the study of essential oils forward. The research work discussed in this paper represents one part of a large research project, which was designed to extensively examine the impact of essential oils on human cells. This study, along with others, will further the understanding of the health benefits of essential oils for a wide research audience. Besides essential oils, we are also interested in studying the health benefits of herbal supplements and skin care products. Dr Han holds a PhD in Biological Sciences and is an elected Fellow of the American College of Nutrition. Dr Parker holds a PhD in Nutritional Sciences.

PUBLIC INTEREST STATEMENT

Essential oils have become more popular globally for skincare purposes. Our study examined the effects of melaleuca essential oil (MEO, also known as tea tree oil) in a human skin disease model. The effects of MEO were determined by measuring the levels of biomarkers that are linked to inflammation, immune function, and wound healing. The effects of MEO on genome-wide gene expression were also studied. MEO showed strong anti-proliferative, tissue remodeling, and immune modulatory activities. Notably, SEO impacted critical genes and pathways that are associated with tissue remodeling and metabolism processes. The findings from this study suggest that SEO may be a good therapeutic candidate for wound care and metabolic conditions. Advanced exploration of the health benefits of MEO may lead to viable options for fighting many of these diseases. Thus, this study provides an important stepping stone for further research on MEO and its health benefits in human beings.

Subjects: Biochemistry; Pharmaceutical Science; Pharmacology; Cell Biology; Human Biology; Immunology

Keywords: melaleuca essential oil; *Melaleuca alternifolia*; terpinen-4-ol; tissue remodeling; metabolism; cell proliferation; tea tree oil

1. Introduction

Melaleuca (*Melaleuca alternifolia*) essential oil (MEO), also known as tea tree oil, is commonly used in skincare products. MEO has been shown to possess antimicrobial, antifungal, antioxidant, anti-inflammatory, immunomodulatory, and pro-wound healing properties (Pazyar, Yaghoobi, Bagherani, & Kazerouni, 2013). However, reports on the biological effects of MEO or its main active components (such as terpinen-4-ol, γ-terpinene, and α-terpinene) on human skin cells are scarce. Homeyer et al. (2015) studied the toxicity of MEO in human fibroblasts and keratinocytes, and reported that the oil has negligible toxicity at concentrations <10%. Furthermore, several small clinical studies found that MEO seems to improve wound healing, presumably due to its antimicrobial, anti-inflammatory, and immunomodulatory activities (Chin & Cordell, 2013; Edmondson et al., 2011; Pazyar et al., 2013).

In the current study, we investigated the biological activity of a commercially available MEO in a human dermal fibroblast system, which was designed to simulate chronic inflammation. We analyzed the effects of MEO on 17 important protein biomarkers that are critically related to the processes of inflammation, immune response, and tissue remodeling. We then studied the effect of MEO on genome-wide gene expression. The study provides important evidence of the biological activity of MEO in human skin cells. Moreover, to the best of our knowledge, it is the first to document the impact of MEO on human genome-wide gene expression. The data provide important insights into the mechanism of action of MEO and will likely stimulate further research.

2. Materials and methods

All experiments were conducted using a Biologically Multiplexed Activity Profiling (BioMAP) system HDF3CGF (Berg et al., 2010; Kunkel, Dea, et al., 2004), which was designed to model the pathology of chronic inflammation in a robust and reproducible manner. The system comprises three components: a cell type, stimuli to create the disease environment, and a set of biomarker (protein) readouts to examine how the treatments affected the disease environment (Berg et al., 2010). The methodologies used in this study were essentially the same as those previously described (Han & Parker, 2017a, 2017b; Kunkel, Plavec, et al., 2004).

2.1. Cell culture

Primary human neonatal fibroblasts were prepared as previously described (Bergamini et al., 2012) and were plated under low serum conditions (0.125% fetal bovine serum) for 24 h. Then, the cell culture was stimulated with a mixture of interleukin (IL)-1β, tumor necrosis factor (TNF)-α, interferon (IFN)-γ, basic fibroblast growth factor (bFGF), epidermal growth factor (EGF), and platelet-derived growth factor (PDGF), for another 24 h. The study agent was added 1 h before stimulation and was present during the entire 24 h stimulation period. The cell culture and stimulation conditions for the HDF3CGF assays have been described in detail elsewhere and were performed in a 96-well plate (Bergamini et al., 2012; R Development Core Team, 2011).

2.2. Protein-based readouts

An enzyme-linked immunosorbent assay (ELISA) was used to measure the biomarker levels of cell-associated and cell membrane targets. Soluble factors in the supernatants were quantified using either homogeneous time-resolved fluorescence detection, bead-based multiplex immunoassay, or capture ELISA. The adverse effects of the test agents on cell proliferation and viability (cytotoxicity) were measured using the sulforhodamine B (SRB) assay. For proliferation assays, the cells were cultured and measured after 72 h, which is optimal for the HDF3CGF system, and the detailed procedure has been described in a previous study (Bergamini et al., 2012). Measurements were performed in triplicate wells, and a glossary of the biomarkers used in this study is provided in Supplementary Table S1.

Quantitative biomarker data are presented as the mean $\log_{10}$ relative expression level (compared to the respective mean vehicle control value) ± standard deviation of triplicate measurements. Differences in biomarker levels between MEO- and vehicle-treated cultures were tested for significance with the unpaired Student's t test. A p-value <0.05, outside of the significance envelope, with an effect size of at least 10% (more than 0.05 $\log_{10}$ ratio units), was considered statistically significant.

2.3. RNA isolation

Total RNA was isolated from cell lysates using the Zymo *Quick-RNA* MiniPrep kit (Zymo Research Corp., Irvine, CA, USA) according to the manufacturer's instructions. RNA concentration was determined using a NanoDrop ND-2000 system (Thermo Fisher Scientific, Waltham, MA, USA). RNA quality was assessed using a Bioanalyzer 2100 (Agilent Technologies, Santa Clara, CA, USA) and an Agilent RNA 6000 Nano kit. All samples had an A260/A280 ratio between 1.9 and 2.1 and a RNA integrity number score greater than 8.0.

2.4. Microarray analysis of genome-wide gene expression

The effect of 0.011% MEO on the expression of 21,224 genes was evaluated in the HDF3CGF system after a 24 h treatment. Samples for microarray analysis were processed by Asuragen, Inc. (Austin, TX, USA) according to the company's standard operating procedures. Biotin-labeled cRNA was prepared from 200 ng of total RNA using an Illumina TotalPrep RNA Amplification kit (Thermo Fisher Scientific) and one round of amplification. The cRNA yields were quantified using ultraviolet spectrophotometry, and the distribution of the transcript sizes was assessed using the Agilent Bioanalyzer 2100. Labeled cRNA (750 ng) was used to probe Illumina human HT-12 v4 expression bead chips (Illumina, Inc., San Diego, CA, USA). Hybridization, washing, staining with streptavidin-conjugated cyanine-3, and scanning of the Illumina arrays were carried out according to the manufacturer's instructions. The Illumina BeadScan software was used to produce the data files for each array; the raw data were extracted using Illumina BeadStudio software.

The raw data were uploaded into R (R Development Core Team, 2011) and analyzed for quality-control metrics using the beadarray package (Dunning, Smith, Ritchie, & Tavare, 2007). The data were normalized using quantile normalization (Bolstad, Irizarry, Astrand, & Speed, 2003), and then re-annotated and filtered to remove probes that were non-specific or mapped to intronic or intragenic regions (Barbosa-Morais et al., 2010). The remaining probe sets comprised the data-set for the remainder of the analysis. The fold-change expression for each set was calculated as the $\log_2$ ratio of MEO to the vehicle control. These fold-change values were uploaded onto Ingenuity Pathway Analysis (IPA, Qiagen, Redwood City, CA, USA, www.qiagen.com/ingenuity) to generate the networks and pathway analyses.

2.5. Reagents

MEO (dōTERRA Intl., Pleasant Grove, UT, USA) was diluted in dimethyl sulfoxide (DMSO) to 8 × the specified concentrations (final DMSO concentration in culture media was no more than 0.1% [v/v]). Then, 25 µL of each 8 × solution was added to the cell culture to obtain a final volume of 200 µL, and DMSO (0.1%) served as the vehicle control. The gas chromatography–mass spectrometry analysis of MEO indicated that its major chemical constitutes (i.e., >5%) were terpinen-4-ol (42%), γ-terpinene (21%), and α-terpinene (11%).

3. Results and discussion

3.1. Bioactivity profile of MEO in a human dermal fibroblast system HDF3CGF

We analyzed the activity of MEO in a dermal fibroblast system, HDF3CGF, which features the microenvironment of inflamed human skin cells with a high inflammation and immune response. Four concentrations (0.011, 0.0037, 0.0012, and 0.00041% v/v) of MEO were initially studied for their effects on cell viability. MEO was not overtly toxic to the cells at any of the concentrations tested; thus, all four concentrations were included in the further analyses. The biomarkers were designated as

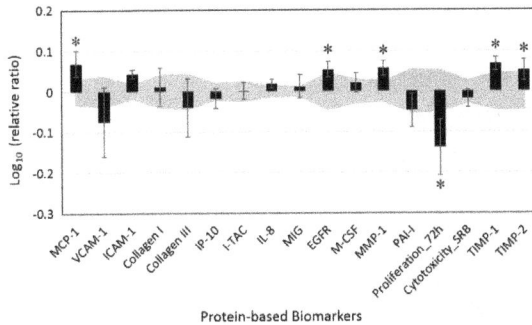

Figure 1. Bioactivity profile of melaleuca essential oil (MEO, 0.011% v/v) in a human dermal fibroblast culture (HDF3CGF).

Notes: The y-axis denotes the relative expression levels of biomarkers compared to vehicle control values, in $\log_{10}$ form. The error bars indicate the standard deviation calculated from triplicate measurements. The vehicle control values are shaded in gray and denote a 95% significance envelope. *indicates a biomarker designated as having "key activity," i.e. the biomarker value is significantly different ($p < 0.05$) from that of the vehicle control, out of the significance envelope, and with an effect size of at least 10% (more than 0.05 log ratio units). MCP-1, monocyte chemoattractant protein 1; VCAM-1, vascular cell adhesion molecule 1; ICAM-1, intracellular cell adhesion molecule 1; IP-10, interferon γ-induced protein 10; I-TAC, interferon-inducible T-cell α chemoattractant; IL-8, interleukin-8; MIG, the monokine induced by gamma interferon; EGFR, epidermal growth factor receptor; M-CSF, macrophage colony-stimulating factor; MMP-1, matrix metalloproteinase 1; PAI-1, plasminogen activator inhibitor 1; SRB, sulforhodamine B; TIMP, tissue inhibitors of metalloproteinase.

having key activity if their expression levels were significantly different ($p < 0.05$) after treatment of the cells with 0.011% v/v MEO with an effect size of at least 10% (more than 0.05 log ratio units) (Figure 1).

MEO showed a significant anti-proliferative activity in the human dermal fibroblasts (Figure 1). In addition, it increased the levels of monocyte chemoattractant protein 1 (MCP-1), epidermal growth factor receptor (EGFR), matrix metalloproteinase 1 (MMP-1), and tissue inhibitor of metalloproteinase (TIMP)-1 and -2 in a slightly significant manner. Other biomarker readouts, including those of several inflammatory biomarkers, were not significantly affected by MEO. These results suggest that MEO may possess tissue remodeling and wound healing properties.

It has been demonstrated that MEO and its major active component terpinen-4-ol show anti-inflammatory and immunomodulatory activities (Hart et al., 2000; Koh, Pearce, Marshman, Finlay-Jones, & Hart, 2002; Low et al., 2015). Some small clinical studies have indicated that the anti-inflammatory, immunomodulatory, and antimicrobial activities of MEO possibly contribute to its pro-wound healing properties (Chin & Cordell, 2013; Edmondson et al., 2011; Pazyar et al., 2013). The results show that MEO affects molecules that are critical to the process of tissue remodeling in the skin, which suggests the involvement of an alternative mechanism in the effect of MEO on the modulation of wound healing.

3.2. Effects of MEO on genome-wide gene expression

We further explored the effects of MEO on human skin cells by studying it at a concentration of 0.011% v/v, which was the highest concentration we studied that was non-cytotoxic to the cells, on the RNA expression of 21,224 genes in the HDF3CGF system. The results showed a diverse effect of MEO on the regulation of these genes (Table S2). Among the 87 genes that were highly regulated (with a $\log_2$ [fold change ratio of expression over vehicle control] $\geq |1.5|$) by MEO, a majority of them (65 genes) were significantly upregulated, whereas the rest were significantly downregulated (Table S2).

The IPA studies showed that the bioactivity of MEO significantly overlapped with many canonical signaling pathways from a literature-validated database (Figure 2, Table S3–S6). It was observed that many of the pathways are closely related to cellular metabolism. For instance, the phosphorylation and dephosphorylation of nicotinamide adenine dinucleotide (NAD), NAD salvage pathways, and biosynthesis of cholesterol were among these most-matched pathways. These findings indicate

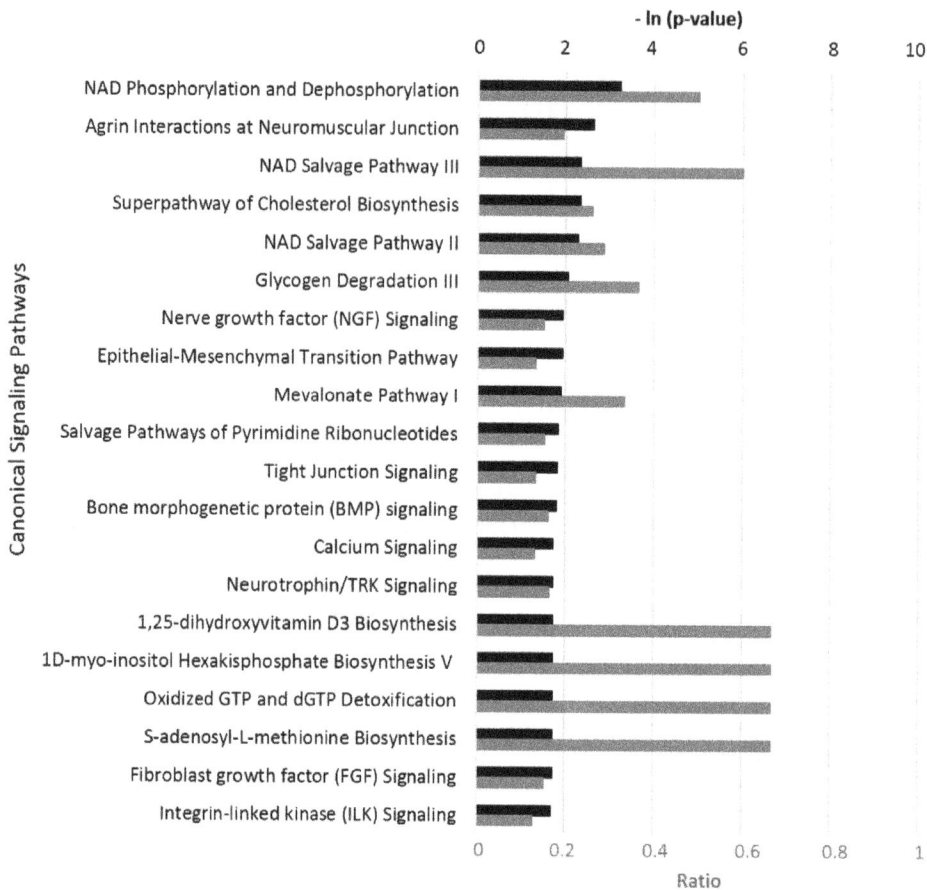

Figure 2. Top 20 canonical pathways matching the bioactivity profile of melaleuca essential oil (MEO, 0.011% v/v) in the HDF3CGF system produced using Ingenuity Pathway Analysis.

Notes: Each p-value was calculated using the right-tailed Fisher's Exact Test. The p-value measures the likelihood that an observed association between a specific pathway and the data-set is due to random chance. A smaller p-value (i.e. bigger - ln [p-value], indicated by the black bars) of a pathway denotes a more significant match with the bioactivity of MEO. A ratio, indicated by each gray bar, was calculated by dividing the number of genes from the MEO data-set that participate in a canonical pathway by the total number of genes in that pathway. NAD, nicotinamide adenine dinucleotide; TRK, tropomyosin-related kinase; GTP, guanosine-5'-triphosphate; dGTP, deoxyguanosine triphosphate.

that MEO possibly plays a role in modulating cellular metabolism and thus, it may be a viable treatment for chronic metabolic conditions.

A literature search conducted by us revealed no published studies on the effects of MEO on metabolism in animal or human models. However, MEO has been found to inhibit the activities of metabolic enzymes in *Candida albicans* (Rajkowska, Kunicka-Styczyńska, Maroszyńska, & Dąbrowska, 2014). Furthermore, given that many metabolic conditions are often associated with inflammation and immune response disorders, it is reasonable to assume that MEO might be able to modulate metabolism via influencing inflammatory and immune mediators.

Collectively, the study, along with existing literature, suggests that MEO may possess tissue-remodeling and metabolism-modulating properties. However, further research into the biological and physiological mechanisms of action of MEO is recommended.

The current study has several limitations. Although the disease model was designed to simulate chronic inflammation and fibrosis, the *in vitro* study results cannot be directly applied to the more complex human system. In addition, the impact of MEO on gene expression was evaluated after short-term treatment of the cells with MEO. Therefore, how MEO impacts global gene expression over a longer term is unclear. Nevertheless, the protein and gene expression data are an evidence of the biological effect of MEO on human skin cells and will likely stimulate further research into the mechanism of action of MEO.

4. Conclusions

MEO significantly inhibited the proliferation of human dermal fibroblasts. Additionally, it slightly increased the levels of MCP-1, EGFR, MMP-1, and TIMP-1 and -2. Furthermore, genome-wide gene expression analysis showed that MEO modulates global gene expression. It was also observed that MEO robustly affected signaling pathways that are critical for cellular metabolism. The data obtained largely support that MEO possesses tissue-remodeling and metabolism-modulating properties.

Funding
This study was funded by dōTERRA (Pleasant Grove, UT, USA) and conducted at DiscoverX (Fremont, CA, USA).n/a

Competing Interests
Xuesheng Han and Tory Parker are employees at dōTERRA, where the study agent MEO was manufactured.

Author details
Xuesheng Han[1]
E-mails: lhan@doterra.com, lawry.han@gmail.com
ORCID ID: http://orcid.org/0000-0003-2720-3011
Tory L. Parker[1]
E-mail: tparker@doterra.com
[1] dōTERRA International, LLC, 389 S. 1300 W., Pleasant Grove, UT 84062, USA.

References
Barbosa-Morais, N. L., Dunning, M. J., Samarajiwa, S. A., Darot, J. F. J., Ritchie, M. E., Lynch, A. G., & Tavaré, S. (2010). A re-annotation pipeline for Illumina BeadArrays: Improving the interpretation of gene expression data. *Nucleic Acids Research, 38,* e17. doi:10.1093/nar/gkp942

Berg, E. L., Yang, J., Melrose, J., Nguyen, D., Privat, S., Rosler, E., & Kunkel, E. J. (2010). Chemical target and pathway toxicity mechanisms defined in primary human cell systems. *Journal of Pharmacological and Toxicological Methods, 61,* 3–15. doi:10.1016/j.vascn.2009.10.001

Bergamini, G., Bell, K., Shimamura, S., Werner, T., Cansfield, A., Müller, K., & Perrin, J. (2012). A selective inhibitor reveals PI3Kγ dependence of TH17 cell differentiation. *Nature Chemical Biology, 8,* 576–582. doi:10.1038/nchembio.957

Bolstad, B. M., Irizarry, R. A., Astrand, M., & Speed, T. P. (2003). A comparison of normalization methods for high density oligonucleotide array data based on variance and bias. *Bioinformatics, 19,* 185–193. doi:10.1093/bioinformatics/19.2.185

Chin, K. B., & Cordell, B. (2013). The effect of tea tree oil (*Melaleuca alternifolia*) on wound healing using a dressing model. *The Journal of Alternative and Complementary Medicine, 19,* 942–945. doi:10.1089/acm.2012.0787

Dunning, M. J., Smith, M. L., Ritchie, M. E., & Tavare, S. (2007). beadarray: R classes and methods for Illumina bead-based data. *Bioinformatics, 23,* 2183–2184. doi:10.1093/bioinformatics/btm311

Edmondson, M., Newall, N., Carville, K., Smith, J., Riley, T. V., & Carson, C. F. (2011). Uncontrolled, open-label, pilot study of tea tree (*Melaleuca alternifolia*) oil solution in the decolonisation of methicillin-resistant Staphylococcus aureus positive wounds and its influence on wound healing. *International Wound Journal, 8,* 375–384. doi:10.1111/j.1742-481X.2011.00801.x

Han, X., & Parker, T. L. (2017a). Anti-inflammatory activity of Juniper (*Juniperus communis*) berry essential oil in human dermal fibroblasts. *Cogent Medicine, 4,* 1306200. doi:10.1080/2331205X.2017.1306200

Han, X., & Parker, T. L. (2017b). Anti-inflammatory, tissue remodeling, immunomodulatory, and anticancer activities of oregano (*Origanum vulgare*) essential oil in a human skin disease model. *Biochimie Open, 4,* 73–77. doi:10.1016/j.biopen.2017.02.005

Hart, P. H., Brand, C., Carson, C. F., Riley, T. V., Prager, R. H., & Finlay-Jones, J. J. (2000). Terpinen-4-ol, the main component of the essential oil of *Melaleuca alternifolia* (tea tree oil), suppresses inflammatory mediator production by activated human monocytes. *Inflammation Research, 49,* 619–626. doi:10.1007/s000110050639

Homeyer, D. C., Sanchez, C. J., Mende, K., Beckius, M. L., Murray, C. K., Wenke, J. C., & Akers, K. S. (2015). *In Vitro* activity of *Melaleuca alternifolia* (tea tree) oil on filamentous fungi and toxicity to human cells. *Medical Mycology, 53,* 285–294. doi:10.1093/mmy/myu072

Koh, K. J., Pearce, A. L., Marshman, G., Finlay-Jones, J. J., & Hart, P. H. (2002). Tea tree oil reduces histamine-induced skin inflammation. *British Journal of Dermatology, 147,* 1212–1217. doi:10.1046/j.1365-2133.2002.05034.x

Kunkel, E. J., Dea, M., Ebens, A., Hytopoulos, E., Melrose, J., Nguyen, D., & Berg, E. L. (2004). An integrative biology approach for analysis of drug action in models of human vascular inflammation. *FASEB Journal, 18,* 1279–1281. doi:10.1096/fj.04-1538fje

Kunkel, E. J., Plavec, I., Nguyen, D., Melrose, J., Rosler, E. S., Kao, L. T., & Wang, Y. (2004). Rapid structure-activity and selectivity analysis of kinase inhibitors by BioMAP analysis in complex human primary cell-based models. *ASSAY and Drug Development Technologies, 2,* 431–442. doi:10.1089/adt.2004.2.431

Low, P., Clark, A. M., Chou, T. C., Chang, T. C., Reynolds, M., & Ralph, S. J. (2015). Immunomodulatory activity of *Melaleuca alternifolia* concentrate (MAC): Inhibition of LPS-induced NF-κB activation and cytokine production in myeloid cell lines. *International Immunopharmacology, 26,* 257–264. doi:10.1016/j.intimp.2015.03.034

Pazyar, N., Yaghoobi, R., Bagherani, N., & Kazerouni, A. (2013). A review of applications of tea tree oil in dermatology. *International Journal of Dermatology, 52*, 784–790. doi:10.1111/j.1365-4632.2012.05654.x

R Development Core Team. (2011). *R: A language and environment for statistical computing.* Vienna: The R Foundation for Statistical Computing. Retrieved from http://www.R-project.org/

Rajkowska, K., Kunicka-Styczyńska, A., Maroszyńska, M., & Dąbrowska, M. (2014). The effect of thyme and tea tree oils on morphology and metabolism of Candida albicans. *Acta Biochimica Polonica, 61*, 305–310.

10

Large-scale integration in tablet screens for blue-light reduction with optimized color: The effects on sleep, sleepiness, and ocular parameters

Masahiko Ayaki[1]*, Atsuhiko Hattori[2], Yusuke Maruyama[2], Kazuo Tsubota[1] and Kazuno Negishi[1]

*Corresponding author: Masahiko Ayaki, Department of Ophthalmology, Keio University, 35 Shinanomachi, Shinjuku, 1608582 Tokyo, Japan
E-mail: mayaki@olive.ocn.ne.jp

Reviewing editor: Jurg Bahler, University College London, UK

Additional information is available at the end of the article

Abstract: We investigated sleep quality and visual symptoms in 30 adults who spent two hours before bedtime using a tablet device with and without the advanced technology of large-scale integration for blue-light reduction and color management. Dry eye- and eye fatigue-related symptom scores were significantly better with than without blue-light reduction. Sleepiness and saliva melatonin during the task were greater with blue-light reduction, however, overnight melatonin secretion and sleep quality parameters were similar in both conditions. In conclusion, tablet devices using large-scale integration for blue-light reduction increased sleepiness and reduced eye fatigue and dryness during tasks before bedtime.

Subjects: Bioscience; Engineering & Technology; Health and Social Care; Medicine, Dentistry, Nursing & Allied Health

Keywords: blue light; sleep; vision; dry eye

1. Introduction

There is growing evidence that exposure to light emitted from a portable device before bedtime disrupts sleep and daytime function (Chang, Aeschbach, Duffy, & Czeisler, 2015; Higuchi, Motohashi, Liu, & Maeda, 2005). In this respect the more recently available portable devices have become a major problem, unlike the conventional displays of desktop computers and ambient lighting, since they are manipulated at a very short distance and even very young users are using them late at night for several hours (Bartel, Gradisar, & Williamson, 2015; Figueiro & Overington, 2015; Gradisar

ABOUT THE AUTHORS

Our research team has three subgroups; refraction, cataract, and sleep research. Currently ongoing major projects include progression of myopia and presbyopia, happiness and eye diseases, and driving performance and visual function. Please visit our HP for details; http://ophthal.med.keio.ac.jp/research/index.html.

PUBLIC INTEREST STATEMENT

This scientific paper describes large-scale integration (LSI) built in tablet PC for blue-light reduction maintained natural sleepiness and reduced eye fatigue and dryness at night. In modern society, many people use portable devices at night and even in bed before sleep. Ordinary tablet emits much blue-light component and may induce sleep disorder and eye symptoms by disturbing biological clock and optical property of the eye. The authors examined 30 healthy adults viewing tablet PC during tasks before bedtime. We found new technology enabled the subjects to do tasks with minimum disadvantages for sleep, dry eye symptoms, and eye fatigue. We believe LSI should be a promising solution for busy mobile user to maintain systemic and eye health.

et al., 2013). Eyewear, software, and films have been introduced to filter and control the blue light emitted from displays (Ayaki et al., 2015; Burkhart & Phelps, 2009; Heath et al., 2014; Ide, Toda, Miki, & Tsubota, 2015; van der Lely et al., 2015); however, these methods have some disadvantages including decreased visual quality with darkness and yellowness that may be potentially linked to eye fatigue, decreased performance and concentration, and increased errors. The recently developed large-scale integration (LSI) IROMI® engine (Dai Nippon Printing Co. Ltd., Tokyo, Japan), designed to provide acceptable image quality even under reduced blue-light conditions, enables the user to choose a blue-light filter and color management to optimize visual quality.

In addition to the well-known sleep and circadian rhythm disorders caused by the use of portable displays at night, there is also a risk of ophthalmological problems. Eye fatigue and dryness are the most common ocular symptoms in patients visiting ophthalmology clinics due to work-related issues or the habitual use of portable devices late at night. Difficulty focusing and reduced blinking frequency are closely associated with eye fatigue during computer display tasks and the symptoms of patients with dry eye disease (DED) worsen in the evening (Tsubota & Nakamori, 1993; Uchino et al., 2013; Walker, Lane, Ousler, & Abelson, 2010). These symptoms may not be detected in adolescents since difficulty focusing and DED are not common in young people.

To date, there has been no ophthalmic assessment of adults using blue-light control technology at night. Therefore, in the present study, we focused on whether innovative display technology could reduce eye fatigue and DED-related symptoms. We used two ocular examinations that can be performed by untrained participants at night in their bedroom. The first test, of functional visual acuity (FVA), reflects an individual's visual performance and detects impairment of visual function in various eye diseases including DED, with normal visual acuity measured by standard examination (Kaido, Dogru, Ishida, & Tsubota, 2007; Kaido et al., 2015). FVA may decrease with disruption of the stable tear-film layer over the surface of the cornea. The second test, maximum blinking interval (MBI), is useful for detecting and quantifying DED because differences in the mean MBI have been reported between normal subjects (8.9 ± 4.0 s) and patients with DED (4.2 ± 2.4 s) (Tsubota et al., 1996).

Here, we describe a sleep and ophthalmological study conducted in adults to explore whether blue-light control using LSI can deliver acceptable visual performance and sleep quality against the effects of blue-light pollution from self-luminous portable devices used at night.

2. Methods

2.1. Participants and ethics approval

Participants were recruited via an advertisement in two research centers. The selection criteria were habitual use of a portable tablet device at night, normal vision, and good general health according to a workplace health check by attending physicians. Normal vision was confirmed by a board-certified ophthalmologist for all participants before they completed the study. Exclusion criteria were a history of shift work, systemic illness, use of medication, and any history of psychiatric illness. Participants maintained their usual weekday lifestyle before the study period and were asked to refrain from drinking alcohol, smoking, and consuming caffeine during the study period. The study was conducted during June (20 subjects), October (3 subjects), and November (7 subjects) 2015 in Tokyo, Japan, where the latitude is 35.68° N and day length varies by 4–6 h over the year, according to the 1981–2010 averages reported by the Japan Meteorological Agency.

The study was approved by the institutional review board of Shinseikai Toyama Hospital and was carried out in accordance with the approved guidelines. All participants provided written informed consent.

2.2. Study design and use of self-luminous portable devices

Participants were asked to stay in a dark room from 22:00 to 24:00 doing tasks and reading Japanese literature of their choice on a portable self-luminous device either with or without LSI for blue-light management. They were asked to go to bed at about midnight and sleep for seven hours in the same dark room. Participants were required to send e-mail alerts and submit three reports during the study to confirm awakeness. A cross-over study design was used, so the use of LSI for blue-light management was changed in the second phase (after two weeks). Participants and investigators were masked either with or without LSI for blue-light management.

The self-luminous devices used were 10.1 inch IROMI® Tablet devices (DNP, Co. Ltd., Tokyo, Japan) emitting approximately 410 cd/m² of visible light; they were held less than 25 cm from the participants' eyes and participants were instructed to set or hold it within their elbow length. The tasks performed using these devices were reading literature chosen by the participant and/or writing about health and habits. Participants were monitored by actigraphy and email alerts to ensure that they stayed awake during the 2-h task period.

2.3. Specification of LSI

Blue-light reduction and color control LSI (IROMI®, Dai Nippon Printing Co. Ltd., Tokyo, Japan) are an image-processing LSI that has functions of color management, contrast enhancement, and blue-light control. It is built into the IROMI® Tablet device and study participants were masked for its function in the present study. The color filter controls the amount of the three primary colors of light, red (R), green (G), and blue (B), in every pixel from the transmitted image to manage the color display. The color resolution is determined by the data width. For example, in the case of each 8-bit RGB (256 × 256 × 256), it is possible to display 16.77 million colors and in the case of each 6-bit RGB (64 × 64 × 64), it is possible to display 262,000 colors. To reduce blue light, the B component can be reduced, but when only B is reduced the color balance collapses and the color turns yellowish. Because the IROMI® engine can adjust RGB individually, IROMI® Tablet devices adjust R and G to display the correct color. The blue-light reduction rate was reduced by 20.1% at a peak wavelength of 434 nm in the present study and reduced by 14.2% in total irradiance (Figure 1). As such, under the blue-light reduction mode, IROMI® LSI controlled the emitted blue light by optimizing color and contrast to acceptable levels (Figure 2).

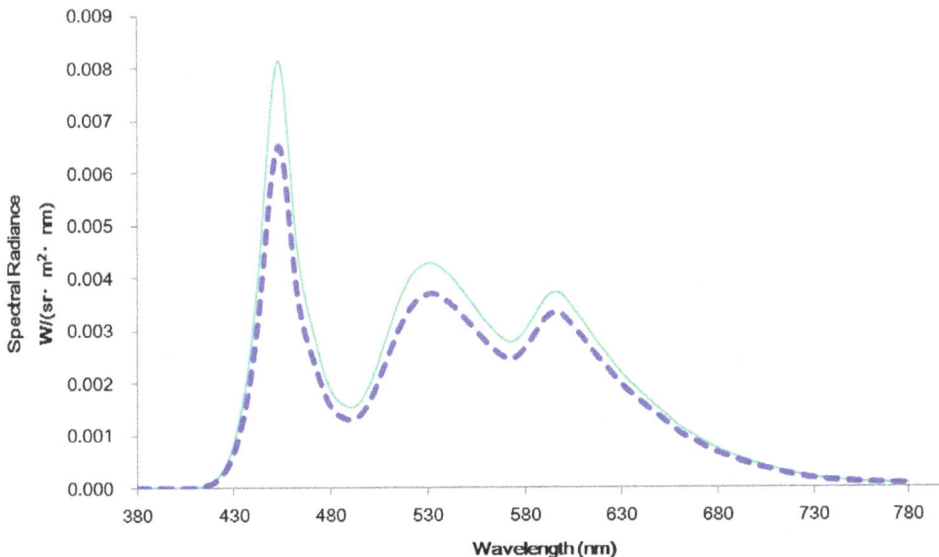

Figure 1. The spectral component of white light emitted from the display of the IROMI® Tablet device with (dotted line) and without large-scale integration for blue-light control (solid line). Blue-light reduction rate is 20.1%
(@ 434 nm).

(a)   (b)   (c)

Figure 2. Screen images from the tablet device with and without large-scale integration (LSI) for blue-light control. (a) Original screen image without LSI for blue-light control; (b) Screen image with LSI for blue-light reduction only; (c) Screen image with LSI for blue-light reduction and color control. Note the display is yellowish with blue-light reduction only whereas the image is acceptable with color control by LSI.

2.4. Ophthalmological examinations

FVA, MBI, and questionnaires regarding visual quality and symptoms related to dry eye were evaluated. The FVA test consisted of a 30–60 s continuous measurement of visual acuity under the best corrected condition, measured at 24:00 with a vision chart on the display downloaded from a free software website to measure FVA (https://itunes.apple.com/jp/app/dryeyekt/id781168068?mt=8). The result was expressed as the mean visual acuity of several measurements during the 30–60 s examination. MBI was expressed as the number of seconds the eyes were kept open, and was measured once at 24:00. Visual quality (clarity, sharpness, eye fatigue, and hue), with or without LSI for blue-light reduction, was measured using a visual analogue scale ranging from 5 (best) to 1 (worst) compared with their own portable device. Dry eye-related symptoms (dryness, irritation, fatigue, blurring) were evaluated as the sum of the symptoms present at 24:00. These symptoms are the most common ocular symptoms included in validated questionnaires (Gulati et al., 2006; Sakane et al., 2013).

2.5. Actigraphy and questionnaires for measurement of sleepiness and sleep quality

Participants completed a validated questionnaire (Karolinska Sleepiness Scale [KSS]) before and after the task (22:00 and 24:00, respectively). Each participant's sleep/wake cycle was monitored by a micro-motion logger (AMI, New York, NY, USA) during the study period. Micro-motion data were analyzed using the Cole–Kripke algorithm (Cole, Kripke, Gruen, Mullaney, & Gillin, 1992).

2.6. Urine melatonin

First morning-void urine samples were collected from all participants at about 7:00 on the day following each intervention. The samples were frozen immediately and stored at −20°C until testing was undertaken. Urinary 6-sulfatoxymelatonin levels were measured by enzyme-linked immunosorbent assay (ELISA) using a Melatonin-Sulfate Urine ELISA kit (IBL International GmbH, Hamburg, Germany). The sensitivity of this assay was 0.41 ng/mL, and the intra-assay and inter-assay coefficients of variation were 1.2 and 5.7%, respectively. To adjust for variation in the dilution of urine, 6-sulfatoxymelatonin concentration was expressed as urine 6-sulfatoxymelatonin/urine creatinine using LabAssay Creatinine (Jaffe method) (Wako Pure Chemical Industries, Ltd, Tokyo, Japan) (Oba, Nakamura, Sahashi, Hattori, & Nagata, 2008).

2.7. Statistical analyses

Data of each phase of the study were analyzed and, where appropriate, given as mean ± standard deviation. Sleep indices, ocular parameters, and melatonin levels obtained during the use of tablet devices with LSI for blue-light reduction (Figure 2(c)) and without LSI for blue-light reduction (Figure 2(a)) were compared. All analyses were performed using StatFlex (Atech, Osaka, Japan) and SPSS version 21 (SPSS Inc, Chicago, IL). A p value less than 0.05 was considered significant.

3. Results

Thirty healthy Japanese adults (age range, 24–58 years; mean age, 35.4 ± 8.9 years; 3 women) participated in the study.

FVA tended to be better with (0.02 ± 0.22, LogMAR scale) than without LSI for blue-light reduction (0.13 ± 0.21, LogMAR scale; $p = 0.08$, paired t test; Figure 3). MBI was significantly greater with (22.2 ± 23.0 s) than without LSI for blue-light reduction (18.1 ± 18.5 s; $p = 0.03$). The dry eye symptom score was improved with the use of LSI for blue-light reduction (0.46 ± 0.80), compared to without it (0.79 ± 0.89; $p = 0.01$). Visual quality, in terms of eye fatigue, was also significantly better when LSI for blue-light reduction was used, rather than not used ($p = 0.04$; Table 1).

The increase in sleepiness from 22:00 to 24:00, measured with the KSS corrected by normalizing to that at the start of the study, was greater with LSI for blue-light reduction (2.26 ± 2.18, mean KSS from 4.2 to 6.5) than without LSI for blue-light reduction (1.52 ± 2.40, mean KSS from 4.8 to 6.3; $p = 0.11$) (Figure 4). Urine melatonin and sleep parameters measured by actigraphy were not different between conditions. Specifically, for control and LSI, respectively, the mean sleep efficacy was 94.3 ± 6.1% and 93.7 ± 7.6%, sleep latency was 5.18 ± 1.54 min and 5.64 ± 1.11 min, and wake-up time after sleep onset was 26.8 ± 12.0 min and 24.7 ± 9.5 min (all $p > 0.05$).

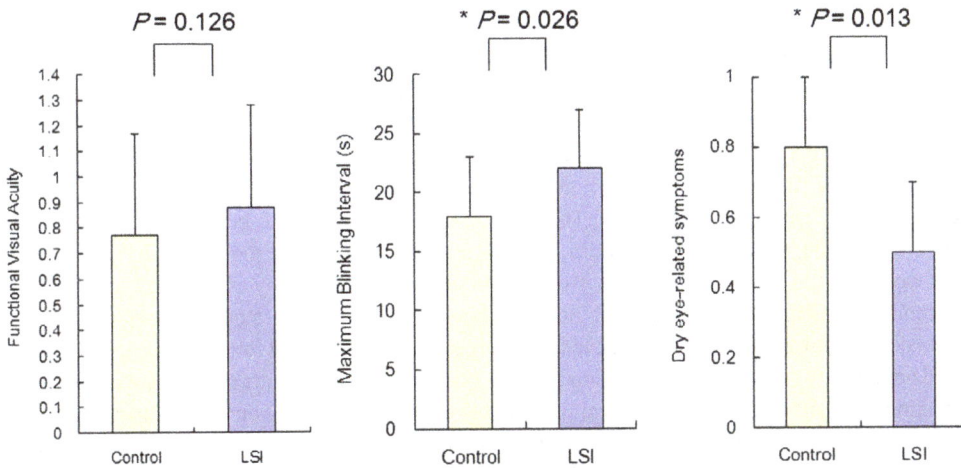

Figure 3. The ocular parameters of functional visual acuity (left panel), maximum blinking interval (center panel), and number of dry eye symptoms (right panel) are shown for participants while they performed tasks for two hours, from 22:00 to 24:00, on self-luminous devices with large-scale integration (LSI) for blue-light reduction and without LSI (control).

Notes: *$p < 0.05$; $n = 30$, two-tail paired t test.

Table 1. Visual quality[a] with and without large-scale integration for blue-light reduction			
	Without LSI	**With LSI**	**p-value[b]**
Clarity	3.39 ± 1.12	4.42 ± 1.04	0.500
Sharpness	3.86 ± 0.83	3.64 ± 0.93	0.102
Less fatigue	2.53 ± 1.34	3.00 ± 1.32	0.048
Hue	3.50 ± 1.45	3.29 ± 1.21	0.055

Notes: LSI, large-scale integration for blue-light reduction.

Values are mean ± standard deviation for 30 adults.

[a]Each element of visual quality was assessed on a visual analog scale from 5 (best) to 1 (worst) compared with their own portable device.

[b]Two-tail paired t test.

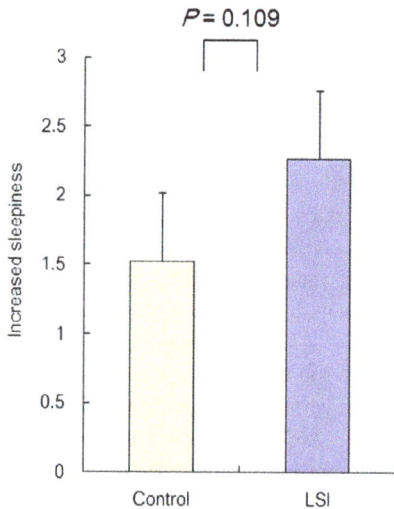

Figure 4. The increase in sleepiness of participants (*n* = 30) measured by the Karolinska Sleepiness Scale at bedtime (24:00) after a two-hour task at night. The increase in sleepiness tended to be greater with large-scale integration (LSI) for blue-light reduction than without LSI (control).

4. Discussion

Our present results demonstrate that incorporating LSI for blue-light control in the newly developed tablet devices preserves acceptable image quality and provides better visual function and ocular comfort. In addition, the LSI addition does not appear to alter sleep quality compared with the same tablet device without LSI for blue-light reduction. However, all the participants in our study were adults and many confounding factors might interact with the results, including light transmission into the eye (Cuthbertson, Peirson, Wulff, Foster, & Downes, 2009; Turner & Mainster, 2008) and light sensitivity (Higuchi, Nagafuchi, Lee, & Harada, 2014). Sleepiness was only measured during the time spent watching the screen at night, whereas daytime activity and many other factors can affect melatonin secretion. Thus, the apparent inconsistency between sleepiness and overnight urine melatonin does not exclude the possible advantages of blue-light control LSI for sleep quality, although we did not analyze saliva melatonin during the task since preliminary results showed wide variations among participants not fit for statistical analysis. The discrepancy between melatonin and sleep might therefore be due to daytime activity, sunlight exposure, chronotype, and individual difference.

One investigation reported that 60-min exposures to a tablet screen before bedtime did not affect sleep even in adolescences (Heath et al., 2014). The blue-light reduction rate in this study was 20% and thus might not be enough to enhance melatonin secretion. Interestingly, the present results were different from our previous results investigating the effect of blue-light filter eyewear on melatonin secretion and sharpness of image with two hours of tablet device use before bedtime (Ayaki et al., 2015). Taken together, eyewear and LSI for blue-light management have unique functions and abilities in reducing the blue-light effect at night in terms of melatonin secretion, sleepiness, vision, and dry eye-related symptoms. LSI for blue-light reduction has no side effects and is not complicated to use. Although further investigations are warranted, this new technology allows users to easily manage blue-light exposure according to their purpose and preference.

Several studies have revealed health and ocular problems caused by the evening use of tablet devices and smartphones. Decreased liquid tear secretion (Srinivasan, Chan, & Jones, 2007) and decreased blinking (Tsubota & Nakamori, 1993) can exacerbate dry eye symptoms in the evening. Blue light emitted from displays can also induce eye fatigue by light scattering (Kaido et al., 2016). Intrinsically photosensitive retinal ganglion cells are implicated in ocular pain when exposed to blue light (Matynia et al., 2015) and younger subjects are more sensitive to light in terms of melatonin suppression and headache (Harle & Evans, 2004; Higuchi et al., 2014; Wilkins, Patel, Adjamian, &

Evans, 2002). Tinted eyewear has been proven effective for blue-light control during day and night (Ide et al., 2015; Kaido et al., 2016), and sleep studies have revealed a distinct effect of blue-light filter eyewear in maintaining healthy sleep (Ayaki et al., 2016; van der Lely et al., 2015).

Blue-light reduction LSI presented better images and less scattering on the display, and these effects induced an increased blinking interval that could be likened to better FVA and less dry eye-related symptoms. MBI, FVA, and dry eye-related questionnaires are clinically proven to be useful for evaluating ocular condition (Gulati et al., 2006; Kaido et al., 2015; Tsubota et al., 1996) and our results favored LSI for maintaining ocular health. Taken together, we speculate that the new LSI could contribute to ocular health with a reduction in the blue light and alteration to the spectral composition.

This study has several limitations. The sample size was small, and the sleep quality data we collected should be extended to a larger number of participants to cover variations in age, level of light sensitivity, and genetic background. Sleep should be evaluated by polysomnography and sleepiness should be measured by saliva melatonin in an environment that is strictly controlled for light level. Also, the study period was short, so longer observation periods should be used in future research to determine the effect of blue-light control LSI on circadian rhythms. A control night during which participants slept the entire night would have been useful because most of the measured values were so variable that the statistical analysis provided few significant findings. Further investigation should be performed with more strict protocols including a control night and experimentally prepared LSI since the LSI used in the present study was a commercial product.

In conclusion, LSI for blue-light reduction was effective for vision and ocular health during a task before bedtime. A longer MBI, reduced dry eye-related symptoms, and the tendency for a better FVA all support the substantial advantages of LSI technology for users. Additionally, hue, sharpness, and clarity of the display using LSI for blue-light reduction were comparable to no blue-light reduction, while eye fatigue was relieved with LSI for blue-light control.

Acknowledgments
We thank Mr Michitaka Yoshimura for his valuable suggestions and Dr Takayuki Abe for help with statistical analyses. We acknowledge Inter-Biotech (http://www.inter-biotech.com) for assistance with the English language editing of this manuscript.

Funding
This work was supported by the Dai Nippon Printing Co. Ltd.

Competing Interests
The authors declare no competing interest.

Author contributions
Masahiko Ayaki, Atsuhiko Hattori, and Yusuke Maruyama collected and analyzed the data. Masahiko Ayaki wrote the manuscript. Masahiko Ayaki, Atsuhiko Hattori, Kazuno Negishi, and Kazuo Tsubota designed the study. Masahiko Ayaki, Atsuhiko Hattori, Yusuke Maruyama, Kazuno Negishi, and Kazuo Tsubota reviewed and approved the final version of the manuscript.

Author details
Masahiko Ayaki[1]
E-mail: mayaki@olive.ocn.ne.jp

Atsuhiko Hattori[2]
E-mail: ahattori.las@tmd.ac.jp

Yusuke Maruyama[2]
E-mail: sobs0930@yahoo.co.jp

Kazuo Tsubota[1]
E-mail: tsubota@z3.keio.jp

Kazuno Negishi[1]
E-mail: fwic7788@mb.infoweb.ne.jp

[1] Department of Ophthalmology, Keio University School of Medicine, 35 Shinanomachi, Shinjuku, 1608582 Tokyo, Japan.

[2] Department of Biology, Tokyo Medical and Dental University, 2-8-30 Kokufudai, Ichikawa, 2720827 Chiba, Japan.

References
Ayaki, M., Hattori, A., Maruyama, Y., Nakano, M., Yoshimura, M., Kitazawa, M., ... Tsubota, K. (2015). Protective effect of blue-light shield eyewear for adults against light pollution from self-luminous devices used at night. *Chronobiology International, 33*, 134–139.
Ayaki, M., Kawashima, M., Negishi, K., Kishimoto, T., Mimura, M., & Tsubota, K. (2016). Sleep and mood disorders in dry eye disease and allied irritating ocular diseases. *Scientific Reports, 6*, 22480. http://dx.doi.org/10.1038/srep22480

Bartel, K. A., Gradisar, M., & Williamson, P. (2015). Protective and risk factors for adolescent sleep: A meta-analytic review. *Sleep Medicine Reviews, 21*, 72–85. http://dx.doi.org/10.1016/j.smrv.2014.08.002

Burkhart, K., & Phelps, J. R. (2009). Amber lenses to block blue light and improve sleep: A randomized trial. *Chronobiology International, 26*, 1602–1612.

Chang, A. M., Aeschbach, D., Duffy, J. F., & Czeisler, C. A. (2015). Evening use of light-emitting eReaders negatively affects sleep, circadian timing, and next-morning alertness. *Proceedings of the National Academy of Sciences, 112*, 1232–1237. http://dx.doi.org/10.1073/pnas.1418490112

Cole, R. J., Kripke, D. F., Gruen, W., Mullaney, D. J., & Gillin, J. C. (1992). Automatic sleep/wake identification from wrist activity. *Sleep, 15*, 461–469.

Cuthbertson, F. M., Peirson, S. N., Wulff, K., Foster, R. G., & Downes, S. M. (2009). Blue light–filtering intraocular lenses: Review of potential benefits and side effects. *Journal of Cataract & Refractive Surgery, 35*, 1281–1297. http://dx.doi.org/10.1016/j.jcrs.2009.04.017

Figueiro M., & Overington D. (2015, May 6). Self-luminous devices and melatonin suppression in adolescents. *Lighting Res Technol.* Article ID: 1477153515584979. doi:10.1177/1477153515584979

Gradisar, M., Wolfson, A. R., Harvey, A. G., Hale, L., Rosenberg, R., & Czeisler, C. A. (2013). The sleep and technology use of Americans: Findings from the National Sleep Foundation's 2011 Sleep in America poll. *Journal of Clinical Sleep Medicine, 9*, 1291–1299.

Gulati, A., Sullivan, R., Buring, J. E., Sullivan, D. A., Dana, R., & Schaumberg, D. A. (2006). Validation and repeatability of a short questionnaire for dry eye syndrome. *American Journal of Ophthalmology, 142*, 125–131. http://dx.doi.org/10.1016/j.ajo.2006.02.038

Harle, D. E., & Evans, B. J. (2004). The optometric correlates of migraine. *Ophthalmic and Physiological Optics, 24*, 369–383. http://dx.doi.org/10.1111/opo.2004.24.issue-5

Heath, M., Sutherland, C., Bartel, K., Gradisar, M., Williamson, P., Lovato, N., & Micic, G. (2014). Does one hour of bright or short-wavelength filtered tablet screenlight have a meaningful effect on adolescents' pre-bedtime alertness, sleep, and daytime functioning? *Chronobiology International, 31*, 496–505. http://dx.doi.org/10.3109/07420528.2013.872121

Higuchi, S., Motohashi, Y., Liu, Y., & Maeda, A. (2005). Effects of playing a computer game using a bright display on presleep physiological variables, sleep latency, slow wave sleep and REM sleep. *Journal of Sleep Research, 14*, 267–273. http://dx.doi.org/10.1111/jsr.2005.14.issue-3

Higuchi, S., Nagafuchi, Y., Lee, S. I., & Harada, T. (2014). Influence of light at night on melatonin suppression in children. *The Journal of Clinical Endocrinology & Metabolism, 99*, 3298–3303. http://dx.doi.org/10.1210/jc.2014-1629

Ide, T., Toda, I., Miki, E., & Tsubota, K. (2015). Effect of blue light–reducing eye glasses on critical flicker frequency. *Asia-Pacific Journal of Ophthalmology, 4*, 80–85. http://dx.doi.org/10.1097/APO.0000000000000069

Kaido, M., Dogru, M., Ishida, R., & Tsubota, K. (2007). Concept of functional visual acuity and its applications. *Cornea, 26*, S29–S35. http://dx.doi.org/10.1097/ICO.0b013e31812f6913

Kaido, M., Kawashima, M., Yokoi, N., Fukui, M., Ichihashi, Y., Kato, H., ... Tsubota, K. (2015). Advanced dry eye screening for visual display terminal workers using functional visual acuity measurement: The Moriguchi study. *British Journal of Ophthalmology, 99*, 1488–1492. http://dx.doi.org/10.1136/bjophthalmol-2015-306640

Kaido, M., Toda, I., Oobayashi, T., Kawashima, M., Katada, Y., & Tsubota, K. (2016). Reducing short-wavelength blue light in dry eye patients with unstable tear film improves performance on tests of visual acuity. *PLoS ONE, 11*, e0152936. http://dx.doi.org/10.1371/journal.pone.0152936

Matynia, A., Parikh, S., Deot, N., Wong, A., Kim, P., Nusinowitz, S., & Gorin, M. B. (2015). Light aversion and corneal mechanical sensitivity are altered by intrinscally photosensitive retinal ganglion cells in a mouse model of corneal surface damage. *Experimental Eye Research, 137*, 57–62.

Oba, S., Nakamura, K., Sahashi, Y., Hattori, A., & Nagata, C. (2008). Consumption of vegetables alters morning urinary 6-sulfatoxymelatonin concentration. *Journal of Pineal Research, 45*, 17–23. http://dx.doi.org/10.1111/jpi.2008.45.issue-1

Sakane, Y., Yamaguchi, M., Yokoi, N., Uchino, M., Dogru, M., Oishi, T., ... Ohashi, Y. (2013). Development and Validation of the Dry Eye–Related Quality-of-Life Score Questionnaire. *JAMA Ophthalmology, 131*, 1331–1338. http://dx.doi.org/10.1001/jamaophthalmol.2013.4503

Srinivasan, S., Chan, C., & Jones, L. (2007). Apparent time-dependent differences in inferior tear meniscus height in human subjects with mild dry eye symptoms. *Clinical and Experimental Optometry, 90*, 345–350. http://dx.doi.org/10.1111/cxo.2007.90.issue-5

Tsubota, K., Hata, S., Okusawa, Y., Egami, F., Ohtsuki, T., & Nakamori, K. (1996). Quantitative videographic analysis of blinking in normal subjects and patients with dry eye. *Archives of Ophthalmology, 114*, 715–720. http://dx.doi.org/10.1001/archopht.1996.01100130707012

Tsubota, K., & Nakamori, K. (1993). Dry eyes and video display terminals. *New England Journal of Medicine, 328*, 584. http://dx.doi.org/10.1056/NEJM199302253280817

Turner, P. L., & Mainster, M. A. (2008). Circadian photoreception: Ageing and the eye's important role in systemic health. *British Journal of Ophthalmology, 92*, 1439–1444. http://dx.doi.org/10.1136/bjo.2008.141747

Uchino, M., Yokoi, N., Uchino, Y., Dogru, M., Kawashima, M., Komuro, A., ... Tsubota, K. (2013). Prevalence of dry eye disease and its risk factors in visual display terminal users: The Osaka study. *American Journal of Ophthalmology, 156*, 759–766. http://dx.doi.org/10.1016/j.ajo.2013.05.040

van der Lely, S., Frey, S., Garbazza, C., Wirz-Justice, A., Jenni, O. G., Steiner, R., ... Schmidt, C. (2015). Blue blocker glasses as a countermeasure for alerting effects of evening light-emitting diode screen exposure in male teenagers. *Journal of Adolescent Health, 56*, 113–119. http://dx.doi.org/10.1016/j.jadohealth.2014.08.002

Walker, P. M., Lane, K. J., Ousler, 3rd., G. W., & Abelson, M. B. (2010). Diurnal variation of visual function and the signs and symptoms of dry eye. *Cornea, 29*, 607–612. http://dx.doi.org/10.1097/ICO.0b013e3181c11e45

Wilkins, A. J., Patel, R., Adjamian, P., & Evans, B. J. (2002). Tinted spectacles and visually sensitive migraine. *Cephalalgia, 22*, 711–719. http://dx.doi.org/10.1046/j.1468-2982.2002.00362.x

11

Differential biochemical response of rice (*Oryza sativa* L.) genotypes against rice blast (*Magnaporthe oryzae*)

P.U. Anushree[1], R.M. Naik[1]*, R.D. Satbhai[1], A.P. Gaikwad[1] and C.A. Nimbalkar[1]

*Corresponding author: R.M. Naik, Department of Statistic, MPKV, Rahuri 413722, Ahmednagar, India
E-mail: rajeevnaik2@rediffmail.com

Reviewing editor: Jurg Bahler, University College London, UK

Additional information is available at the end of the article

Abstract: The disease-free (control) and blast infected leaf samples of 11 rice genotypes were evaluated for activity profile of defense-related and antioxidative enzymes. The amplification genomic DNA with two SSR markers RM124 and RM224 were also performed for identification of blast resistance and susceptible genotypes. The activity of chitinase, PAL and β-glucosidase of post pathogen-infected leaf samples increased significantly in all rice genotypes, thought the increase was comparable less in to blast susceptible genotypes Chimansal and EK-70. The activity of antioxidative enzymes was comparatively higher in the infected leaf of blast resistant genotypes recording highest increase in NLR-20104 and KJT-5. The activity of defense-related and antioxidative enzymes in the disease-free leaf samples differed among the genotypes and was even higher in the two blast susceptible genotypes. RM144 and RM224 SSR primers clearly amplified in blast resistant KJT-5, NLR-20104, KJT-2, Tetep genotypes whereas RM144 missing in susceptible Chimansal but prominently present in susceptible genotype EK-70. This study revealed that higher level of induction of defense-related and antioxidative enzymes and presence of specific

ABOUT THE AUTHORS

We at the department of Biochemistry are undertaking research work to identify biochemical and molecular markers to screen the available germplasm of crop plants for major abiotic and biotic stresses particularly in pigeon pea, chickpea for wilt and sterility mosaic, rice for blast and blight, sorghum for shoot fly. The work is also being undertaken for abiotic stresses particularly drought and salinity and also for combinational stress. The present work is similar attempt which was carried out in collaboration with rice pathologist working at ARS, Lonavala, India which is a hot spot for blast disease of rice. Disease-free and infected leaf samples of 11 rice genotypes were collected and biochemical analysis of some defense-related and antioxidative enzymes was carried out. Flanking markers RM144 and RM224 also amplified for validation of blast resistant and susceptible rice genotypes in the present study. These efforts are helping the plant breeder to understand breeding program and developing mapping population is being utilized for marker trait association.

PUBLIC INTEREST STATEMENT

Rice blast disease is a serious fungal disease caused by *Magnaporthe oryzae*. The use of chemical is expensive and not environment-friendly; hence utilization of host resistance has been the best way to manage the disease. To identify biochemical and molecular markers selectively differentiating the germplasm which can be environmentally sustainable plan for developing rice resistant genotypes. In this report, disease-free and infected leaf samples of 11 rice genotypes were screened against fungal blast disease for biochemical analysis of some defense-related and antioxidative enzymes. Amplification genomic DNA with two SSR markers RM124 and RM224 were also performed for identification of blast resistance and susceptible genotypes. This study revealed that higher level of induction of defense-related and antioxidative enzymes and presence of specific amplified fragments with RM144 and RM224 could be useful for screening the resistant and susceptible rice genotypes against *M. oryzae*.

amplified fragments with RM144 and RM224 could be useful for screening the resistant and susceptible rice genotypes against *Magnaporthe oryzae*.

Subjects: Bioscience; Built Environment; Environment & Agriculture; Environmental Studies & Management

Keywords: rice blast (*M. oryzae*); SSR marker; defense-related enzymes; antioxidants

1. Introduction

Rice blast disease is caused by the filamentous ascomycete fungus *Magnaporthe oryzae*, is the most devastating fungal disease in the rice growing world thus resulting in huge yield losses (Samalova, Meyer, Gurr, & Fricker, 2014). The disease symptoms appear on the aerial parts of the plant. Most infections occur on the leaves during vegetative phase and on panicle and neck during reproductive phase of the crop. Plant diseases are often severe during periods of warm temperatures and high moisture. Generally, rice blast is favored by moderate temperatures 24°C and periods of high moisture that is 12 h or longer, conditions readily attainable in flooded rice fields.

In India, management of rice blast diseases is highly dependent on chemical fungicides, and due to low levels of host plant resistance in many of the cultivated rice varieties. The expensive use of fungicide is not an environment-friendly for disease control and hence utilization of host resistance has been the best way to manage the disease, for which identification of sources of resistance are necessary (Bonman, Khush, & Nelson, 1992). A better understanding of the mechanisms involved in defense to *M. oryzae* infection and responsible for damage to the host plant may provide new methods to control this disease. The need for a better understanding of this disease becomes clear if we consider the poor durability of many blast resistant cultivars of rice, which have a typical field life of only 2–3 growing seasons before disease resistance is overcome. Rice blast control strategies that can be deployed as part of an environmentally sustainable plan for increasing the efficiency are therefore urgently required.

Plants defend themselves against pathogen challenge by the activation of defense response pathways (Staskawicz, Ausubel, Baker, Ellis, & Jones, 1997). The systemic resistance induction process increases enzymatic activity of peroxidase (POX) and polyphenol oxidase (PPO) which are responsible for catalyzing lignin formation, and phenylalanine ammonia lyase (PAL) which is involved in the biosynthesis of phytoalexins and phenols. The pathogenesis-related proteins (PRPs) β-1,3-glucanase and chitinase, enzymes that belong to PR-2 and PR-3 families, respectively (van Loon, Rep, & Pieterse, 2006) have been related more often to Systemic acquired resistance (SAR) and sometimes to Induced systemic resistance (ISR). All these enzymes have been shown to be involved in plant defense against pathogens in several pathosystem (Kini, Vasanthi, & Shetty, 2000). The activation and the expression levels of defense genes vary in different plant-pathogen interactions. Plants have also developed complex antioxidant defense systems that respond to biotic and abiotic stresses and mitigate the deleterious effects of reactive oxygen species (Panda, 2007). The levels of ROS and the extent of oxidative damage depend largely upon the level of coordination among ROS-scavenging enzymes (Liang, Chen, Liu, Zhang, & Ding, 2003). In transgenic rice plant, phenolic compounds and activity profile of some enzymes such as SOD, POX, APX and hydrolytic enzyme such as chitinase, β-glucosidase have shown to play active role in resistant mechanism of plant disease.

A combination of major resistance genes and defensive response genes form the basis for durable resistance. It has been reported that DNA markers that co-segregate with the resistant gene are a powerful method to accelerate development of a resistant cultivar (Fjellstrom et al., 2004a, 2004b). The availability of different molecular markers allows characterization of genes of interest. Single sequence repeat (SSR) can be applied to identify markers tightly linked to blast resistance genes and to detect genes and QTLs on rice chromosomes (Fjellstrom et al., 2004a; Liu, Wang, Chen, Lin, & Pan, 2005; Zhu, Wang, & Pan, 2004) which also extensively used in diversity analyses (Baraket et al., 2011; Swapna, Sivaraju, Sharma, Singh, & Mohapatra, 2010), marker-assisted selection (Zhu et al., 2009) and inheritance studies (Campoy et al., 2011). During the last few years, genetics of blast resistance in rice has been extensively studied and many dominant R genes conferring complete resistance to *M. oryzae* have been identified. In this study, the genomic DNA of four blast resistant and two blasts susceptible (Chimansal and EK-70) genotypes were amplified using these two flanking SSR markers. RM 224/RM 144 are flanking SSR markers for blast resistance located on chromosome 11 of rice crop which were used to distinguish resistance and susceptible cultivars of rice in present study.

Biochemical studies on defense-related and antioxidative enzymes can also be applied to identify markers tightly linked to blast resistance genes and QTLs on rice chromosomes. This information also helped in understanding the nature and mechanisms of resistance and aid in screening for disease resistant genotypes. The constitutive and induced biochemical defense of rice genotypes with amplification of identified two SSR markers RM-124 and RM-224 against rice blast was therefore undertaken at Lonavala region of Pune district in India which is a hot spot of rice blast disease (Krishnaveni et al., 2012).

2. Materials and methods

2.1. M. oryzae *inoculation and disease rating*

Five week old *M. oryzae* infected and disease-free leaf samples (Control) of rice seedlings of same plant from different blast resistant (KJT-2, TeTep, NLR 20104, KJT-5, Rp-Biopatho-3, Swarnadhan, RAU-631-9-10, CN-1447-9-4-2 and CB-06-555) and susceptible (EK-70 and Chimansal) genotypes were collected from Agricultural Research Station, Lonavala during rainy season of 2014. For screening of rice genotypes against leaf blast, a pot culture experiment was carried out in green house condition (20°C) at ARS, Lonavala, The inoculum load of the pathogen used was 10^6 spores ml^{-1} by spraying the leaves until run off. After 10 days of inoculum, spraying samples was collected immediately frozen in liquid nitrogen for biochemical analysis. The rating scale used for different lesion types on the basis of disease severity symptomatic leaf blades was measured according to the severity of the lesions described in Table 1 (Anonymous, 2002).

The chitinase (EC.3.2.1.14) activity was assayed by the method of Giri et al. (1998). For enzyme extraction 0.5 g of infected and control leaf samples were weighed separately and macerated with 2 ml of 0.1 M sodium citrate buffer in precooled mortar and pestle. The homogenate was centrifuged at 10,000 rpm for 10 min at 10°C and the supernatant was used as crude source of chitinase. 0.5 ml of supernatant was added to 2 ml of chitin suspension containing 7.5 mg of BSA and was incubated in water bath at 37°C for 3 h. From that an aliquot of 0.1 ml was taken for the estimation of N-acetyl glucosamine as per the method of Nelson-Smogyi. The chitinase activity was expressed in terms of μg N-acetyl glucosamine released per min per mg of protein. The PAL activity (E.C.4.1.3.5) was assayed by the method of Campos, Nonogaki, Suslow, and Saltveit (2004). Infected and control seedlings were separately weighed and 0.5 g were macerated with 2 ml of 50 mM borate buffer (pH 8.5) containing 5 mM of 2-mercaptoethanol and 0.4 g polyvinylpyrrolidone (PVP). The homogenate was

Table 1. Grade description for screening against *M. oryzae*	
0	No lesions observed
1	Small brown specks of pinpoint size or larger brown specks without sporulating center
2	Small roundish to slightly elongated, necrotic gray spots, about 12 mm in diameter with a distinct brown margin
3	Lesion type is the same as in scale 2, but a significant number of lesions are on the upper leaves
4	Typical susceptible blast lesions 3 mm or longer, infecting less than 4% of the leaf area
5	Typical blast lesions infecting 4–10% of the leaf area
6	Typical blast lesions infection 11–25% of the leaf area
7	Typical blast lesions infection 26–50% of the leaf area
8	Typical blast lesions infection 51–75% of the leaf area
9	More than 75% leaf area affected

centrifuged at 20,000 rpm at 4°C for 20 min. The collected supernatant was used as an enzyme source. The assay mixture containing 1 ml aliquots of supernatant and 110 µl of 100 mM L-phenylalanine were incubated at 40°C for 30 min. 1 ml of 4% tri-chloro acetic acid (TCA) was added in it to terminate the reaction. Similarly, the TCA was added in one of the test tubes at zero min to serve as blank. The assay mixture was incubated with TCA for 5 min at room temperature and the absorbance was read at 290 nm. PAL activity was calculated as µ moles of trans-cinnamic acid released per min per mg proteins under the specific condition.

The β-glucosidase activity was assayed by the modified method of Agrawal and Bahl (1969). Enzyme extraction was done using 0.5 g of infected and control seedlings and samples were macerated with 0.05 M of sodium acetate buffer (pH 4.6). The homogenate was centrifuged at 23,000 rpm at 41°C for 20 min. The supernatant was used as an enzyme source. A reaction mixture was prepared by adding 100 µl of a solution of p-nitro phenyl-β-D glucopyranoside to 350 µl of 0.05 m sodium buffer (pH 4.6), followed by initial incubation at 30°C for 5 min. After the addition of 50 µl of enzyme extract, the mixture was further incubated at 30°C for 15 min. The reaction was stopped by adding 700 µl of 0.2 M sodium carbonate. The yellow color formed was measured at 420 nm by spectrophotometer. Enzyme activity was calculated as µ moles of p-nitro phenol released per min per mg of protein. The activity was calculated based on molar extinction coefficient $(U) = 1.12 \times 10^4$ M^{-1} cm^{-1}.

Ascorbate peroxidase (APX) (EC.1.11.1.11) activity was assayed as per the method described by Nakona and Asada (1981). Enzyme extract for APX was prepared by grinding 0.5 g of controlled and infected leaf samples separately with 2 ml of 100 mM Potassium phosphate buffer (pH = 7.5). The homogenate was centrifuged at 15,000 rpm for 15 min at 4°C and the supernatant used as the enzyme source. The reaction mixture contained 2.3 ml phosphate buffer, 0.2 ml ascorbic acid, 0.2 ml (EDTA, 50 µl enzyme extract, 50 µl H$_2$O$_2$ and 0.3 ml distilled water. The reaction was started with addition of 0.2 ml of hydrogen peroxide. Decrease in absorbance after 30 s. was measured at 290 nm in UV–visible spectrophotometer. The activity was determined using molar extinction coefficient $U = 2.8$ mM^{-1} cm^{-1}.

Superoxide dismutase (SOD) (EC.1.15.1.1) activity was measured immediately in fresh extract as described by Dhindsa, Plumb-Dhindsa, and Thorpe (1981). Enzyme extract for SOD was prepared by grinding 0.5 g of controlled and infected leaf samples separately with 2 ml of 100 mM Potassium phosphate buffer (pH 7.5). The homogenate was centrifuged at 15,000 rpm for 15 min at 4°C and the supernatant was used as the enzyme source. The reaction mixture contained, 1.5 ml phosphate buffer, 0.2 ml methionine, 0.1 ml Ethylene-diaminetetraacetic acid (EDTA), 0.1 ml sodium carbonate, 0.1 ml enzyme extract, 0.1 ml NBT, 0.9 ml distilled water, and 0.1 ml riboflavin. The reaction was started by adding 0.1 ml of riboflavin and placing the tubes under two 15 W fluorescent lamps for 15 min. A complete reaction mixture without enzyme, which gives the maximal color, served as control. Switching off the lights and putting the tubes into dark stopped the reaction. The non-irradiated complete reaction mixture served as blank.

The assay of peroxidase activity (EC.1.11.1.7) was performed as described by Sadasivam and Manickam (1996). Enzyme extract for peroxidase was prepared by grinding 0.5 g of controlled and inoculated leaf samples separately with 2 ml of 100 mM potassium phosphate buffer (pH 7.5). The homogenate was centrifuged at 15,000 rpm for 15 min at 4°C and the supernatant was used as the enzyme source. For the peroxidase assay, the reaction mixture was prepared by adding 2.85 ml 0.1 M phosphate buffer (pH 7.0), 50 µl of 20 mM guaiacol solution, 50 µl of 12.3 mM hydrogen peroxide and 50 µl of enzyme extract. The reaction was allowed to proceed for 3 min. Absorbance at 470 nm were measured 30 s after adding the enzyme extract to the substrate, and change in the absorbance was recorded up to 3 min. The peroxidase activity was determined using molar absorption coefficient $U = 26.6$ mM^{-1} cm^{-1}. The protein content in the crude enzyme extract was estimated according to the method of Lowry, Rosebrough, Farr, and Randall (1951).

2.2. DNA Isolation, purification and amplification of PCR product

Isolation of genomic DNA from young seedlings was carried out by modified cetyltrimethyl ammonium bromide (CTAB) method described by Keim, Olson, and Shoemaker (1988). Fresh young leaves of rice, about 0.5 g were taken and cut into small pieces (about 10 mm²) with blade, powdered in liquid nitrogen with mortar and pestle. The powder was homogenized in prechilled 1.5 ml CTAB buffer with mortar and pestle and transferred in 2 ml eppendorf tubes. The tubes were incubated for 60 min at 65°C in thermostatic water bath. The contents in the tubes were mixed after every 15 min by inversion during incubation and the tubes were allowed to cool at room temperature. An equal volume of CI (24:1) was added into the contents of fresh tube. The tubes were then centrifuged at 10,000 rpm for 10 min at 4°C in a high speed refrigerated centrifuge (Kubota 6,500, Japan). Then aqueous phase was carefully recovered and transferred to a fresh tube for precipitation of DNA. The tubes were again centrifuged at 12,000 rpm for 10 min and collect upper layer in 1.5 ml tube. About 500 µl, ice cold 70% (v/v) ethanol was added into the tubes and 50 µl of 7.5 M ammonium acetate was added to it and kept it for precipitation for 1 h at −20°C. Then the pellets were washed with 70% ethanol. The DNA pellet was air dried till the last traces of ethanol was evaporated. The pellets were resuspended in suitable volume of TE (10/1) buffer for PCR use.

For purification, 500 µl of DNA sample was taken in a fresh eppendorf tube with 10 µl of RNase A (10 mg/ml) and incubated at 37°C for 1 h with occasional gentle shaking. After incubation, 20 µl of Proteinase K and 100 µl of 3 M sodium acetate (pH 4.8) was added followed by incubation for 60°C for 1 h. Further steps was followed as per above procedure. DNA amplification was carried out in a 0.2 ml PCR tubes having 25 µl reaction volume as described by Mahatma, Bhatnagar, Solanki, Mittal, and Shah (2011) with some modification. The reaction mixture containing 17 µl of sterile distilled water with 10x reaction buffer 2 µl with 15 mM Mgcl$_2$, 1 µl of dNTPs (10 mM), 0.5 µl(3U) of Tag DNA polymerase, 2 µl (1 µl forward and 1 µl reverse primer form 10 pmol stock and 2 µl (20 ng/µl) template DNA (Table 2).

The 25 µl reaction mixture was gently vortexed and spinned down. The DNA amplification was carried out on a thermal cycler (Eppendorf Master Cycle gradient, Germany). PCR reaction performed as follows: The first step consist of 1 cycle for initial denaturation at 94°C for 5.0 min, annealing as per the TM values of primers for 45 s and primer extention at 72°C for 1.0 min and 30 s. A final extention at 72°C for 7 min was given at the end of the cycles and the samples were held at 4°C till retrieval. The amplification of PCR product was subjected to 3% agarose gel electrophoresis. For this 3 g of metaphore agarose gel was added to 100 ml of 1x TBE buffer and melted by heating the solution in a microwave oven. The solution was cooled to about 50°C and 5 µl ethidium bromide was added in it. In prepared gel 25 µl PCR product were analyzed by mixing 2 µl of tracking dye and loaded carefully in the wells of gel and amplification was performed. The amplified PCR products were observed under UV transilluminater in gel imaging and spot picking work station documentation system and image was captured.

All biochemical parameters were analyzed in three replication. The data obtained by biochemical constituents and enzymes determination were subjected to factorial completely randomized design (CRD) for the significance of various data analyzed and means were compared by Tukey test ($p = 0.05$) using SAS (2002).

Table 2. Chromosome location (CL), size range (SR), primers sequence and annealing temperature (AT) for two SSR markers used in the study

S. no	Marker SSR	CL	Forward sequence	Reverse sequence	AT (°C)	SR (bP)
1	RM144	11	5'TGCCCTGGCGCAAATTT-GATCC-3'	5'-GCTAGAGGAGATCAGATG GTAGTGCATG-3'	55	214–255
2	RM224	11	5'-ATCGATCGATCTTCAC-GAGG-3'	5'TGCTATAAAAGGCATTC-GGG-3'	55	119–155

Source: Jia and Moldenhauer (2010).

3. Results

3.1. Disease severity

Five-week old disease-free rice seedlings when challenge with pathogen (10^6 spores ml^{-1}) by spraying and after 10 significant differences was observed for disease severity. From the Table 3, it is observed that out of 11 genotypes, NLR 20104, KJT-5-1-10-22-38-13, CB-06-555, TeTep, Swarnadhan, RAU-631-9-10 and RP-Biopatho-3 were categorized as a resistant genotypes to leaf blast, whereas CN-1447-9-4-7 and KJT-2 were found to be moderately resistant. On the contrary, Ek70 and Chimansal, which are the susceptible check, recorded highly susceptible reaction to leaf blast. There was no any infection on any genotypes/entry before inoculation.

3.2. Activity of antioxidative enzymes

The activity of SOD recorded in range from 3.91 to 14.15 U mg^{-1} protein in the control leaf samples of 11 rice genotypes and in infected it was ranged between 8.06 and 27.12 U mg^{-1}protein (Figure 1). The activity increased significantly highest in NLR-20104 followed by KJT-5. The induction levels of SOD in term of fold increase after inoculum spraying was higher (2.96 fold) in KJT-5 followed 2.63 fold in RAU-631-9-10 resistant and 1.64 fold in CN-1447-9-4-2 moderately blast resistant and least in EK-70 (0.33 fold) and Chimansal (0.41 fold) the two blast susceptible genotypes of rice.

Table 3. Leaf blast severity were recorded by following 0–9 SES scale as per IRRI, Philippines

Genotypes	Score	Percent disease severity	Reaction
KJT-5-1-10-22-38-13	2	2.5	Resistant
KJT-2	3	3.87	Moderately resistant
Tetep	2	2.77	Resistant
Swarnadhan	2	2.70	Resistant
CN-1447-9-4-7	3	3.33	Moderately resistant
NLR-20104	1	1.01	Resistant
RAU-631-9-10	2	1.93	Resistant
CB-06-555	2	2.11	Resistant
RP-Biopatho-3	2	2.00	Moderately resistant
EK-70	9	77	Highly susceptible
Chimansal	9	80	Highly susceptible

Source: Anonymous (2002).

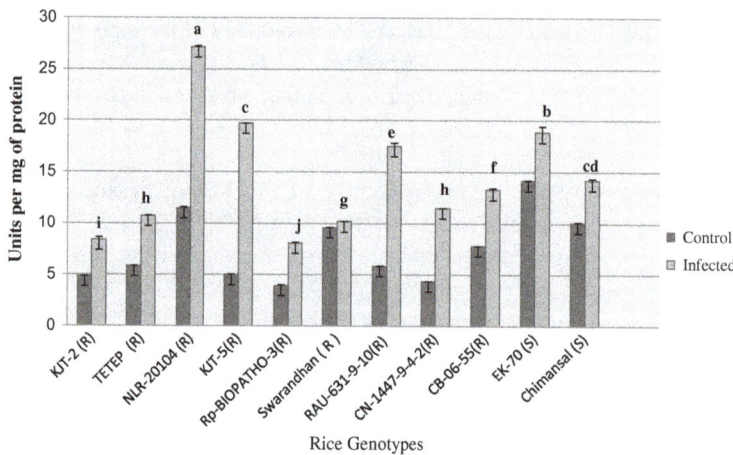

Figure 1. Activity profile of superoxide dismutase (SOD).

Notes: The vertical bar in the figures indicates activity profile of antioxidative enzymes from disease free (control) and after challenge to inoculation with *M. oryzae* (infected) leaf samples of 11 rice genotypes. The bar indicates SE ± of mean (*n* = 3) at *p* < 0.05 probability.

The peroxidase activity profile ranged between 5.67 and 132.61 n moles H_2O_2 oxidized min^{-1} mg^{-1} protein in the infected leaf samples of all 11 rice genotypes (Figure 2). The activity profile of peroxidase (POX) in the infected leaf samples ranged between 51.16 and 725 n moles H_2O_2 oxidized min^{-1} mg^{-1} protein. After 10 days of inoculum spraying, the POX activity significantly induced in blast resistant genotypes by 27.96 fold followed by 21.82 fold in NLR-20104 and KJT-5 followed by TETEP (21.61 fold) ($p < 0.05$). The least level of induction was recorded in Chimansal (3.92 fold), a blast susceptible rice. It thus appears that the level of induction higher than 20 fold over the uninfected control (constitutive) may be considered as a criterion for selection.

APX activity profile ranged between 42.48 and 296.59 n moles ascorbate oxidized/min/mg protein in the uninfected leaf samples of 11 rice genotypes (Figure 3). Whereas in the infected leaf samples, it was ranged between 184.43 and 587.67 n moles ascorbate oxidized min^1 mg^{-1} protein. Significantly ($p < 0.05$) highest APX activity was recorded after inoculum sprayed leaf samples of NLR-20104 (587.67 n moles ascorbate oxidized min^{-1} mg^{-1} protein) followed by KJT-5 (476.67 n moles ascorbate oxidized min^{-1} mg^{-1} protein) followed by Swarandhan (463.23 n moles ascorbate oxidized min^{-1} mg^{-1} protein), respectively. EK-70 but these blast susceptible genotypes recorded constitutive higher levels of APX as against other resistant genotypes but level of induction was nonsignificant. After 10 days of inoculum spraying the APX activity significantly induced in blast resistant genotypes by 8.20 fold in CN-1447-9-4-2(R) followed by 5.09 fold in KJT-5 and least in EK-70 (1.14 fold).

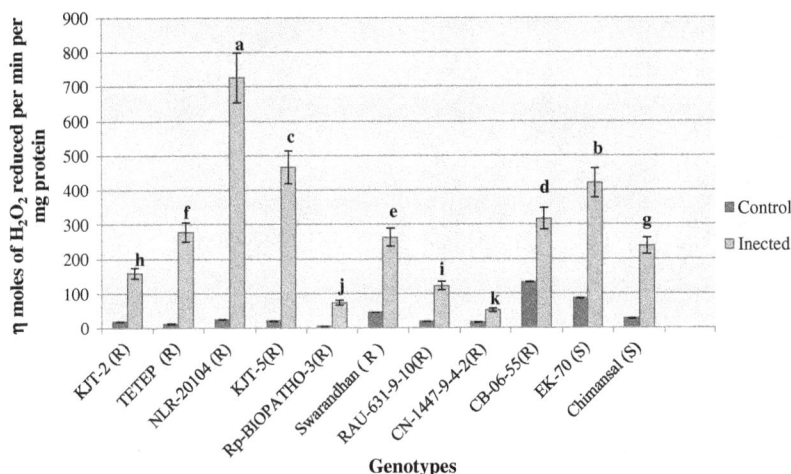

Figure 2. Activity profile of peroxidase (POX).

Notes: The vertical bar in the figures indicates activity profile of antioxidative enzymes from disease free (control) and after challenge to inoculation with *M. oryzae* (infected) leaf samples of 11 rice genotypes. The bar indicates SE ± of mean (*n* = 3) at *p* < 0.05 probability.

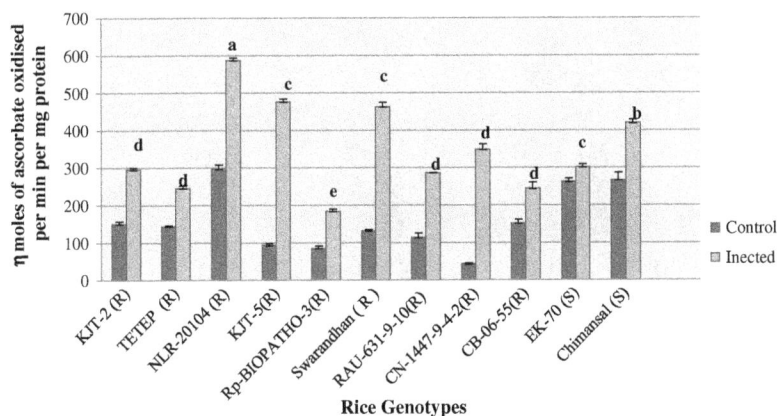

Figure 3. Activity profile of ascorbate peroxidase (APX).

Notes: The vertical bar in the figures indicates activity profile of antioxidative enzymes from disease free (control) and after challenge to inoculation with *M. oryzae* (infected) leaf samples of 11 rice genotypes. The bar indicates SE ± of mean (*n* = 3) at *p* < 0.05 probability.

3.3. Activity profile of defense-related enzymes

The PAL activity profile 11 rice genotypes ranged between 3.21 and 4.34 µ moles cinnamic acid released/min/mg protein in the disease-free leaf samples whereas it was ranged between 3.39 and 5.01 µ moles cinnamic acid released/min/mg protein in the infected leaf samples (Figure 4). Highest increase of PAL activity recorded in KJT-2 (R) (4.67 µ moles cinnamic acid released min^1 mg^{-1} protein). The order of induction of PAL activity on infection with the pathogen was in the order KJT-2(R) followed by NLR-20104 and RAU-631-9-10 and the least level of induction was recorded in Chimansal a blast susceptible rice genotype. From Figure 4, it was also observed that the PAL activity constitutively recorded higher in both group of genotypes.

The leaf chitinase activity profile ranged between 11.53 and 33.35 µg NAG released min^1 mg^{-1} protein in the disease-free leaf samples of 11 rice genotypes. As compared to control samples chitinase activity significantly increased after infection of *M. oryzae* from 23.21 to 77.15 µg NAG released/min/mg protein (Figure 5). The leaf chitinase activity was significantly higher both in the uninfected (constitutive) and infected (induced) recorded in RAU-631-9-10 that is 33.35 to 77.15 µg NAG released min^{-1} mg^{-1} protein followed by Swarandhan and in susceptible EK-70. Chimansal recorded constitutively higher activity after infection but it was low as compared to other resistant rice genotypes. The order of induction level of chitinase activity significantly ($p < 0.05$) higher in NLR-20104 and CN-1447-9-4-2 (R) (4.17) and was at par in Swarandhan and RAU-631-9-10 and least level of induction was recorded in Rp-Biopath-3 (R).

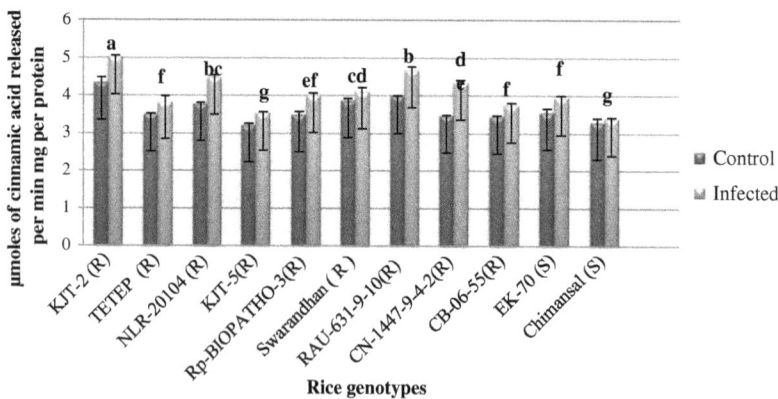

Figure 4. Activity of phenylalanine ammonia lyase (PAL).

Notes: The vertical bar in the figures indicates activity profile of defense-related enzymes from disease free (control) and after challenge to inoculation with *M. oryzae* (infected) leaf samples of 11 rice genotypes. The bar represent standard error of mean ($n = 3$) at $p < 0.05$ probability.

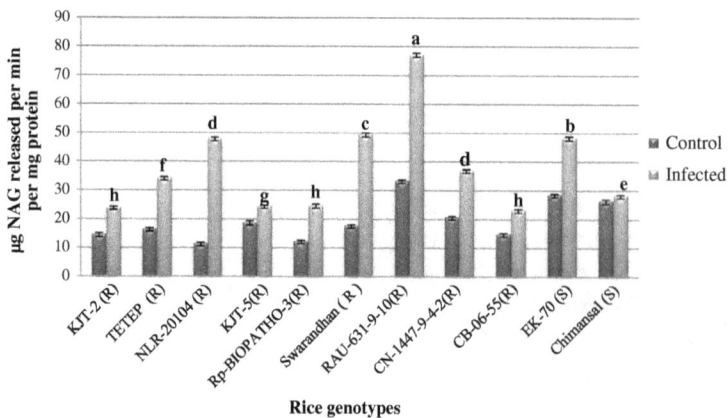

Figure 5. Activity of Chitinase.

Notes: The vertical bar in the figures indicates activity profile of defense-related enzymes from disease free (control) and after challenge to inoculation with *M. oryzae* (infected) leaf samples of 11 rice genotypes. The bar represent standard error of mean ($n = 3$) at $p < 0.05$ probability.

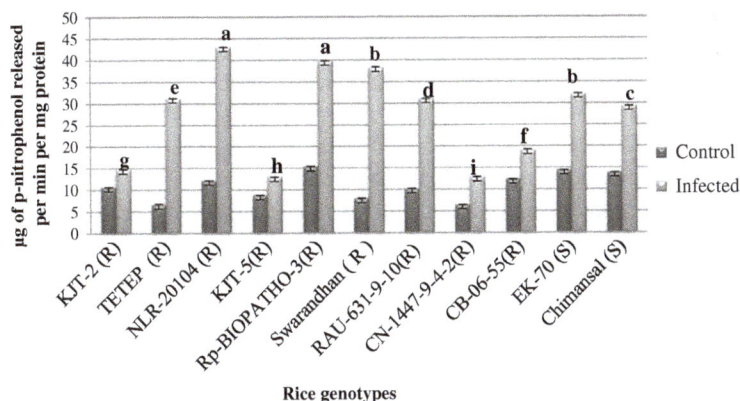

Figure 6. Activity of β-glycosidase.

Notes: The vertical bar in the figures indicates activity profile of defense-related enzymes from disease free (control) and after challenge to inoculation with *M. oryzae* (infected) leaf samples of 11 rice genotypes. The bar represent standard error of mean ($n = 3$) at $p < 0.05$ probability.

In case of β-glucosidase, as compared to control samples (6.45 to 15.42 n moles p-nitro phenol released min^{-1} mg^{-1} protein) the activity profile significantly induced in all genotypes and it was ranged from 12.93 to 43.08 n moles p-nitro phenol released min^{-1} mg^{-1} protein in infected leaf samples of all 11 rice genotypes (Figure 6). Highest β-glucosidase activity in the infected leaf samples was recorded in NLR-20104 and Rp-BIOPATHO-3R of 43.08 and 39.87 n moles p-nitro phenol released/ min/mg protein respectively followed by Swarandhan (R) 38.34 n moles p-nitro phenol released min^{-1} mg^{-1} protein and EK-70(S) (n moles p-nitro phenol released min^{-1} mg^{-1} protein) and Swarandhan with. Constitutive higher β-glucosidase activity recorded in both group of rice genotypes.

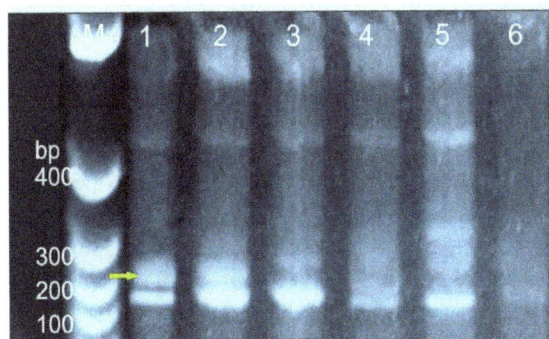

(a)

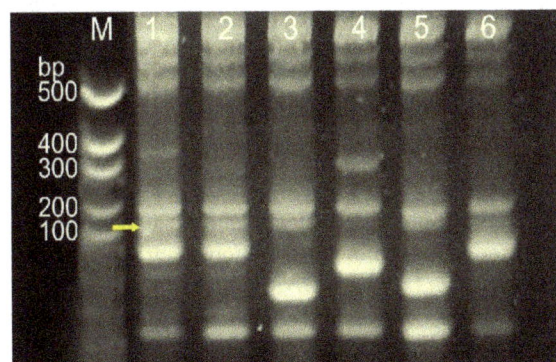

(b)

Figure 7. The purified genomic DNA was quantified and equal amount of DNA was used for PCR amplification. (a) Amplified genomic DNA of blast resistant and susceptible genotypes with SSR primer RM 224 and (b) with SSR RM144. The Lane M:100 bp, lane 1:1-KJT-2 (R),Lane 2:KJT-5 (R), Lane 3-TETEP (R), Lane 4:Chimansal(S), Lane 5:NLR- 20104 (R),Lane 6:-EK-70 (S).

3.4. Genomic DNA amplification using SSR primers

The genomic DNA of four blast resistant (KJT-2, KJT-5, Tetep and NLR-20104) and two blast susceptible (Chimansal and EK-70) genotypes were amplified using these two SSR markers, RM 224 with an
expected allele size fragment of 122 bp, corresponding to *PiL* locus and RM 144 with an expected
allele size of 254 bp corresponding to *Pik* locus. It is observed that a band corresponding ~122 bp
with RM 224 is clearly visible in all the four blast resistant genotypes, while the same is not very
prominent in the two blast susceptible genotypes (Figure 7(a)). As regard to the amplification pattern using RM144, a band corresponding to a allelic size of ~254 bp was distinctly visible in line 1, 2
and 5 corresponding to KJT-2, KJT-5 and NLR-20104 a three blast resistant genotypes, which was
missing in the susceptible genotype Chimansal, however, the same was prominently present in another susceptible genotype EK-70 (Figure 7(b)).

4. Discussion

In plant-pathogen interactions formation of new proteins that have direct or indirect effect on plant
resistance to pathogen include a heterogeneous group of proteins collectively called as pathogenesis-related (PR) proteins (Jones & Dangl, 2006). PR protein such as PAL, β-glucosidase and chitinase
has been suggested to be involved in plant resistance against fungal pathogens (Kini et al., 2000).
The ROS can be second messengers in resistance mechanisms leading to activation of defense-related genes and interferes with other important signaling molecules (Chen, Kidd, Carvalhais, &
Schenk, 2014). Uncontrolled accumulation of ROS results in spreading cell death which in some
cases can enhance plant susceptibility (Torres, Jones, & Dang, 2006). Thus, ROS production and elimination are tightly controlled during plant-pathogen interactions. Enzymatic antioxidants such as
SOD and peroxidase (POX) are participating in scavenging various types of ROS (Barna, Fodor,
Harrach, Pogány, & Király, 2012). The enzyme SOD constitutes the first line of defense against ROS
by catalyzing the dismutation of O^{2-} to O_2 and H_2O_2 (Alscher, Erturk, & Heath, 2002). In our study, SOD
activity was significantly induced in infected leaf of both resistant and susceptible rice genotypes
(Figure 2). POX is one of the most important enzymes active in elimination of ROS and catalyzes the
oxidoreduction of various substrates using hydrogen peroxide. Depending upon physiological condition peroxidase may acts as either H_2O_2 scavanger or generator (Almagro et al., 2009). Many reports
have suggested that POX plays a role in resistance to pathogens (Kawaoka et al., 2003). In our study,
SOD and POX activities induced significantly over controls suggesting their activation might be due
to generation of ROS there by activating other defense cascade (Rahman, Uddin, & Wenner, 2014) or
display direct toxicity toward invading pathogen (Torres, 2006), thus this two enzymes might be
more efficiently control ROS (Figures 1 and 2).

APX is an enzymatic antioxidant present in practically all sub cellular compartments. In rice,
Agrawal, Jwa, Han, Agrawal, and Rakwal (2003) reported cytosolic APX genes are up regulated upon
wounding suggesting that the cytosolic APX isozymes play a protective role against stressful conditions. In our study, constitutive higher APX activity has been recorded in both group of rice genotypes and it was induced on post infection (Figure 3). This indicates their participation in removal of
H_2O_2 produced by induced SOD activity. Similar increase of APX activity with glutathione reductase
activity is reported in pearl millets *Sclerospora graminicola* by Kumar, Naik, Satbhai, and Patil (2015).
Overall, this study revealed that priming in ROS production and the activity of antioxidant enzymes
such as SOD, POX and APX occurred during interaction with the pathogen more in resistant rice cultivar compared to susceptible cultivar. Furthermore, the greater resistance of NLR-20104 (R), KJT-5
(R), Swarnadhan (R) genotypes might be associated with greater lignin contents in this genotypes.
Together, these findings suggest the critical role of early O^{2-} and H_2O_2 accumulation, also SOD and
POX and APX dependent lignifications as a defense mechanism involved in basal resistance in our
pathosystem. The induced levels of antioxidants after post infection in rice plants might be due to
pathogen-associated molecular pattern triggered immunity and effectors triggered immunity that
accompanied by ROS generation (Filippi et al., 2011; Taheri, Irannejad, Goldani, & Tarighi, 2014) and
that also strengthening of plant cell wall to limit proliferation of pathogen under control condition as
a basal defense and during the pathogen attack of *M. oryzae* (Chisholm, Coaker, Day, & Staskawicz,
2006; Jones & Dangl, 2006). In addition to the enzymatic H_2O_2 scavenging system, phenolics are

strong non-enzymatic antioxidants due to availability of their phenolic hydrogen. Some phenolics are constituents of lignin and these phenolics are oxidized by POX using H_2O_2 (Nikraftar, Taheri, Falahati Rastegar, & Tarighi, 2013; Sharma, Jha, Dubey, & Pessarakli, 2012).

PAL catalyses the first committed step for biosynthesis of the phenyl propanoid pathway in higher plants and involved in the synthesis of both phytoalexin and lignin. These PR protein prevent cell wall penetration by the pathogen (Dixon, 2001). In this study PAL also recorded higher constitutive level in both group of rice genotypes with similar trend of increment in infected samples (Figure 4). Earlier positive correlation of PAL in six rice cultivars differing in resistance to M. oryzae with the degree of resistance has been reported (Zhang, Duan, & Yu, 1987). PAL activity in highly resistant cultivars was 63.5% higher than in susceptible cultivars which cause hardening of infection sites, thus preventing pathogen entry into the host plant reported by Hsieh, Ma, Yang, and Lee (2010).

In this study, the chitinase activity constitutively higher at control in all 11 rice genotypes that indicate their importance and involvement in basal defense response in rice. Early research showed low constitutive expression of chitinases in healthy plants and induced to much higher levels upon infection or wounding (Boller, 1988). Shimizu et al. (2010) reported that Slp1, a novel effector secreted by M. oryzae competes for chitin binding with the rice pattern recognition receptor (PRR) chitin elicitor binding protein (CEBiP), which is required for chitin triggered immunity in rice, acting in cooperation with the LysM receptor-like kinase Os-CERK1. Transgenic over expression lines, using family 19 chitinases from bean, tobacco, and rice confirmed that higher constitutive expression of some chitinases indeed does contribute to increased fungal resistance (Datta et al., 2001).

β-glucosidase a enzymes with a broad substrate specificity is a microbial cell wall degrading enzymes and have relationship to pathogenicity (Takeda et al., 2010). In our study, KJT(5), CN-1447-9-4-2 and CB-06-55 a resistant rice genotypes recorded lowest activity than susceptible EK-70 and Chimansal both at control and post infection indicates these enzymes showed more activation to avoid infection but their action might insufficient to limit the spread of M. oryzae than the resistant (Figure 6). Yang, Jiang, Yan, and Zhu (2008) reported microbial and fungal β-glucosidase (EC 3.2.1.21) are produced extra cellularly and intra cellularly, and are thought to play a significant role in saccharifying cellulosic materials and acquiring nutrients by producing glucose. Earlier Whetten, MacKay, and Sederoff (1998) reported that the plant β-glucosidase may be involved in the processing and release of fungal glucan elicitors, triggering a chain of reactions in the host, including phytoalexin formation and the biosynthesis of phenylpropanoids and lignin-like phenol aglucones by hydrolyzing B-phenyl glucosides. These aglucones are basically fungi toxic and fungi static in action and may limit the spread of M. oryzae in resistant plants. A similar result in accumulation of β-1-3 glucanase, PAL and chitinase enzymes in incompatible interactions of pearl millets has been reported recently by Kumar et al. (2015). Suppression of blast infection M. oryzae has been reported with induction of this enzymes by soil drenching of rhizobacteria in rice field by Filippi et al.(2011). This induced resistance (IR) may control the pathogens or damaging factors, completely or partially (Chen et al., 2014; Kuc, 1982). Several studies have shown that genes expressed during IR responses produce proteins with chitinase, glucanase and other enzymatic activities that are involved in defense reactions to a wide spectrum of pathogens (van Loon, Rep, & Pieterse, 2006). In resistant rice genotypes, the activity of defense-related and antioxidative enzymes increased might associated with hypersensitive response (HR) and or M. oryzae has robust the defense system. Our results are in agreement with the results of above quotation.

PCR-based microsatellite markers have been widely used to screen, characterize and evaluate genetic diversity in cereal species. In particular, microsatellite based methods offer an high through put and non labour intensive way to tag resistance genes in breeding programs. RM 224/RM 144 are flanking SSR markers for blast resistance loci located on chromosome 11 of rice crop which were used to detect allelic differences in resistance and susceptible cultivars of rice. In present study all the four resistant genotypes KJT-2, KJT-5, Tetep and NLR-20104 could amplify the fragment 139 and 254 bp using RM244 and RM144 primers, respectively (Figure 7(a) and (b)). These genotypes also recorded lower score of blast infection. Suh et al. (2009) reported the presence or absence of the eight major blast-resistance genes (*Pib, Pia, Piz, Piz-t, Pi9, Pi5, Pita,* and *Pi40*) was validated using foreground selection with gene specific DNA markers in the parents. According to report by *Liu, Pi-l* is located 6.8 cM and *Pi-k* is located 1.2 cM away from the RM-144 microsatellite. More recently, there has been a report concerning the identification of rice blast resistance using RM 144. Markers RM 144 and RM224 co-segregated with both *Pi-kh* and *Pi-ks* resistance factors found at this locus (Fjellstrom et al., 2004a, 2004b). The presence of the *Pi-kh* and *Pi-ks* alleles can be differentiated by RM-224 and RM-144. The *Pi-kh* gene is associated with the RM-224 = 139 nucleotide (nt) and RM-144 = 255 nt alleles, whereas the *Pi-ks* allele is associated with the RM-224 = 120 nt and RM-144 = 255 nt alleles. It is reported that RM-144 an allelic fragment size of 254 is expected in resistant genotypes corresponding to *Pik* locus. However, in the present finding, a band corresponding to 200 bp was observed both in susceptible and resistant genotype. These markers are ideally suited for marker assisted selection for blast resistance in rice because of their tight linkage with resistance genes and ease of use through analysis of amplification products (Fjellstrom et al., 2004a, 2004b). Fjellstrom, McClung, and Shank (2006) reported, SSR markers linked to the *Pi* gene were useful for selection of resistance genes at the *Pi* gene locus in rice germplasm. Eizenga, Agrama, Lee, Yan, and Jia (2006) reported marker RM225 to be close to the blast resistant genes *Pi22* and *Pi27*, where as Temnykh et al. (2001) mapped the marker RM225 on short arm of chromosome 6 at 1.1 cM away from the marker RM204. It suggests that marker RM224 identified in present study is in close vicinity to the gene.

5. Conclusions and future prospects

The activity of defense-related enzymes i.e. pathogenesis related protein was induced differentially in different genotypes upon pathogen infection and appreciable differences was observed in both resistant and susceptible rice genotypes. However, the activity profile of these enzymes in the disease-free leaf samples of some of the resistant genotypes was less than the activity recorded in blast susceptible genotypes (Chimansal). The resistant genotypes NLR-20104, KJT-5 (R), Swaranadhan, KJT-2 recorded higher levels of defense-related and antioxidative enzymes both at constitutive and post infection be considered as most promising blast resistant genotypes. Whereas significant higher induction levels of chitinase and with higher disease severity in susceptible EK-70 than the other resistant genotypes needs further validation and needs to be tested at different stages of post inoculation. In addition to two SSR markers RM-224, RM-144 some already reported SSR markers on different linkage group needs to be attempted for further study. This study revealed that, the post infection induction in activity level of defense-related and antioxidant enzymes with least disease severity and validation of genomic DNA with RM144 and RM224 could be effectively exploited to screen the rice genotypes for blast resistance.

Funding

The authors received no direct funding for this research.

Author details

P.U. Anushree[1]

E-mail: anusree.uthaman94@gmail.com

R.M. Naik[1]

E-mail: rajeevnaik2@rediffmail.com

R.D. Satbhai[1]

E-mail: ravimh17@gmail.com

ORCID ID: http://orcid.org/0000-0002-7577-5965

A.P. Gaikwad[1]

E-mail: apgaikwad@rediffmail.com

C.A. Nimbalkar[1]

E-mail: canimbalkar@gmail.com

ORCID ID: http://orcid.org/0000-0002-7651-9832

[1] Department of Statistic, MPKV, Rahuri 413722, Ahmednagar, India.

Cover image

Source: A.P Gaikwad.

References

Agrawal, G. K., Jwa, N. S., Han, K. S., Agrawal, V. P., & Rakwal, R. (2003). Isolation of a novel rice PR4 type gene whose mRNA expression is modulated by blast pathogen attack and signaling components. *Plant Physiology and Biochemistry, 41,* 81–90. http://dx.doi.org/10.1016/S0981-9428(02)00012-8

Agrawal, K. M. L., & Bahl, O. P. (1969). Glycosidases of *Phaseolus vulgaris. The Journal of Biological Chemistry, 243,* 103–111.

Almagro, L., Gomez Ros, L. V., Belchi-Navarro, S., Bru, R., Ros Barcelo, A., & Pedreno, M. A. (2009). Class III peroxidases in plant defence reactions. *Journal of Experimental Botany, 60,* 377–390. http://dx.doi.org/10.1093/jxb/ern277

Alscher, R. G., Erturk, N., & Heath, L. S. (2002). Role of superoxide dismutases (SODs) in controlling oxidative stress in plants. *Journal of Experimental Botany, 53*, 1331–1341. http://dx.doi.org/10.1093/jexbot/53.372.1331

Anonymous. (2002). Find out how the qualities of rice are evaluated and scored in this authoritative sourcebook. *Standard evaluation system for rice* (p. 1518) IRRI, Philippines.

Baraket, G., Chatti, K., Saddoud, O., Abdelkarim, A. B., Mars, M., & Trifi, M. (2011). Comparative assessment of SSR and AFLP markers for evaluation of genetic diversity and conservation of fig, *Ficus carica* L., genetic resources in Tunisia. *Plant Molecular Biology Reporter, 29*, 171–184. http://dx.doi.org/10.1007/s11105-010-0217-x

Barna, B., Fodor, J., Harrach, B. D., Pogány, M., & Király, Z. (2012). The Janus face of reactive oxygen species in resistance and susceptibility of plants to necrotrophic and biotrophic pathogens. *Plant Physiology and Biochemistry, 59*, 37–43. http://dx.doi.org/10.1016/j.plaphy.2012.01.014

Boller, T. (1988). Ethylene and the regulation of antifungal hydrolases in plants. *Oxford Surveys of Plant Molecular and Cell Biology, 5*, 145–174.

Bonman, J. M., Khush, G. S., & Nelson, R. J. (1992). Breeding rice for resistance to pests. *Annual Review of Phytopathology, 30*, 507–528. http://dx.doi.org/10.1146/annurev.py.30.090192.002451

Campos, R., Nonogaki, H., Suslow, T., & Saltveit, M. S. (2004). Isolation and characterization of a wound inducible phenylalanine ammonia-lyase gene (LsPAL1) from Romaine lettuce leaves. *Physiologia Plantarum, 121*, 429–438. http://dx.doi.org/10.1111/ppl.2004.121.issue-3

Campoy, J. A., Ruiz, D., Egea, J., Rees, D. J. G., Celton, J. M., & Martínez-Gómez, P. (2011). Inheritance of flowering time in apricot (*Prunus armeniaca* L.) and analysis of linked quantitative trait loci (QTLs) using simple sequence repeat (SSR) markers. *Plant Molecular Biology Reporter, 29*, 404–410. http://dx.doi.org/10.1007/s11105-010-0242-9

Chen, Y. C., Kidd, B. N., Carvalhais, L. C., & Schenk, P. M. (2014). Molecular defense responses in roots and the rhizosphere against *Fusarium oxysporum. Plant Signaling & Behavior, 9*(12), e977710. doi:10.4161/15592324.2014.977710

Chisholm, S. T., Coaker, G., Day, B., & Staskawicz, B. J. (2006). Host-microbe interactions: Shaping the evolution of the plant immune response. *Cell, 124*, 803–814. http://dx.doi.org/10.1016/j.cell.2006.02.008

Datta, K., Tu, J., Oliva, N., Ona, I. I., Velazhahan, R., Mew, T. W., Muthukrishnan, S., & Datta, S. K. (2001). Enhanced resistance to sheath blight by constitutive expression of infection-related rice chitinase in transgenic elite indica rice cultivars. *Plant Science, 160*, 405–414. http://dx.doi.org/10.1016/S0168-9452(00)00413-1

Dhindsa, R. S., Plumb-Dhindsa, P., & Thorpe, T. A. (1981). Leaf senescence: Correlated with increased levels of membrane permeability and lipid peroxidation, and decreased levels of superoxide dismutase and catalase. *Journal of Experimental Botany, 32*, 93–101. http://dx.doi.org/10.1093/jxb/32.1.93

Dixon, R. A. (2001). Natural products and plant disease resistance. *Nature, 411*, 843–847. http://dx.doi.org/10.1038/35081178

Eizenga, G. C., Agrama, H. A., Lee, F. N., Yan, W., & Jia, Y. (2006). Identifying novel resistance genes in newly introduced blast resistant rice germplasm. *Crop Science, 46*, 1870–1878. http://dx.doi.org/10.2135/cropsci2006.0143

Filippi, M. C. C., da Silva, G. B., Silva-Lobo, V. L., Côrtes, M. V. C. B., Moraes, A. J. G., & Prabhu, A. S. (2011). Leaf blast (*Magnaporthe oryzae*) suppression and growth promotion by rhizobacteria on aerobic rice in Brazil. *Biological Control, 58*, 160–166. http://dx.doi.org/10.1016/j.biocontrol.2011.04.016

Fjellstrom, R., Conaway-Bormans, C. A., McClung, A., Marchetti, M. A., Shank, A. R., & Park, W. D. (2004a). Development of DNA markers suitable for marker assisted selection of three genes conferring resistance to multiple pathotypes. *Crop Science, 44*, 1790–1791. http://dx.doi.org/10.2135/cropsci2004.1790

Fjellstrom, R., Conaway-Bormans, C. A., McClung, A. M., Marchetti, M. A., Shank, A. R., & Park, W. D. (2004b). Development of DNA markers suitable for marker assisted selection of three Pi. *Genetics, 138*, 1251–1274.

Fjellstrom, R., McClung, A. M., & Shank, A. R. (2006). SSR markers closely linked to the Pi-z locus are useful for selection of blast resistance in a broad array of rice germplasm. *Molecular Breeding, 17*, 149–157. http://dx.doi.org/10.1007/s11032-005-4735-4

Giri, A. P., Harsulkar, A. M., Patankar, A. G., Gupta, V. S., Sainani, M. N., Deshpande, V. V., & Ranjekar, P. K. (1998). Association of induction of protease and chitinase in chickpea roots with resistance to *Fusarium oxysporum f. sp. ciceri. Plant Pathology, 47*, 693–699. http://dx.doi.org/10.1046/j.1365-3059.1998.00299.x

Hsieh, L. S., Ma, G. T., Yang, C. C., & Lee, P. D. (2010). Cloning, expression, site-directed mutagenesis and immunolocalization of phenylalanine ammonia-lyase in *Bambusa oldhamii. Phytochemistry, 71*, 1999–2009. http://dx.doi.org/10.1016/j.phytochem.2010.09.019

Jia, Y., & Moldenhauer, K. (2010). Development of monogenic and digenic rice lines for blast resistance genes. *Journal of Plant Registrations, 4*, 163–166. http://dx.doi.org/10.3198/jpr2009.04.0223crmp

Jones, J. D. G., & Dangl, J. (2006). The plant immune system. *Nature, 444*, 323–329. http://dx.doi.org/10.1038/nature05286

Kawaoka, A., Matsunaga, E., Endo, S., Kondo, S., Yoshida, K., Shinmyo, A., & Ebinuma, H. (2003). Ectopic expression of a horseradish peroxidase enhances growth rate and increases oxidative stress resistance in hybrid aspen. *Plant Physiology, 132*, 1177–1185. http://dx.doi.org/10.1104/pp.102.019794

Keim, P., Olson, T. C., & Shoemaker, R. C. (1988). A rapid protocol for isolating soybean DNA. *Soybean Genetics Newsletter, 15*, 150–152.

Kini, K. R., Vasanthi, N. S., & Shetty, H. S. (2000). Induction of β-1,3-glucanase in seedlings of pearl millet in response to infection by *Sclerospora graminicola. European Journal of Plant Pathology, 106*, 267–274. http://dx.doi.org/10.1023/A:1008771124782

Krishnaveni, D., Laha, G. S., Prasad, M. S., Ladha Lakshmi, D., Mangrauthia, S. K., Prakasam, V., & Viraktamath, B. C. (2012). *Sources of resistance in "Disease resistance in rice"* (p. 35). Hyderabad: Directorate of Rice Research.

Kuc, J. (1982). Induced immunity to plant diseases. *BioScience, 32*, 854–860.

Kumar, S., Naik, R., Satbhai, R., & Patil, H. (2015). Activity profile of defense related enzymes in pearl millet against downy mildew (*Sclerospora graminicola*). *Journal of Pure and Applied Microbiology, 9*, 1465–1474.

Liang, Y. C., Chen, Q., Liu, Q., Zhang, W. H., & Ding, R. X. (2003). Exogenous silicon (Si) increases antioxidant enzyme activity and reduces lipid peroxidation in roots of salt-stressed barley (*Hordeum vulgare* L.). *Journal of Plant Physiology, 160*, 1157–1164. http://dx.doi.org/10.1078/0176-1617-01065

Liu, X. Q., Wang, L., Chen, S., Lin, F., & Pan, Q. H. (2005). Genetic and physical mapping of Pi36(t), a novel rice blast resistance gene located on rice chromosome 8. *Molecular Genetics and Genomics, 274*, 394–401. http://dx.doi.org/10.1007/s00438-005-0032-5

Lowry, O. H., Rosebrough, N. J., Farr, A. L., & Randall, R. J. (1951). Protein measurement with the folin phenol reagent. *The Journal of Biological Chemistry, 193*, 265–275.

Mahatma, M. K., Bhatnagar, R., Solanki, R. K., Mittal, G. K., & Shah, R. R. (2011). Characterisation of downy mildew resistant and susceptible pearl millet (*Pennisetum glaucum* (L.) R.Br.) genotypes using isozyme, protein, randomly amplified polymorphic DNA and inter-simple sequence repeat markers. *Archives of Phytopathology and Plant Protection, 44*, 1985–1998. http://dx.doi.org/10.1080/03235408.2011.559038

Nakona, Y., & Asada, K. (1981). Hydrogen peroxide is scavenged by ascorbic specific peroxidase in spinach chloroplasts. *Plant and Cell Physiology, 22*, 868–880.

Nikraftar, F., Taheri, P., Falahati Rastegar, M., & Tarighi, S. (2013). Tomato partial resistance to *Rhizoctonia solani* involves antioxidative defense mechanisms. *Physiological and Molecular Plant Pathology, 81*, 74–83. http://dx.doi.org/10.1016/j.pmpp.2012.11.004

Panda, S. K. (2007). Chromium-mediated oxidative stress and ultrastructural changes in root cells of developing rice seedlings. *Journal of Plant Physiology, 164*, 1419–1428. http://dx.doi.org/10.1016/j.jplph.2007.01.012

Rahman, A., Uddin, W., & Wenner, N. G. (2014). Induced systemic resistance responses in perennial ryegrass against *Magnaporthe oryzae* elicited by semi-purified surfactin lipopeptides and live cells of *Bacillus amyloliquefaciens*. *Mole Plant Pathology*. doi:10.1111/mpp.12209

Sadasivam, S., & Manickam, A. (1996). *Biochemical methods* (2nd ed.) New Delhi: New Age International.

Samalova, M., Meyer, A. J., Gurr, S. J., & Fricker, M. D. (2014). Robust anti-oxidant defences in the rice blast fungus *Magnaporthe oryzae* confer tolerance to the host oxidative burst. *New Phytologist, 201*, 556–573. http://dx.doi.org/10.1111/nph.12530

SAS, Institute. (2002). *Version 9.2*. Cary, NC: Author.

Sharma, P., Jha, A. B., Dubey, R. S., & Pessarakli, M. (2012). Reactive oxygen species, oxidative damage, and antioxidative defense mechanism in plants under stressful conditions. *Journal of Botany, 2012*, 1–26.

Shimizu, T., Nakano, T., Takamizawa, D., Desaki, Y., Ishii-Minami, N., Nishizawa, Y., ... Shibuya, N. (2010). Two LysM receptor molecules, CEBiP and OsCERK1, cooperatively regulate chitin elicitor signaling in rice. *The Plant Journal, 64*, 204–214.
http://dx.doi.org/10.1111/tpj.2010.64.issue-2

Staskawicz, B. J., Ausubel, F. M., Baker, B. J., Ellis, J. G., & Jones, J. D. C. (1997). Molecular genetics of plant disease resistance. *Science, 268*, 661–667.

Suh, J. P., Roh, J. H., Cho, Y. C., Han, S. S., Kim, Y. G., & Jena, K. K. (2009). The Pi40 gene for durable resistance to rice blast and molecular analysis of Pi40 - Advanced backcross breeding lines. *Phytopathology, 99*, 243–250. http://dx.doi.org/10.1094/PHYTO-99-3-0243

Swapna, M., Sivaraju, K., Sharma, R. K., Singh, N. K., & Mohapatra, T. (2010). Single-strand conformational polymorphism of EST-SSRs: A potential tool for diversity analysis and varietal identification in sugarcane. *Plant Molecular Biology Reporter*. 11105-010-0254.

Taheri, P., Irannejad, A., Goldani, M., & Tarighi, S. (2014). Oxidative burst and enzymatic antioxidant systems in rice plants during interaction with *Alternaria alternata*. *European Journal of Plant Pathology, 140*, 829–839. http://dx.doi.org/10.1007/s10658-014-0512-8

Takeda, T., Takahashi, M., Nakanishi-Masuno, T., Nakano, Y., Saitoh, H., Hirabuchi, A., Fujisawa, S., & Terauchi, R. (2010). Characterization of endo-1,3–1,4-β-glucanases in GH family 12 from *Magnaporthe oryzae*. *Applied Microbiology and Biotechnology, 88*, 1113–1123. http://dx.doi.org/10.1007/s00253-010-2781-2

Temnykh, S., DeClerck, G., Lukashova, A., Lipovich, L., Cartinhour, S., & McCouch, S. (2001). Computational and experimental analysis of microsatellites in rice (*Oryza sativa* L.): Frequency, length variation, transposon associations, and genetic marker potential. *Genome Research, 11*, 1441–1452. http://dx.doi.org/10.1101/gr.184001

Torres, M. A., Jones, J. D. G., & Dang, J. L. (2006). Reactive oxygen species signaling in response to pathogens. *Plant Physiology, 141*, 373–378. http://dx.doi.org/10.1104/pp.106.079467

van Loon, L. C., Rep, M., & Pieterse, C. M. (2006). Significance of inducible defense-related proteins in infected plants. *Annual Review of Phytopathology, 44*, 135–162. http://dx.doi.org/10.1146/annurev.phyto.44.070505.143425

Whetten, R. W., MacKay, J. J., & Sederoff, R. R. (1998). Recent advances in understanding lignin biosynthesis. *Annual Review of Plant Physiology and Plant Molecular Biology, 49*, 585–609. http://dx.doi.org/10.1146/annurev.arplant.49.1.585

Yang, S., Jiang, Z., Yan, Q., & Zhu, H. (2008). Characterization of a thermostable extracellular β-glucosidase with activities of exoglucanase and transglycosylation from *Paecilomyces thermophila*. *Journal of Agricultural and Food Chemistry, 56*, 602–608. http://dx.doi.org/10.1021/jf072279+

Zhang, J. T., Duan, G. M., & Yu, Z. Y. (1987). Relationship between *Phenylalanine ammonia lyase* (PAL) activity and resistance to rice blast. *Plant Physiology Communications, 6*, 34–37.

Zhu, M., Wang, L., & Pan, Q. H. (2004). Identification and characterization of a new blast resistance gene located on rice chromosome 1 through linkage and differential analyses. *Phytopathology, 94*, 515–519. http://dx.doi.org/10.1094/PHYTO.2004.94.5.515

Zhu, W., Lin, J., Yang, D., Zhao, L., Zhang, Y., Zhu, Z., Chen, T., & Wang, C. (2009). Development of chromosome segment substitution lines derived from backcross between two sequenced rice cultivars, indica recipient 93-11 and japonica donor nipponbare. *Plant Molecular Biology Reporter, 27*, 126–131. http://dx.doi.org/10.1007/s11105-008-0054-3

12

Antibiotic profiling of Methicillin Resistant *Staphylococcus aureus* (MRSA) isolates in stray canines and felines

Rajwin Raja Kanagarajah[1], David Charles Weerasingam Lee[1], Daniel Zhi Fung Lee[1], Khatijah Yusoff[2], Sharmini Julita Paramasivam[3], Wai Yee Low[4,5], Kamalan Jeevaratnam[3] and Swee Hua Erin Lim[1,4,6*]

*Corresponding author: Swee Hua Erin Lim, Perdana University – Royal College of Surgeons in Ireland, Perdana University Hall D Level 1, MAEPS Building, MARDI Complex, Jalan MAEPS Perdana, Universiti Putra Malaysia, Serdang 43400, Selangor, Malaysia; Perdana University – Centre of Bioinformatics, Perdana University Hall D Level 1, MAEPS Building, MARDI Complex, Jalan MAEPS Perdana, Universiti Putra Malaysia, Serdang 43400, Selangor, Malaysia; Abu Dhabi Women's College, Higher Colleges of Technology, Abu Dhabi 41012, United Arab Emirates

E-mail: erinlimsh@gmail.com

Reviewing editor: Yusuf Akhter, Central University of Himachal Pradesh, India

Additional information is available at the end of the article

Abstract: *Background*: Methicillin resistant *Staphylococcus aureus* (MRSA) is an important Gram positive pathogen that has raised concerns due to its increasing prevalence despite pharmaceutical and technological advances. It does not only cause infections in humans but it can also be zoonotic in nature. The aim of this study is to investigate the prevalence of MRSA from presumably healthy shelter animals, in particular, canines and felines. *Methods*: Fifty-two faecal samples from canines and felines were collected from an animal shelter for the isolation of MRSA. This was carried out using the ChromMRSA (Oxoid, United Kingdom) media, followed by antibiotic susceptibility testing using the Kirby–Bauer disk diffusion test in accordance to the Clinical and Laboratory Standards Institute (CLSI) for ceftazidime, enrofloxacin, methicillin, oxacillin and vancomycin. Principal component analysis (PCA) was then employed to identify the variables in antibiotic sensitivity and emphasise any patterns between isolates and the antibiotic profile from both animal samples types. This was then summarised using descriptive statistics. *Results*: 283 and 169 *S.aureus*

ABOUT THE AUTHORS

This work is the core research interest of Animal Neighbours, an initiative led by Sharmini Julita Paramasivam, focusing on microbiological and behavioral zooanthroponotic and anthropozoonotic transmissions of antimicrobial resistant bacteria in non-clinical settings. In this particular work, the prevalence of methicillin resistant *Staphylococcus aureus* (MRSA) among canines and felines housed in animal shelters in Malaysia and the antibiotic profiling of these isolates relating to their host types were carried out. As there is limited data on the prevalence of MRSA among stray animals in South-East Asia in non-conventional, non-clinical settings, this research aims to reveal the prevalence of MRSA amongst these animals in the region, hence, the sample collection from mainly local breeds of canines and felines in Malaysia, which may differ in prevalence compared to other regions of the world due to varied prescribing practices and different existing guidelines for shelters as well as adoption processes. This research also aims to highlight the importance of practising antibiotic stewardship when treating companion animals.

PUBLIC INTEREST STATEMENT

This research looks into the prevalence of MRSA, which is a type of bacteria that is resistant to many commonly used antibiotics. The research involves taking stool samples and isolates MRSA from these samples. The MRSA isolates are then further analysed to differentiate the bacterial strains. The zoonotic nature of MRSA is a matter of high concern due to which it can spread from animals to humans and animal shelters act as a potential hub where this contact can take place. We hope that this research can provide a public health awareness about the importance of biosecurity and sewage or waste management from animal shelters.

isolates were obtained respectively from the canine and feline samples on selective media. Of this, 33/283 (11.66%) and 13/169 (7.70%) were MRSA when grown on ChromMRSA. Canine MRSA isolates exhibited resistance in decreasing order of methicillin (100%), ceftazidime (81.82%), enrofloxacin (78.79%), oxacillin (60.61%) and vancomycin (0%). On the same note feline MRSA isolates indicated resistance to methicillin (100%), ceftazidime (100%), enrofloxacin (92.31%), oxacillin (84.62%) and vancomycin (0%). 51.51% of the canine and 84.62% of feline MRSA isolates indicated resistance to four out of five antibiotics tested. PCA of antibiotic resistance profiles revealed that canine and feline formed distinct groups. However, one of the feline MRSA isolates resembled more of the canine group; although likelihood of cross-transmission between the animals may be low due to separate enclosures for the canines and felines: cross-transmission may have occurred when animals are brought into the main building for vaccination, neutering and consulting procedures. *Conclusion*: The majority of MRSA isolates obtained were not only resistant to methicillin alone but to other antibiotics too. Vancomycin proved to be the only effective antibiotic. This will pose a greater risk of resistance developing in empirical antibiotics in the future if proper antibiotic stewardship practices are not in place.

Subjects: Microbiology; Pharmacology; Bioinformatics; Infectious Diseases; Health Education and Promotion

Keywords: *Staphylococcus aureus*; MRSA; canines; felines; antibiotic resistance

1. Introduction

The discovery of antibiotics in 1928 by Sir Alexander Fleming revolutionised the treatment of infectious diseases and the quality of life of many patients, such as those suffering from Tuberculosis, Lyme disease, etc. Penicillin for instance, saved approximately 12–15% of the lives of Allied soldiers during the second World War and reduced the numbers of amputations as it prevented septicaemia from bacterial infections (Aldridge, Parascandola, & Sturchio, 1999). In tandem with this development, however, is the emergence of antibiotic-resistant bacteria. Healthcare-related problems such as failure to respond to treatment, increase in morbidity and mortality, increase in incidence of complications and increased treatment costs due to lengthened hospital stay and higher use of novel antibacterial drugs are repercussions of this resistance phenomenon (Livermore, 2003). This phenomenon, however in part, is caused by the inappropriate and rampant use of antibiotics by medical practitioners, both in the clinical and agricultural settings.

Overprescription of antibiotics exerts selective pressure on bacteria and those that can resist the harmful effects will be favoured for survival, culminating to the emergence of the fittest strain. The most common multidrug resistant bacteria includes methicillin resistant *Staphylococcus aureus* (MRSA); they are ubiquitous, often present asymptomatically in domestic animals such as canines and felines despite being known as a nosocomial pathogen for the past 30 years (Gardam, 2000). Recent studies show that MRSA development in companion animals is increasing (Loeffler et al., 2005; Strommenger et al., 2006) which can be a public health concern especially in situation where owners are immunocompromised.

The mechanism of resistance of MRSA is associated with the methicillin resistance gene (*mecA*) which codes for a protein that resists binding by methicillin, preventing the drug from exerting its effects. This gene is carried on the mobile genetic element known as staphylococcal chromosomal cassette mec (*SCCmec*) enabling them to become resistant to all β-lactam antibiotics by coding for the penicillin binding protein PBP2a (Hiramatsu, Cui, Kuroda, & Ito, 2001; Hiramatsu, Katayama,

Yuzawa, & Ito, 2002). MRSA is involved in the development of clinical syndromes such as bacterae-mia, pneumonia, cellulitis, osteomyelitis, endocarditis and septic shock (Klevens et al., 2007).

Several research papers have shown that canine faeces contain gastrointestinal tract pathogens that can cause diarrhoea in humans such as *Campylobacter, Salmonella, Yersinia* and *E. coli* (Beutin, 1999; Chaban, Ngeleka, & Hill, 2010; Lefebvre, Reid-Smith, Boerlin, & Weese, 2008). Moreover, the faeces may facilitate the diffusion of protozoa including *Giarda* and *Crytosporidium* and roundworms such as *Toxocara canis* (El-Tras, Holt, & Tayel, 2011). Furthermore, there is evidence to suggest that the stool samples of pets may have the potential to harbour antibiotic-resistant bacteria such as MRSA (Rich & Roberts, 2004). In addition, MRSA strains found in canine faeces and infected pets have been shown to be similar to those isolated from the faecal samples of human subjects with nosoco-mial infections in hospital settings (Cefai, Ashurst, & Owens, 1994). This could be a result of the direct transfer of bacteria between these two hosts, further amplifying the development of antibiotic re-sistance in the host gut that facilitates the hidden selection of resistant bacteria, manifesting in criti-cal illnesses and nosocomial infections caused by resistant bacteria. Furthermore, cross contamination as a direct cause of this transfer is becoming increasingly rampant, adding to the complexity of the problem (Carlet, 2012).

Human–animal relationships shared a long history from the time when humans first used animals such as buffaloes in the agricultural sector in rural settings to modern day context wherein surveil-lance dogs assist the police to sniff out illegal substances and explosives, etc. Other examples of human–animal interaction in non-clinical settings can further be expanded to pet shop owners, vet-erinarians and animal shelter workers. Cases of animal-to-human transmissions of MRSA were re-ported and MRSA is gradually emerging as an important veterinary pathogen (Weese et al., 2006). The focus of this work is primarily on the prevalence of MRSA from shelter animals in nonclinical settings.

2. Methods

2.1. Sample origin and collection
The study was approved by the Institutional Review Board at Perdana University (PUIRB0055). A to-tal of 52 faecal samples obtained from 28 canines and 24 felines of different breeds were collected from a local animal shelter in the Klang Valley, Malaysia during the months of May and June 2014 based on an inclusion criteria such as healthy, vaccinated and neutered animals that were ready for adoption. The samples were placed in faecal containers labelled with details of the animal, before being stored at −20 °C in the freezer at the Virology Laboratory of the Institute of Bioscience, Universiti Putra Malaysia (UPM) until further use.

2.2. Isolation and enumeration of bacteria
0.1 g of each faecal sample was weighed and placed into Eppendorf tubes followed by addition of 900 µL of 0.9% sodium chloride, NaCl (Kollin, USA). The mixture was homogenised briefly using a vortex mixer. Ten-fold serial dilutions were carried out up to dilution of 10^{-9}. Then 100 µL of each dilution was spread onto Tryptone Soya Agar, TSA (Oxoid, United Kingdom) in triplicates. This was done within 30 min of the preparation of the dilutions. The plates were incubated for 48 h at 37°C. Countable number of colonies between 30 and 300 were proceeded to replica plating onto Mannitol Salt Agar, (Oxoid, United Kingdom) for selection of *S. aureus*. To isolate MRSA another set of replica plating was performed onto ChromMRSA, (Chromagar, Oxoid, United Kingdom). The number of iso-lates on Mannitol Salt Agar and ChromMRSA were enumerated after incubation at 37°C for 24 h. *S. aureus* ATCC 25,923 was used as a control.

2.3. Disk diffusion testing
Positively tested MRSA isolates from ChromMRSA were then tested for antibiotic resistance using a range of antibiotic classes which included penicillins, cephalosporins, glycopeptides and fluoroqui-nolones namely methicillin, oxacillin, ceftazidime, vancomycin and enrofloxacin. All antibiotics are

Table 1. Antibiotics, solvent and amount impregnated per disk

Antibiotic	Solvent	Disc potency (µg)	Inhibition zone size (mm)		
			R	I	S
Ceftazidime[a]	Distilled water	30	≤14	15–17	≥18
Enrofloxacin[b]	Distilled water	5	<22	–	≥22
Methicillin[a]	Distilled water	5	≤9	10–13	≥14
Oxacillin[a]	Distilled water	1	≤10	10–12	≥13
Vancomycin[a]	Distilled water	30	–	–	≥15

Notes: *R* = Resistant; *I* = Intermediate; *S* = Susceptible.

[a]Classification adapted from National Committee for Clinical Laboratory Standard (2001).

[b]Classification adapted from Attili et al. (2016).

soluble in water. Volume of 20 µL of varying amounts of antibiotics as shown in Table 1 were impregnated into the 6 mm disks (Oxoid, United Kingdom). The size classification of average inhibition zones (Cockerill et al., 2012) is also given. Antibiotic susceptibility was carried out using Mueller-Hinton Agar which had been spread with putative MRSA isolates adjusted to 0.5 MacFarland (1.5×10^8 CFU/ML) prior to placement of the disks. The inhibition zones for each antibiotic were measured after 24 h of incubation at 37°C. Distilled water was used as a negative control and the experiment was carried out in triplicates.

2.4. Statistical analysis

Descriptive statistics that are used to summarise the raw data were done using Microsoft Excel and R (Team R.C, 2012). Principal component analysis (PCA) was done using the prcomp() function of the R stats package and visualization of the plot were calculated using the ggbiplot package.

3. Results

3.1. ChromMRSA and antibiotic susceptibility results

Not all collected faecal samples yield *S. aureus*. In the search of *S. aureus*, 283 isolates in canines and 169 isolates in felines were obtained. 33 out of 283 (11.66%) isolate for canines and 13 out of 169 (7.70%) isolate for felines were confirmed to be MRSA. These 46 isolates were then subjected to antibiotic susceptibility testing using disk diffusion and the results are listed in Tables 2 and 3.

Feline resistance to ceftazidime, enrofloxacin and oxacillin is higher compared to canine resistance. Canine MRSA isolates exhibited resistance in decreasing order of methicillin (100%), ceftazidime (81.82%), enrofloxacin (78.79%), oxacillin (60.61%) and vancomycin (0%) (Figure 1A). Feline MRSA isolates' resistance are as follows, methicillin (100%), ceftazidime (100%), enrofloxacin (92.31%), oxacillin (84.62%) and vancomycin (0%) (Figure 1B). 51.51% of the canine and 84.62% of feline MRSA isolates indicated resistance to four out of five antibiotics tested (Figure 2).

Table 2. The resistance classification for 33 MRSA isolated from canines

Antibiotic	No. of isolates		
	R	I	S
Ceftazidime[a]	27	2	4
Enrofloxacin[b]	26	0	7
Methicillin[a]	33	0	0
Oxacillin[a]	20	2	11
Vancomycin[a]	0	0	33

Notes: *R* = Resistant; *I* = Intermediate; *S* = Susceptible.

[a]Classification adapted from National Committee for Clinical Laboratory Standard (2001).

[b]Classification adapted from Attili et al. (2016).

Table 3. The resistance classification for 13 MRSA isolated from felines

Antibiotic	No. of isolates		
	R	I	S
Ceftazidime[a]	13	0	0
Enrofloxacin[b]	12	0	1
Methicillin[a]	13	0	0
Oxacillin[a]	11	0	2
Vancomycin[a]	0	0	13

Notes: R = Resistant; I = Intermediate; S = Susceptible.

[a]Classification adapted from National Committee for Clinical Laboratory Standard (2001).

[b]Classification adapted from Attili et al. (2016).

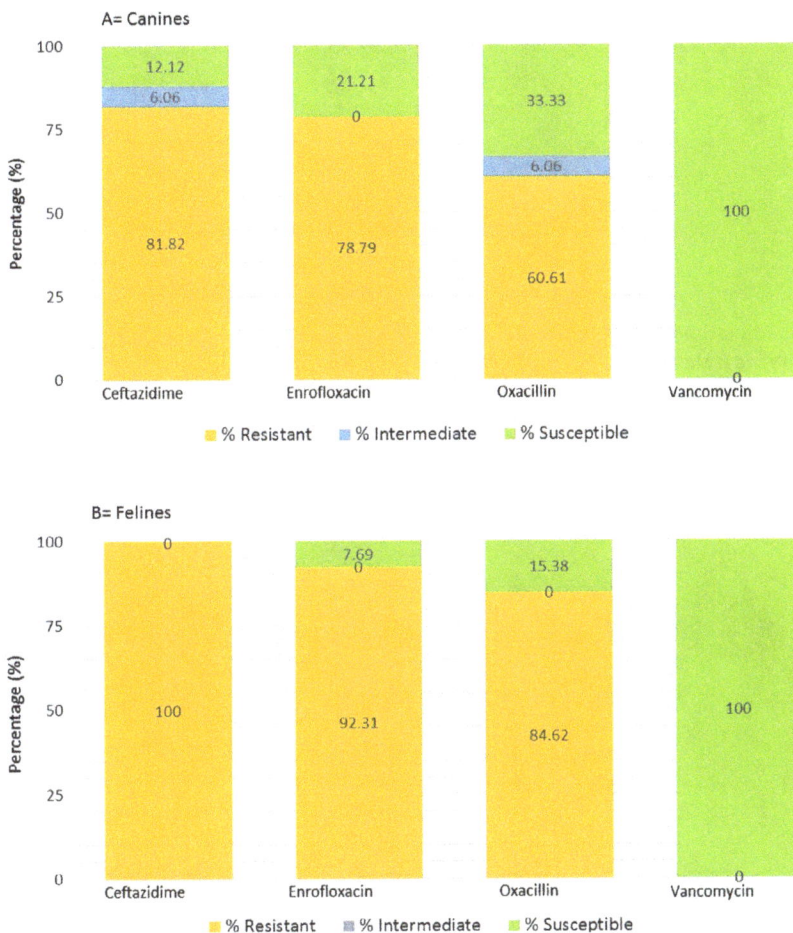

Figure 1. Antibiotic resistance profile for 33 canine (A) and 13 feline (B) MRSA isolates based on level of resistance towards different antibiotics in percentage.

Notes: Colour codes – Golden yellow = resistant, Sky-blue = intermediate Pear-green = Susceptible.

The distribution of the raw data of antibiotic profiling of each MRSA isolate was summarised in boxplots comparing canine and feline groups (Figure 3). From the boxplots, clear differences in the distribution of MRSA isolate of canine vs. feline could be observed and there was more variation amongst canine isolates when compared to feline. From the PCA analysis, the cumulative proportion of variance explained by the first two components was 83%. There was a clear demarcation between the antibiotic profile of the canine and feline isolates (Figure 4). Interestingly, one of the feline isolates indicated an antibiotic profile that was more similar to the canine group.

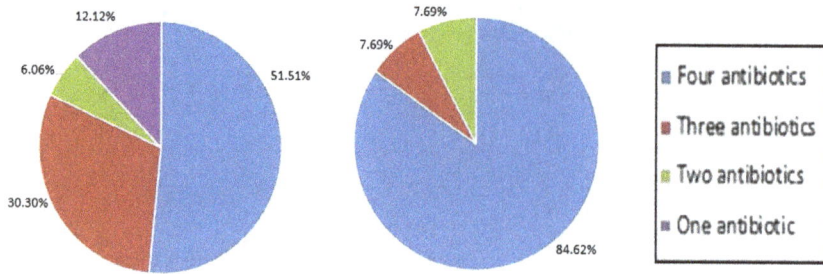

Figure 2. Resistance of the 33 canine (A) and 13 feline (B) MRSA isolates to number of antibiotics.

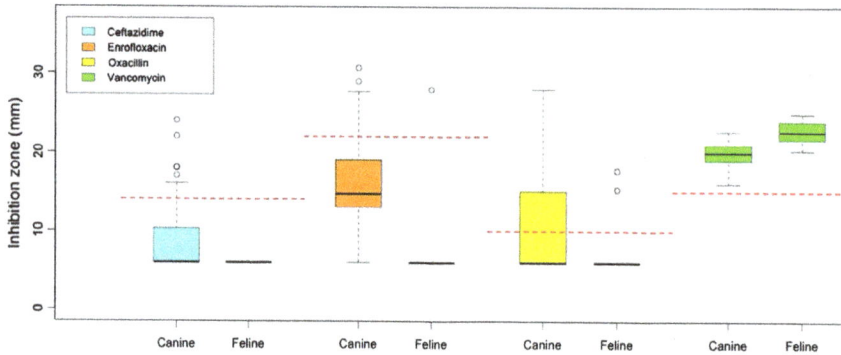

Figure 3. Boxplots of antibiotics resistance profiles, which are measured by inhibition zone (mm), of canine and feline isolates.

Notes: Values in the boxplots below the dashed red lines indicate resistance for the particular antibiotic.

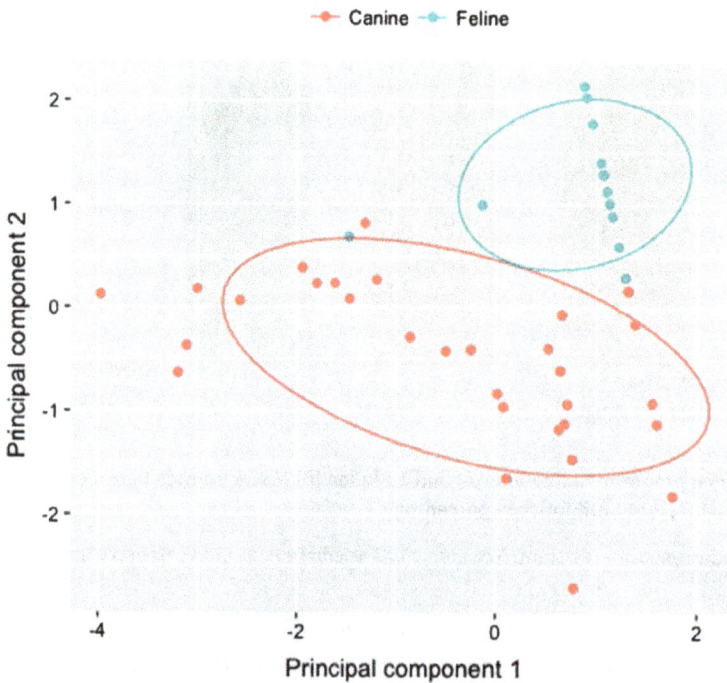

Figure 4. Principal components analysis plot of canine and feline isolates as groups and the four antibiotics, ceftazidime, enrofloxacin, oxacillin and vancomycin, as variables.

Notes: The actual values for the antibiotic variables are derived from the inhibition zone measurements (mm) for all isolates. The antibiotic methicillin is not included because all samples are resistant to it.

4. Discussion

4.1. Handling of shelter animals infected with MRSA

An animal shelter is potentially a major reservoir of MRSA as there is an influx and efflux of large number of animals. The nature of animal shelters that keep large number of animals in a confined space increases the possibility of bacterial transmission. Furthermore, in Malaysia, many animal shelters double up their holding area for excess of stray animals. Generally, these stray animals are immunocompromised and needed to be kept for a limited time before they can be humanely euthanised. This type of management enhances the spreading of pathogens such as MRSA in the healthy animals too. Proper preventive care remains a challenge not only to the resident animals but also to the people working and visiting there. In addition to excretions, bacteria festering in the anterior nares or open wounds of animals can cause boils, impetigo and rashes.

The control and prevention of infectious animals being taken into the shelters is difficult to mitigate. Although many shelters have stringent adoption policies, infectious animals which are in their clinically latent incubation phase may not be identified and housed anyway. MRSA is a commonly found microbe in shelter animals and is becoming more resistant towards available antibiotics (Tacconelli, De Angelis, Cataldo, Pozzi, & Cauda, 2008). Overcrowding and lack of isolation units in shelters cause MRSA to spread to previously uninfected animals thus becoming an important source of community acquired MRSA (CA-MRSA) today (Steneroden, Hill, & Salman, 2011).

4.2. Use of antibiotics to counter MRSA

In our study, 11.66% isolates for canines and 7.70% for felines were identified to be MRSA. Methicillin was initially used to treat penicillin-resistant *S. aureus* and by 1961, there were already reports of the emergence of methicillin resistant strains (Enright et al., 2002). MRSA is an emerging medical and veterinary problem as it can cause complications in immunocompromised shelter animals such as surgical site infection, blood stream infection and pneumonia (Oztoprak et al., 2006). In addition, compared to *S. aureus*, MRSA is not only more difficult to eradicate, but also it is associated with higher mortality due to the higher increased virulence of MRSA, the reduced effectiveness of vancomycin and the delay in administering appropriate antimicrobial therapy (Cosgrove et al., 2003).

Ceftazidime is more effective against MRSA isolates in felines than canines. Ceftazidime is a third generation cephalosporin and generally used for Gram-negative microbes (Tambekar, Dhanorkar, Gulhane, & Dudhane, 2007). β-lactamase enzyme can degrade Ceftazidime, hence MRSA which express this enzyme are resistant to Ceftazidime. It explains the relatively high rates of resistance found in the tested canines and felines. Overall, feline resistance to ceftazidime, enrofloxacin and oxacillin is higher compared to canine resistance. The higher resistance for enrofloxacin in felines may be due to mutation in the DNA gyrase (Lysnyansky et al., 2013). Enrofloxacin, a synthetic antibacterial from the fluoroquinolone class works by preventing the synthesis and supercoiling of the bacteria DNA. This is done by inhibiting DNA gyrase, a type 2 topoisomerase (Ramli, Neoh, Aziz, & Hussin, 2012). Vancomycin is commonly used clinically in MRSA infections by inhibiting the synthesis of peptidoglycan, a major component in the bacterial cell wall by forming complexes with D-Ala-D-ala in various phases of cell wall synthesis (Perkins & Nieto, 1973, 1974). There were no MRSA isolates in this study that were resistant or intermediately resistant to vancomycin. This confirms that there are no vancomycin resistant or intermediate *S. aureus* yet detected in animals in Malaysia (Ng et al., 2011), although they have been reported in many countries worldwide (Tiwari & Sen, 2006).

4.3. Preventive measures against MRSA

It was also noted that most of the isolates were resistant to four of the five antibiotics tested. It is widely accepted that MRSA can colonise healthy animals but do not cause any harm or disease. However, when an animal is immunocompromised or if there is an underlying pathological condition, the organism can cause persistent infections. It is important to note that MRSA cannot be eradicated and thus proper personal hygiene practice remains fundamental. Proper containment measures such as stricter antibiotic policies should be adopted to minimise the spread and development of these resistant strains. Proper aseptic and antisepsis procedures must be employed by all who work with these animals. The use of gloves is especially pertinent when there is a risk of contact with body fluids. Infected animals should be placed in singly housed enclosures.

4.4. Main findings of this study

It is interesting to note that PCA analysis is distinguishing the canine group from the feline group on the basis of antibiotic resistance profile. It suggests that there is no widespread cross-transmission of MRSA isolates between the two different animal specimens. Separate enclosures of the tested animals may be a reason for this finding. Another possible explanation for this observation is that different MRSA isolates may adapt to specific hosts and if cross-transmission occurs, the new host is not suitable for the growth of the bacteria. It is noteworthy that one feline isolate is similar to the canine group in the PCA analysis for the antibiotic resistance profile, suggesting the possibility of feline-canine transmission. However, since evidence of cross-transmission is based on a single isolate, it is inconclusive to measure the rate of cross-transfer between species. Future work should include testing of larger sample sizes from both animal species to better elucidate the possibility of cross-transmission.

5. Conclusion

The close contact between humans and animals have led to the zoonotic spread of community acquired MRSA (CA-MRSA). From the study, it is clear that MRSA was present in both animals. However, the pattern of colonisation may vary in both animals. There were slightly higher MRSA isolates detected in canines compared to the felines. A difference in host enterotype may result in different antibiotic profiling of the isolates. Majority of the isolates were resistant to four antibiotics of the five tested. Furthermore, MRSA colonisation in animals is a cause of concern as they have become established in various animal species other than canines and felines such as pigs, horses and cattle. In addition to careful administration of antibiotics, newer drugs must be developed and more effective molecular methods should be employed to detect these microorganisms earlier in animals and humans.

Funding

The authors are grateful to the Malaysian Medical Association Foundation for funding this work and to Perdana University as well as Universiti Putra Malaysia for providing research facilities.

Competing Interests

The authors declare no competing interest.

Author details

Rajwin Raja Kanagarajah[1]
E-mail: rajwin.raja@perdanauniversity.edu.my
David Charles Weerasingam Lee[1]
E-mail: davidcharles@perdanauniversity.edu.my
Daniel Zhi Fung Lee[1]
E-mail: daniellee@perdanauniversity.edu.my
Khatijah Yusoff[2]
E-mail: kyusoff@gmail.com
Sharmini Julita Paramasivam[3]
E-mail: sharminijp@gmail.com
Wai Yee Low[4,5]
E-mail: lloydlow@perdanauniversity.edu.my
Kamalan Jeevaratnam[3]
E-mail: kamalan@perdanauniversity.edu.my
Swee Hua Erin Lim[1,4,6]
E-mail: erinlimsh@gmail.com
ORCID ID: http://orcid.org/0000-0001-5177-0257
[1] Perdana University – Royal College of Surgeons in Ireland, Perdana University Hall D Level 1, MAEPS Building, MARDI Complex, Jalan MAEPS Perdana, Universiti Putra Malaysia, Serdang 43400, Selangor, Malaysia.
[2] Faculty of Biotechnology and Biomolecular Sciences, Department of Microbiology, Universiti Putra Malaysia, Serdang 43400, Selangor, Malaysia.
[3] Faculty of Health and Medical Sciences, VSM Building, University of Surrey, Guildford, Surrey GU2 7AL, UK.
[4] Perdana University – Centre of Bioinformatics, Perdana University Hall D Level 1, MAEPS Building, MARDI Complex, Jalan MAEPS Perdana, Universiti Putra Malaysia, Serdang 43400, Selangor, Malaysia.
[5] The Davies Research Centre, School of Animal and Veterinary Sciences, University of Adelaide, Roseworthy, SA 5371, Australia.
[6] Abu Dhabi Women's College, Higher Colleges of Technology, Abu Dhabi 41012, United Arab Emirates.

References

Aldridge, S., Parascandola, J., & Sturchio, J. (1999). *The discovery and development of penicillin 1928–1945*. Washington, DC: American Chemical Society.

Attili, A., Preziusoa, S., Ngu, V. N., Cantalamessaa, A., Moriconia, M., & Cuteri, V. (2016). Clinical evaluation of the use of enrofloxacin against *Staphylococcus aureus* clinical mastitis in sheep. *Small Ruminant Research, 136*, 72–77. https://doi.org/10.1016/j.smallrumres.2016.01.004

Beutin, L. (1999). *Escherichia coli* as a pathogen in dogs and cats. *Veterinary Research, 30*(2–3), 285–298.

Carlet, J. (2012). The gut is the epicentre of antibiotic resistance. *Antimicrobial Resistance and Infection Control, 1*(1), 39. https://doi.org/10.1186/2047-2994-1-39

Cefai, C., Ashurst, S., & Owens, C. (1994). Human carriage of methicillin-resistant *Staphylococcus aureus* linked with pet dog. *The Lancet, 344*(8921), 539–540. https://doi.org/10.1016/S0140-6736(94)91926-7

Chaban, B., Ngeleka, M., & Hill, J. E. (2010). Detection and quantification of 14 Campylobacter species in pet dogs reveals an increase in species richness in feces of diarrheic animals. *BMC Microbiology, 10*, 73. https://doi.org/10.1186/1471-2180-10-73

Cockerill, F., Wikler, M., Alder, J., Dudley, M., Eliopoulos, G., Ferraro, M., ... Zimmer, B. (2012). *Performance standards for antimicrobial disk susceptibility tests. Approved standard* (Vol. 32, 11th ed., p. 58). Wayne, PA: CLSI (Clinical and Laboratory Standards Institute).

Cosgrove, S., Sakoulas, G., Perencevich, E. N., Schwaber, M. J., Karchmer, A. W., & Carmeli, Y. (2003). Comparison of mortality associated with methicillin-resistant and methicillin-susceptible *Staphylococcus aureus* Bacteremia: A meta-analysis. *Clinical Infectious Diseases, 36*, 53–59. https://doi.org/10.1086/cid.2003.36.issue-1

El-Tras, W. F., Holt, H. R., & Tayel, A. A. (2011). Risk of Toxocara canis eggs in stray and domestic dog hair in Egypt. *Veterinary Parasitology, 178*(3–4), 319–323. https://doi.org/10.1016/j.vetpar.2010.12.051

Enright, M. C., Robinson, D. A., Randle, G., Feil, E. J., Grundmann, H., & Spratt, B. G. (2002). The evolutionary history of methicillin-resistant *Staphylococcus aureus* (MRSA). *Proceedings of the National Academy of Sciences, 99*(11), 7687–7692. https://doi.org/10.1073/pnas.122108599

Gardam, M. A. (2000). Is methicillin-resistant *Staphylococcus aureus* an emerging community pathogen? A review of the literature. *Canadian Journal of Infectious Diseases, 11*(4), 202–211. https://doi.org/10.1155/2000/424359

Hiramatsu, K., Cui, L., Kuroda, M., & Ito, T. (2001). The emergence and evolution of methicillin-resistant *Staphylococcus aureus*. *Trends in Microbiology, 9*(10), 486–493. https://doi.org/10.1016/S0966-842X(01)02175-8

Hiramatsu, K., Katayama, Y., Yuzawa, H., & Ito, T. (2002). Molecular genetics of methicillin-resistant *Staphylococcus aureus*. *International Journal of Medical Microbiology, 292*(2), 67–74. https://doi.org/10.1078/1438-4221-00192

Klevens, R. M., Morrison, M. A., Nadle, J., Petit, S., Gershman, K., Ray, S., ... Fridkin, S. K. (2007). Invasive methicillin-resistant *Staphylococcus aureus* infections in the United States. *JAMA, 298*(15), 1763–1771. https://doi.org/10.1001/jama.298.15.1763

Lefebvre, S. L., Reid-Smith, R., Boerlin, P., & Weese, J. S. (2008). Evaluation of the risks of shedding salmonellae and other potential pathogens by therapy dogs fed raw diets in Ontario and Alberta. *Zoonoses Public Health, 55*(8–10), 470–480. https://doi.org/10.1111/jvb.2008.55.issue-8-10

Livermore, D. M. (2003). Bacterial resistance: Origins, epidemiology, and impact. *Clinical Infectious Diseases, 36*(Suppl 1), S11–S23. https://doi.org/10.1086/344654

Loeffler, A., Boag, A. K., Sung, J., Lindsay, J. A., Guardabassi, L., Dalsgaard, A., ... Lloyd, D. H. (2005). Prevalence of methicillin-resistant *Staphylococcus aureus* among staff and pets in a small animal referral hospital in the UK. *Journal of Antimicrobial Chemotherapy, 56*(4), 692–697. https://doi.org/10.1093/jac/dki312

Lysnyansky, I., Gerchman, I., Mikula, I., Gobbo, F., Catania, S., & Levisohn, S. (2013). Molecular characterization of acquired enrofloxacin resistance in *Mycoplasma synoviae* field isolates. *Antimicrobial Agents and Chemotherapy, 57*(7), 3072–3077. https://doi.org/10.1128/AAC.00203-13

National Committee for Clinical Laboratory Standard. (2001). *Performance standards for antimicrobial disc susceptibility tests (M2-A7 and M7-A5)* (6th ed., Vol. 21, No. 1). 11th informational supplement 2001 (M100-S11), 1997.

Ng, S. T., Lim, C. Y., Tan, C. S., Karim, A. A., Haron, H., Ahmad, N. S., & Murugaiyah, V. (2011). Emergence of vancomycin-resistant *Staphylococcus aureus* (VRSA). *WebMed Central*, 1–11.

Oztoprak, N., Cevik, M. A., Akinci, E., Korkmaz, M., Erbay, A., Eren, S. S., ... Bodur, H. (2006). Risk factors for ICU-acquired methicillin-resistant *Staphylococcus aureus* infections. *American Journal of Infection Control, 34*(1), 1–5. https://doi.org/10.1016/j.ajic.2005.07.005

Perkins, H. R., & Nieto, M. A. N. U. E. L. (1973). The significance of D-alanyl-D-alanine termini in the biosynthesis of bacterial cell walls and the action of penicillin, vancomycin and ristocetin. *Pure and Applied Chemistry, 35*(4), 371–381.

Perkins, H. R., & Nieto, M. (1974). The chemical basis for the action of the vancomycin group of antibiotics. *Annals of the New York Academy of Sciences, 235*(1), 348–363. https://doi.org/10.1111/nyas.1974.235.issue-1

Ramli, S. R., Neoh, H. M., Aziz, M. N., & Hussin, S. (2012). Screening and detection of heterogenous vancomycin intermediate *Staphylococcus aureus* in Hospital Kuala Lumpur Malaysia, using the glycopeptide resistance detection etest and population analysis profiling. *Infectious Disease Reports, 4*(1), 20. https://doi.org/10.4081/idr.2012.3836

Rich, M., & Roberts, L. (2004). Methicillin-resistant *Staphylococcus aureus* isolates from companion animals. *Veterinary Record, 45*, 591–597.

Steneroden, K. K., Hill, A. E., & Salman, M. D. (2011). Zoonotic disease awareness in animal shelter workers and volunteers and the effect of training. *Zoonoses Public Health, 58*(7), 449–453. https://doi.org/10.1111/jvb.2011.58.issue-7

Strommenger, B., Kehrenberg, C., Kettlitz, C., Cuny, C., Verspohl, J., Witte, W., & Schwarz, S. (2006). Molecular characterization of methicillin-resistant *Staphylococcus aureus* strains from pet animals and their relationship to human isolates. *Journal of Antimicrobial Chemotherapy, 57*(3), 461–465. https://doi.org/10.1093/jac/dki471

Tacconelli, E., De Angelis, G., Cataldo, M., Pozzi, E., & Cauda, R. (2008). Does antibiotic exposure increase the risk of methicillin-resistant *Staphylococcus aureus* (MRSA) isolation? A systematic review and meta-analysis. *Journal of Antimicrobial Chemotherapy, 61*(1), 26–38.

Tambekar, D., Dhanorkar, D., Gulhane, S., & Dudhane, M. (2007). Prevalence and antimicrobial susceptibility pattern of methicillin resistant *Staphylococcus aureus* from healthcare and community associated sources. *African Journal of Infectious Disease, 1*(1), 52–56.

Team R.C. (2012). *R: A language and environment for statistical computing*. Vienna: Author.

Tiwari, H. K., & Sen, M. R. (2006). Emergence of vancomycin resistant Staphylococcus aureus (VRSA) from a tertiary care hospital from northern part of India. *BMC Infectious Diseases, 6*, 520. https://doi.org/10.1186/1471-2334-6-156

Weese, J., Dick, H., Wiley, B. M., Mcgeer, A., Kreisweirth, B. N., Innis, B., & Low, D. E. (2006). Suspected transmission of methicillin-resistant *Staphylococcus aureus* between domestic pets and humans in veterinary clinics and in the household. *Veterinary Microbiology, 115*(1–3), 148–155. https://doi.org/10.1016/j.vetmic.2006.01.004

13

Mutual modulation of femarelle and vitamin D analog activities in human derived female cultured osteoblasts

D. Somjen[1]*, S. Katzburg[1], O. Sharon[1], E. Knoll[1], D. Hendel[1] and G.H. Posner[2]

*Corresponding author: D. Somjen, The Sackler Faculty of Medicine, Institute of Endocrinology, Metabolism and Hypertension and Bone Diseases Unit, Tel-Aviv Sourasky Medical Center, Tel-Aviv University, 6 Weizman Street, Tel Aviv 64239, Israel
E-mail: dsomjen10@gmail.com
Reviewing editor: Steve Winder, University of Sheffield, UK
Additional information is available at the end of the article

Abstract: *Introduction:* Pre-, post-menopausal, and male human osteoblasts (hObs) express estradiol-17β (E$_2$) receptors α and β (ERα, ERβ), vitamin D receptor (VDR), and 25-hydroxy vitamin D$_3$ 1-α hydroxylase (1OHase) and produce 1,25 (OH)$_2$D$_3$ (1,25D). Pre-treatment with JKF (JKF 1624F$_{2-2}$) up-regulated estrogenic responsiveness, via modulation of ERs expression and E$_2$ binding. These estrogens induce VDR and 1OHase expression and 1,25D production. *Purpose:* Compared the effects of femarelle (F) to daidzein (D) and E$_2$ by themselves and their modulation by JKF. *Methods:* hObs derived from human bones at different ages and sex were cultured, treated with different hormones and analyzed for the different parameters. *Results:* (1) F, D and E$_2$ stimulated ^{3}[H] thymidine incorporation (DNA) and creatine kinase specific activity (CK) in female but not in male hObs. The responses to E$_2$ and to D unlike to F were up-regulated by JKF. (2) All hormones increased ERα and decreased ERβ mRNA in all hObs. (3) JKF modulated in different ways the expressions of ERα and ERβ mRNA in all hObs. (4) JKF did not significantly modulate the expressions of ERα and ERβ mRNA in all hObs. (5) JKF increased intracellular competitive binding of E$_2$ by all hormones only in female hObs. (6) Pre-treatment with all hormones increased VDR and 1OHase expression and 1,25D formation in pre- and post-, but only JKF modulated it in male hObs. *Conclusions:* F, D and E$_2$ increase different parameters in hObs. However, pre-treatment with JKF modulates the effect of E$_2$ and D but not of F. F reciprocally modulates mRNA expression and activity of 1OHase

ABOUT THE AUTHORS

The authors of the paper are members of the Tel-Aviv medical center and other institutions as reported. We are interested in bone and vascular endocrinology and especially in synthetic estrogens and vitamin D analogs as reported in here and in other previous studies and papers. We have shown their importance and their use with much less side effects as other ones, we believe in their use in human health and hope to promote it, without any commercial benefits to us.

PUBLIC INTEREST STATEMENT

In the last years a lot of interest is in the use of phytoestrogens and their derivatives in human biology and medicine. Since there are a lot of health problems in the use of either native estrogen or native phytoestrogens, it is important to use new compounds. We therefore decided to analyze our new compound femarelle which is a synthetic new phytoestrogens in our experiments. We used it before in animal models showing its importance, and now in human bone cells. We can see how unique it is in its estrogenic effects and its activity which is not affected by other hormones as the estrogen itself.

We therefore believe in its special importance for bone biology and other effects as shown before.

which in turn up-regulate ERs expression and activity. These finding may contribute to F's beneficial role in treatment of post-menopausal bone loss even in vitamin D deficiency.

Subjects: Environment & Agriculture; Bioscience; Environmental Studies & Management

Keywords: human osteoblasts; femarelle; DNA; CK; phytoestrogens; JKF; ER; VDR

1. Introduction

Phytoestrogens are heterogeneous group of plant-derived compounds, some of which are selective estrogen receptor modulators (SERMs). They are all polyphenolic compounds with structural similarities to natural and synthetic estrogens and they bind to estrogen receptors (ERs) with much lower affinity compared to E_2 (Miksicek, 1994). Phytoestrogens have estrogenic activity in bone and in the cardiovascular system but have anti-estrogenic activity in the breast and in the uterus (Brzezinski & Debi, 1999). They have been proposed to prevent bone resorption and to promote bone formation and increased bone density (Setchell & Lydeking-Olsen, 2003; Tham, Gardner, & Haskell, 1998; Yoles et al., 2003). Femarelle (F), a chemical derivative of the phytoestrogen daidzein (D), which is a standardized extract derived from soybeans was shown to increase bone mineral density in post-menopausal women (Yoles et al., 2004) and to relieve vasomotor symptoms with no effect on sex hormone levels or endometrial thickness (Somjen & Yoles, 2003a). It has also properties as selective estrogen receptor modulator (SERM) as was shown previously both in rat *in vivo* and in different cultured cells *in vitro* (Eriksen & Glerup, 2002; Somjen, Katzburg, Lieberherr, Hendel, & Yoles, 2006; Somjen & Yoles, 2003b). It has an estrogen-like activity and activates human-derived female-cultured bone cells (hObs) which express receptors for E_2 both ERα and ERβ and for vitamin D (VDR). Estrogens and Vitamin D metabolites and their analogs regulate cell proliferation (DNA) and energy metabolism through modulation of the specific activity of creatine kinase (CK). Pre-treatment with vitamin D less-calcemic analog: JKF 1624F_{2-2} (JKF) up-regulated responsiveness to E_2 and to different estrogens, via modulation of ERs mRNA expression. Estrogens, in turn, induce VDR and 25-hydroxy vitamin D_3 1-α hydroxylase (1OHase) expression and 1,25(OH)$_2D_3$ (1,25D) synthesis.

Optimal bone growth and prevention of osteoporosis in post-menopausal women, requires adequate concentrations of vitamin D_3 as well as estrogens (Somjen, Waisman, Weisman, & Kaye, 2000). We have studied previously the interaction between the vitamin D_3 metabolite; 1,25D and its less-calcemic analogs with estrogens in a rat model (Reiss & Kaye, 1981; Somjen, Waisman, Lee, Posner, & Kaye, 2001) using the increase in CK as a response marker for hormonal treatment in cells containing biologically active ERs (Sömjen, Harell, Jaccard, Weisman, & Kaye, 1990; Somjen, Weisman, Harell, Berger, & Kaye, 1989).

Moreover, pre-treatment with 1,25D up-regulated sex-specific responsiveness and sensitivity to E_2 and to several SERMs in osteoblast-like cell lines and rat bone, as measured by both the stimulation of CK and the increase in DNA synthesis (Fournier, Haring, Kaye, & Somjen, 1996; Reiss & Kaye, 1981; Sömjen, Waisman, Weisman, & Kaye, 1998; Sömjen et al., 1990). This mutual interaction between E_2 4 and vitamin D was also manifested by an increase in the expression of ERs after treatment with 1,25D (Kanis et al., 1979). However, the use of vitamin D_3 metabolites is restricted by their hypercalcemic effects (Posner et al., 1998). We reported that multiple treatments with "less-calcemic" analogs of vitamin D, particularly JKF (Posner, 2004), stimulated osteoblast-like cells (Katzburg et al., 2005; Posner, 2004) and pre-treatment of skeletal-derived cells with these analogs, up-regulated both responsiveness and sensitivity to E_2 (Fournier et al., 1996; Katzburg et al., 1999, 2005).

The present study was undertaken to measure age and sex specificities in the responses of cultured human female and male osteoblastss (hObs) to E_2, to the phytoestrogen daidzein; D and to the synthethic phytoestrogen femarelle; F and their modulation by JKF. We sought to correlate these, VDR as β and ERαresponse changes with the expression of mRNA for ER well as 1OHase expression and activity resulting in the synthesis of the active metabolite 1,25D (Somjen et al., 2005), as well as the intracellular binding of E_2.

2. Materials and methods

2.1. Reagents
Estradiol-17β (E_2), Daidzein (D) and creatine kinase (CK) assay kit were purchased from Sigma Chemicals Co. (St. Louis, MO). Femarelle (F) was purchased from Se-Cure Pharmaceuticals (Dalton, Israel). JKF 1624F$_{2-2}$ (JKF) was synthesized and donated by Dr G.H. Posner et al. (1998). All other reagents were of analytical grade.

2.2. Cell Cultures
Human female osteoblasts (hObs) from pre- (age up to 45 years) and post-menopausal (age from 55 years) women or males (age 45–65 years) were prepared from bone explants of healthy individuals which were not taking any medicines, by a non-enzymatic method as described previously (Katzburg et al., 1999). Each culture was prepared from each donor separately. Shortly: samples of the trabecular surface of the iliac crest or long bones were cut into 1 mm³ pieces and repeatedly washed with phosphate buffered saline (PBS) to remove blood components. The explants were incubated in DMEM medium without calcium (to avoid fibroblastic growth) containing 10% fetal calf serum (FCS) and antibiotics. First passage cells were seeded at a density of 3 × 105 cells/35 mm tissue culture dish, in phenol red- free DMEM with 10% charcoal stripped FCS, and incubated at 37oC in 5% CO_2.

2.3. Hormonal treatments
Cells were treated either for 24 h with the different hormones or daily with vehicle (0.01% ethanol in medium) or JKF at 1nM final concentration (Reiss & Kaye, 1981), for 3 days, starting on day 1 after seeding and on day 4 after seeding, the cultures were then treated for additional 24 h with 200 ng/ml F, 300nM D or 300nM E_2, followed by harvesting for CK assay or DNA synthesis. In some experiments, cells were pre-treated with the different estrogenic compounds in similar pattern and the changes in the vitamin D system was assayed (Somjen et al., 2005).

2.4. Creatine kinase extraction and assay in hObs
Cells were treated for 24 h with the various agents as specified, scraped off and homogenized by freezing and thawing three times in an extraction buffer, and CK was extracted and assayed as previously described (Reiss & Kaye, 1981; Somjen et al., 1989, 2001). Protein was determined by Coomasie blue dye binding, using bovine serum albumin (BSA) as the standard.

2.5. DNA synthesis assay in hobs
Cells were grown until sub-confluence and then treated with various hormones or agents as indicated. Twenty-two hours following the exposure to these agents, ³[H]thymidine was added for 2 h and the ³[H]thymidine incorporation into DNA was determined (Katzburg et al., 1999).

2.6. Determination of ERα and ERβ mRNA by real time PCR in hObs
RNA was extracted from cultured human bone cells, shown previously (Eriksen & Glerup, 2002; Somjen et al., 2005) and subjected to reverse transcription as previously described (Katzburg et al., 1999; Somjen et al., 2005). ER standard controls and compared to RNAse P as internal control for mRNA.

2.7. Competitive binding assay for intracellular estrogenic binding sites in hObs
Cells with and without pre-treatment with JKF, were incubated for 60 min at 37oC with ³[H]E$_2$ with and without excess of unlabelled compounds as described. Binding was terminated by four successive washes with ice-cold binding medium, and cellular content of ³[H]E$_2$ was measured in a scintillation counter (Katzburg et al., 1999; Somjen et al., 2005).

2.8. Determination of vitamin D receptor (VDR) and 25-hydroxy vitamin D_3 1-hydroxylase (1OHase) mRNA by real time PCR in hObs

RNA was extracted from cultured cells and subjected to reverse transcription as previously described (Katzburg et al., 1999; Somjen et al., 2005). RNAse P expression served as an internal control.

2.9. Assay of 1OHase activity in hObs

1OHase activity was measured by the level of 1,25D generated in hObs within 60 min after the addition of 25(OH)D_3 (200 ng/ml) to culture, using the 1,25D ^{125}I RIA kit from Dia Sorin, Mn, USA (Somjen et al., 2005). Protein of the cellular layer was assayed by the Bradford method.

2.10. Statistical analysis

Differences between the mean values of experimental and control groups were evaluated by analysis of variance (ANOVA); p values less than 0.05 were considered significant.

3. Results

3.1. The effects of F, D, or E_2 with and without JKF on stimulation of DNA synthesis in hObs

All cells analyzed pre-, post- and male hObs have basal DNA synthesis with lower activity in post-hObs (4,250 + 300 in pre- vs. 3,000 + 375 in post- and 4,450 + 400 dpm/well in male hObs) (Figure 1). Addition of different hormones except JKF stimulated significantly [3][H] thymidine incorporation into all female but not in male hObs, with higher responses in pre-hObs compared to post-hObs 305 + 11% vs. 155 + 10% by E_2, 167 + 11% vs. 150 + 14% by D and 173 + 14% vs. 159 + 12% by F (Figure 2). Pre-treatment with JKF did not change the stimulation of DNA synthesis by F 159 + 5% vs. 141 + 12% in pre- and 159 + 4% vs. 141 + 5% in post-hObs, but did up-regulate the stimulation by E_2 (165 + 6% vs. 188 + 12% in pre- and 135 + 4% vs. 176 + 12% in post-hObs) and by D (175 + 12% vs. 200 + 13% in pre- and 141 + 12% vs. 176 + 13% in post-hObs) (Figure 3). No significant changes in the age dependent responsiveness to F unlike to E_2 or to D were observed (Figures 2 and 3).

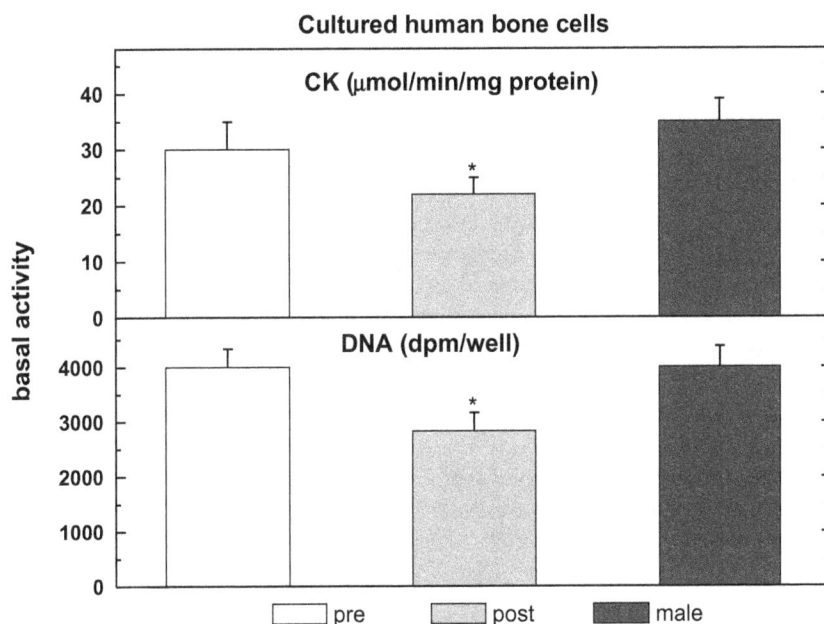

Figure 1. Basal levels of DNA synthesis and CK specific activity in pre-menopausal (pre), in post-menopausal human osteoblasts (post) and male human cultured bone cells.

Notes: Bone cells were cultured, treated, and analyzed for DNA synthesis and CK specific activity as described in Materials. Results are means ± SEM for triplicate cultures from 5 humans/group. Experimental means were compared to control (vehicle alone) means: *p < 0.05, **p < 0.01.

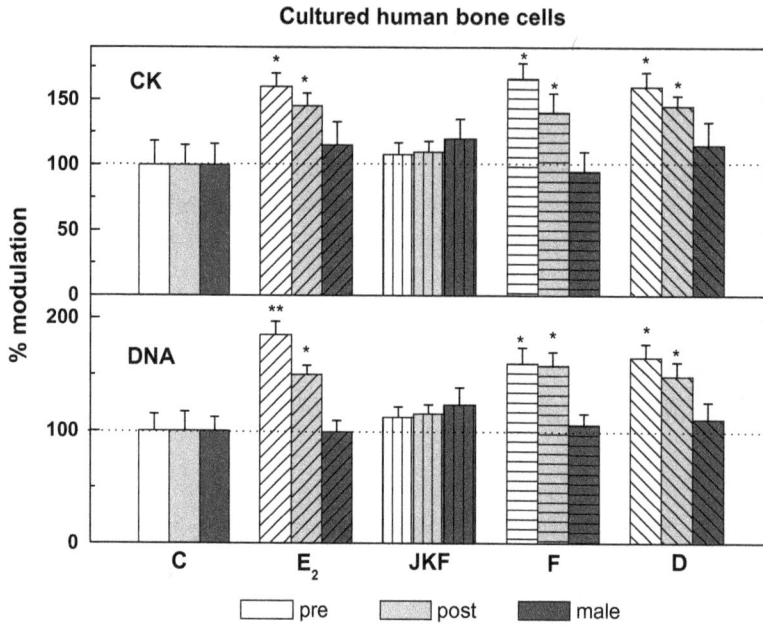

Figure 2. Stimulation of DNA synthesis or CK specific activity by F, D or E$_2$ as well as JKF in hObs from pre- and post-menopausal women or male.

Notes: Cells were obtained, cultured, treated, and assayed as described in Materials and Methods. They were treated for 24 h with vehicle (C), 200 ng/ml F, 300nM D or 30nM E$_2$ as well as 1nM JKF. Results are means ± SEM for cultures obtained from 3–5 women or male. Control means were 6,286 ± 675 and 5,824 ± 460 and 5,964 ± 575 dpm/well, for pre- and post-menopausal women and for male respectively. Experimental means compared to control means: *$p < 0.01$. For CK control means were 28.6 ± 6.5 and 24.6 ± 4.0 and 26.2 ± 5.7 nmol/min/mg protein, for pre- and post-menopausal women and for male respectively. Experimental means compared to control means: *$p < 0.01$.

3.2. The effects of F, D or E$_2$ with and without JKF on stimulation of creatine kinase (CK) specific activity in hObs

All cells analyzed pre-, post- and male hObs have basal CK specific activity with lower activity in post-hObs (32.5 + 0.5 in pre- vs. 24.5 + 3.0 in post- and 32.5 + 0.5 μmol/min/mg protein in male hObs) (Figure 1). Addition of the hormones except JKF stimulated significantly CK in female hObs at different ages but not in male, with higher responses in pre- compared to post-hObs (155 + 9% vs. 141 + 9% by E$_2$, 155 + 12% vs. 141 + 9% by D and 161 + 10% vs. 136 + 11% by F) (Figure 2). Pre-treatment with JKF did not change the stimulation of CK by F (150 + 6% vs. 138 + 10% in pre- and 150 + 5% vs. 144 + 12% in post-hObs) but did up-regulate the stimulation by E$_2$ (150 + 5% vs. 169 + 11% in pre- and 144 + 5% vs. 200 + 25% in post-hObs) and by D (156 + 11% vs. 175 + 12% in pre- and 138 + 5% vs. 175 + 12% in post-hObs) (Figure 3). No significant changes in the age dependent responsiveness to F unlike to E$_2$ or to D were observed (Figures 2 and 3).

3.3. The expression and its modulation by JKF, F, D or E$_2$ of ERα and ERβ in hObs

All cells analyzed; pre-, post- and male hObs expressed mRNA for both ERα and ERβ as measured by real time PCR, corrected for RNAse P mRNA. In pre-ERα expression was 0.094 + 0.007 vs. 0.076 + 0.006 in post- and 0.100 + 0.008 $2^{-\Delta CT}$ in male hObs, whereas ERβ in pre- 0.00085 + 0.00012 vs. 0.00098 + 0.00011 in post- and 0.00094 + 0.00008 $2^{-\Delta CT}$ in male hObs. The ratio of ERα to ERβ was 121:1 in pre-, 78:1 in post- and 105:1 in male hObs (Figure 4). Pre-treatment with the different hormones as well as JKF modulated the expression of ERα and ERβ to different extents in both female age groups and to less extent in male hObs (in ERα 117 + 20% in pre- and 167 + 10% in post-hObs and 117 + 20% in male hObs and in ERβ 50 + 9% in pre-, 49 + 15% in post- and 67 + 17% in male hObs) (Figure 5). On the other hand E$_2$ increased ERα by 156 + 11 in pre- but not in post- and in male hObs (67 + 15% and 99 + 12% respectively). D increased ERα (by 167 + 12% in pre-, by 133 + 5% in post- and 101 + 21% in male hObs). F increased ERα by 178 + 22% in pre-, but not in post- (78 + 5%)

Cultured human female bone cells

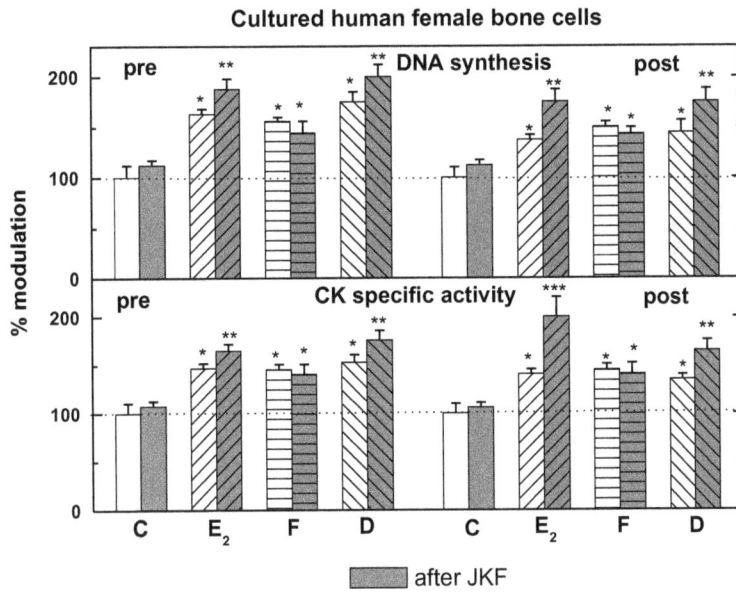

Figure 3. **Stimulation of DNA synthesis or CK specific activity by F, D, or E$_2$ in hObs from pre- and post-menopausal women after pre-treatment with JKF (light gray bars).**

Notes: Cells were obtained, cultured, treated, and assayed as described in Materials and Methods. Cells were treated for 3 days by daily addition of 1nM JKF and then treated for 24 h with 200 ng/ml F, 300nM D or 30nM E$_2$. Results are means cultures obtained from 5 women/group for JKF treated cultures and 10 women for control cultures. Control means were 6,286 ± 675 and 5,824 ±4 60 dpm/well, for pre- and post-menopausal women, respectively. Experimental means compared to control means: $*p < 0.01$; experimental means with JKF compared to experimental means with control: $\#p < 0.01$; experimental means with JKF compared to experimental means with control: $\#p < 0.01$.

cultured human bone cells

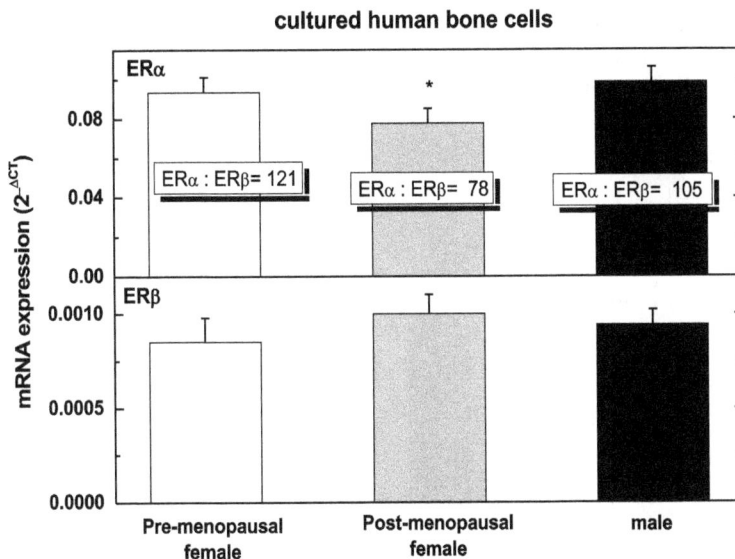

Figure 4. **The expression of estrogen receptors ERα and ERβ in pre-menopausal human osteoblasts (pre), in post-menopausal human osteoblasts (post) and human males (male).**

Notes: Bone cells were cultured, treated, and extracts prepared for mRNA expression as described in Materials and Methods. Results are means ± SEM for cultures from 5 humans/group.

and in male (106 + 10%) hObs. D increased ERβ by 267 + 30% in pre-, by 167 + 18% in post- and by 100 + 10% in male hObs. E$_2$ did not increase ERβ in pre- (50 + 15%), or in post- (33 + 5%) and in male (67 + 10%) hObs, JKF increased ERα in pre- and post-hObs and decreased ERβ in pre- and post-hObs with no effect on male hObs (Figure 5).

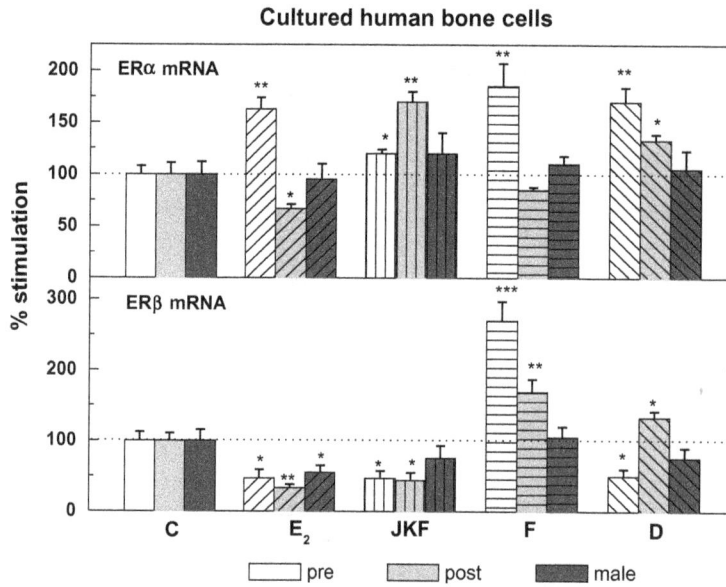

Figure 5. Modulation by pre-treatment with E$_2$, D, F or JKF on the expression of ER mRNA in hObs obtained from pre- and post-menopausal women.

Notes: Bone cells were obtained, cultured, treated daily for 3 days with 20 ng/ml F, 30nM D or 3nME$_2$ or 1nM JKF, and extracts prepared for analysis as described in Materials and Methods. Results are SEM for triplicate cultures from 5 donors for each group. Means of hormonal treated compared to means of vehicle-treated cells: *$p < 0.01$.

3.4. The intracellular binding of E$_2$ and its modulation by JKF, F, D or E$_2$ in hObs

Both pre- and post-menopausal female but not male-derived hObs demonstrated E$_2$ specific binding of ^{3}[H]E$_2$ (Figure 6), presumably predominantly nuclear under these conditions (37° for 60 min). All compounds tested for competition with ^{3}[H]E$_2$ i.e. E$_2$, F and D, but not JKF showed significant binding in both age groups but not in male hObs (Figure 6). Pre-treatment of hObs with JKF increased only slightly non significantly the specific binding of ^{3}[H]E$_2$ in female cells from both age groups by all hormones studied (Figure 7).

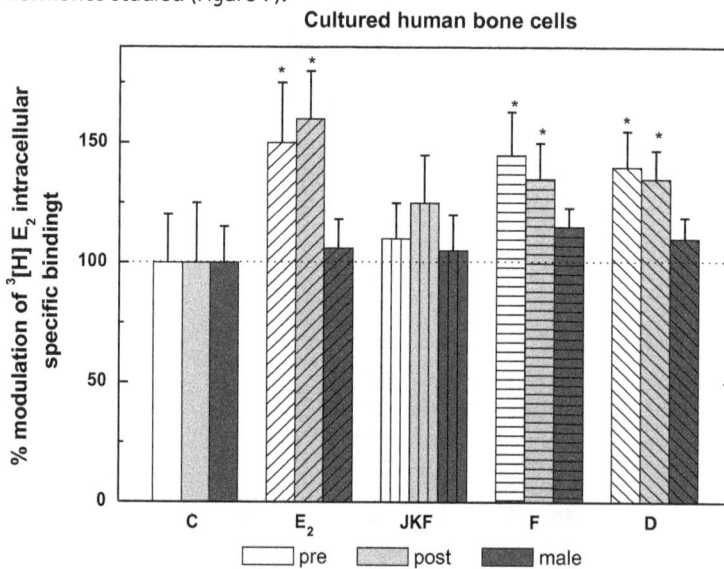

Figure 6. The effect of E$_2$, D, F or JKF on intracellular binding of E$_2$ in pre- or post-menopausal female hObs, as measured by competition of the binding of ^{3} [H] E$_2$. The effects of treatment with JKF (3 days with 1nM) or E$_2$, D and F on intracellular binding of E$_2$ in pre- or post-menopausal female hObs, as measured by competition of the binding of ^{3} [H] E$_2$.

Notes: Results are means of 4 donors for each group, and are expressed as % modulation of the binding in hormone treated cells compared to untreated cells. **$p < 0.01$.

Cultured human female bone cells

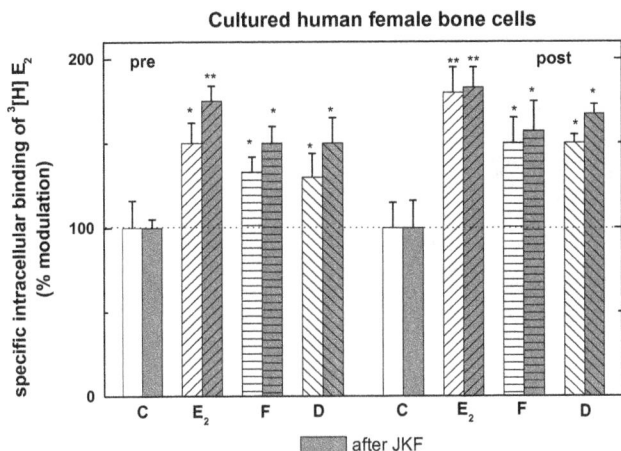

Figure 7. Modulation of the specific intracellular binding of ^{3} [H] E$_2$ binding in pre-menopausal human osteoblasts (pre-), in post-menopausal human osteoblasts (post-) and males (male) by 30nM E$_2$, 3,000 nM D or 200 ng/ml F after three daily treatments with 1nM JKF.

Notes: Bone cells were cultured, treated, and extracts prepared for binding activity as described in Materials and Methods. Results are means of 5 donors for each group and are means of experimental compared to control (vehicle alone) means: *$p < 0.01$.

3.5. The modulation by JKF, F, D or E$_2$ of VDR and 1OHase mRNA expression in hObs

Female-derived bone cells from both ages and male hObs expressed mRNA for VDR and 1OHase as measured by real time PCR, corrected for RNAse P mRNA (VDR 0.8 + 0.033 in pre-, 0.41 + 0.067 in post- and 0.50 + 0.133 $2^{-\Delta CT}$ in male hObs, 1OHase 0.188 + 0.025 in pre-, 0.25 + 0.02 in post- and 0.075 + 0.002 $2^{-\Delta CT}$ in male hObs) (Figure 8). Treatment with all hormones tested increased the expression of 1OHase and VDR by about 35–165%, in both age groups but only JKF increased the expressions in male hObs by about 50% (Figure 9).

3.6. The production of 1,25D and its modulation by JKF, F, D or E$_2$ in hObs

Female-derived bone cells from both ages as well as male hObs produced 1,25D as measured by radio-immunoassay 53.7 + 7.1 in pre-, 72.8 + 14.3 in post- and 35.7 + 7.1 pg/ml in male hObs (Figure 8). F, D and E$_2$ treatment increased the activity of 1OHase by increasing the production of 1,25D by about 80–150%, in both age groups but not in male hObs (Figure 9). JKF increased 1,25D production also in male hObs by about 40% as well as by about 180% in pre- and about 160% in post-hObs (Figure 9).

Cultured human bone cells

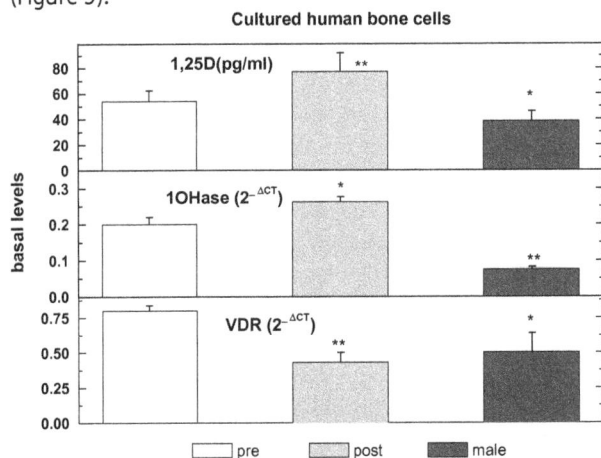

Figure 8. The expression of VDR or 1OHase and 1,25D production in pre-menopausal human osteoblasts (pre-), in post-menopausal human osteoblasts (post-) and in human male osteoblasts (male).

Notes: Bone cells were cultured, treated, and extracts prepared for mRNA expression as described in Materials ± SEM for cultures from 5 humans group.

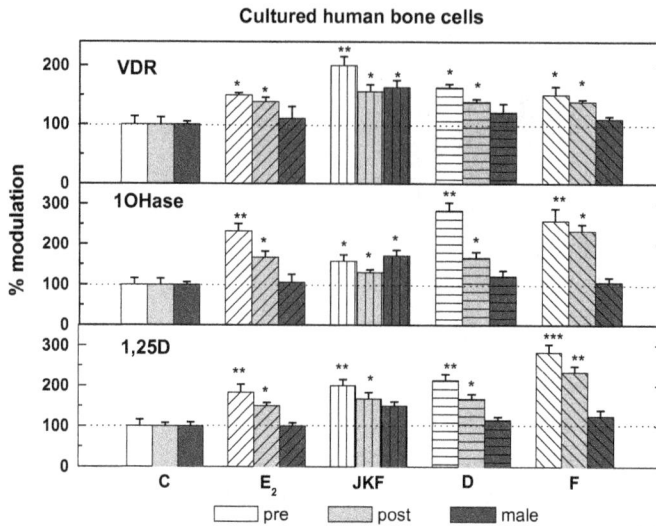

Figure 9. The effects of treatment with E₂, D, F, or JKF on 1,25D production (lower panel) and 1OHase mRNA expression (middle panel), as well as on VDR mRNA expression (upper panel) in pre- and post-menopausal and male hObs.

Notes: Cells were incubated for 24 h for the production of 1,25D or for 3 days for VDR and 1OHase mRNA expressions with F (20 ng/ml), D (30nM) or E₂ (3nM) or JKF (1nM). Results are expressed as % change in the concentration of 1,25D (pg/mg protein) or in 1OHase and VDR mRNA expression quantified by real time PCR (n = 4–8 human/group). $*p < 0.05$; $**p < 0.01$.

4. Discussion

In our previous studies as well as in the present one, we found that F similar to D (its precursor) or E₂ stimulates—age- and sex-dependently—both DNA synthesis and CK specific activity in primary cultures of human female hObs but not in human male hObs (Eriksen & Glerup, 2002). Pre-treatment with JKF emphasizes the difference between F and D or E₂, whereas the stimulation of D or E₂ on both parameters and in both female age groups were up-regulated by the vitamin D analog JKF, the effects of F were not significantly changed in either age groups by JKF pre-treatment (Figures 3, 7 and 9). We showed that up-regulation of the stimulation by vitamin D₃ less-calcemic analogs occurs by modulation of ERα and ERβ mRNA in these cells. We have previously demonstrated, using western blot analysis (Kanis et al., 1979) that JKF increased the protein levels of ERα and ERβ with a greater increase of ERα than ERβ in pre-menopausal-derived cells and the opposite occurred in post-menopausal-derived cells (Sömjen, Weisman, & Kaye, 1995; Sömjen et al., 1990). The effects of JKF on nuclear binding (Katzburg et al., 1999; Somjen et al., 2005) are consistent with the increased responsiveness to E₂ after JKF pre-treatment as well as changes in ER proteins (Fournier et al., 1996). In contrast, the membranal binding of E₂ and all the phytoestrogens tested was abolished by JKF pre-treatment (Katzburg et al., 1999). Membranal binding is therefore not correlated with the up-regulation of DNA synthesis or CK specific activity stimulated by estrogens after JKF pre-treatment. This indicates that membranal processes are not involved in CK stimulation by estrogenic compounds tested in this study or others (Katzburg et al., 1999; Somjen & Yoles, 2003b; Somjen et al., 2006).

It is important to notice that these results are seen in vitamin D deficient rats in which the response of bone to E₂ is attenuated and is restored after treatment with sufficient amounts of vitamin D (Somjen et al., 2007).

ERα and ERβ mRNA were found in both ages of female-derived osteoblasts but while JKF increased ERα expression and decreased ERβ expression, E₂ and D significantly down regulated ERβ without affecting ERα suggesting a negative effect of E₂ on estrogenic responsiveness in cells from both age groups. On the other hand, F increased both ERs suggesting its positive effect on estrogenic responsiveness in human female-derived osteoblasts from both age groups.

The synthesis of 1,25D from its precursor 25-hydroxyvitamin D_3 (25(OH) D_3), is catalyzed by 1OHase in epithelial cells comprising various parts of the human nephron (Sömjen, Kaye, Harell, & Weisman, 1989; Zehnder, Bland, Walker, et al., 2001). Renal 1OHase is subject to tight systemic metabolic control by parathyroid hormone (PTH), calcium, phosphate and vitamin D_3 metabolites, predominantly 1,25D itself (Reichel, Koeffler, & Norman, 1987). Renal 1OHase is the major source of circulating 1,25D, which controls systemic calcium homeostasis; nevertheless, external-renal expression of 1OHase and 1,25D is now well documented in various tissues and cell types such as prostate cells (Reichel et al., 1987; Sömjen et al., 1989; Zehnder, Bland, Walker, et al., 2001). Molecular studies indicate that the enzyme is expressed in both the kidney and non-renal tissues (Sömjen et al., 1989; Zehnder, Bland, Walker, et al., 2001) and is differentially regulated at least in some of these tissues (Feldman, Malloy, Goldschmidt, & Gross, 2001; Pryke, Duggan, White, Posen, & Mason, 1990; Zehnder, Bland, Williams, et al., 2001). In contrast to circulating 1,25D which controls systemic calcium homeostasis through its action on the intestinal mucosa, bone and kidney, accumulating data indicate that the 1,25D produced by extra-renal 1OHase in various tissues does not contribute to circulatory levels but rather appears to act in an autocrine and/or paracrine fashion by modulating cell proliferation, differentiation, apoptosis, immunoregulation and other functions at a local level (Holick, 2003; Walters, 1992) .

We have shown that human female osteoblasts express 1OHase mRNA and produces 1,25D both of which are modulated by different hormones including estrogenic compounds (Somjen et al., 2007). The synthesis of 1,25D in hObs is quantitatively significant at basal production rate of ~1.5 pmol/mg protein/hr, reaching ~4 pmol/mg protein/hr under saturating concentrations of its substrate 25(OH)D_3 (Somjen et al., 2005), which might be important for the physiology of bone as was shown for other parameters.

The effects of F as well as D and E_2 on 1OHase expression and activity remain obscure. The current study demonstrates that F modulates local 1OHase expression in hObs which might, in turn, modulate the effects of other estrogenic compounds on the cells. It is not known, however, whether this happens also *in vivo*. If it does, it might be used as a mixed new treatment for bone formation in post-menopausal osteoporosis.

The effect of the locally formed 1,25D might be involved in the modulation by vitamin D of the hormonal responses of hObs to the different estrogenic compounds. This might be due to the changes in the expression of estrogen receptors, which were reported to be changed with external vitamin D treatments as described previously (Holick, 2003; Somjen et al., 2005, 2007). This might be a defence mechanism in the pathophysiological conditions of absence of vitamin D in the body. The different parameters changed in hObs response to estrogens by vitamin D analogs show a high increase in CK, and a slight increase in intracellular estrogen binding as well as increased mRNA for ERα. On the other hand, both binding of membranal receptors and mRNA for ERβ were reduced. Whether this indicates a differential interaction of F with the different binding sites through which it exerts its biological activity, needs to be studied.

In conclusion, the present study provides evidence for the mutual interaction between F or other estrogenic compounds and the vitamin D_3 system. The potential role of this system as an autocrine/paracrine mechanism to modulate bone cell metabolism and physiology warrants further investigation. Of importance is the fact that since F is not up-regulated by vitamin D-like estrogens, it is also active in vitamin D deficient conditions, unlike estrogens. This might add to its advantage in hormone replacement therapy treatment over the use of E_2 and other estrogens.

Funding

The authors received no direct funding for this research.

Competing Interests

The authors declare no competing interest.

Author details

D. Somjen[1]
E-mail: dsomjen10@gmail.com
S. Katzburg[1]
E-mail: skatzburg@gmail.com
O. Sharon[1]
E-mail: osharon@gmail.com
E. Knoll[1]
E-mail: eknoll@gmail.com
D. Hendel[1]
E-mail: dhendel@gmail.com
G.H. Posner[2]
E-mail: ghp@jhu.edu

[1] The Sackler Faculty of Medicine, Tel-Aviv Sourasky Medical Center, Institute of Endocrinology, Metabolism and Hypertension and Bone Diseases Unit, Tel-Aviv University, 6 Weizman Street, Tel Aviv 64239, Israel.

[2] Department of Orthopedic Surgery, Sharei-Zedek Medical Center, Jerusalem, Israel.

References

Brzezinski, A., & Debi, A. (1999). Phytoestrogens: The "natural" selective estrogen receptor modulators? *European Journal of Obstetrics & Gynecology and Reproductive Biology, 85,* 47–51. https://doi.org/10.1016/S0301-2115(98)00281-4

Eriksen, E. F., & Glerup, H. (2002). Vitamin D deficiency and aging: Implications for general health and osteoporosis. *Biogerontology, 3,* 73–77. https://doi.org/10.1023/A:1015263514765

Feldman, D., Malloy, P. J., Goldschmidt, D., & Gross, C. (2001). Vitamin D: Biology, action and clinical implications. In R. Marcus, D. Feldman, & J. Kelsey (Eds.), *Osteoporosis* (Vol 1, pp. 230–257). San Diego, CA: Academic Press.

Fournier, B., Haring, S., Kaye, A. M., & Somjen, D. (1996). Stimulation of creatine kinase specific activity in human osteoblast and endometrial cells by estrogens and antiestrogens and its modulation by calciotropic hormones. *Journal of Endocrinology, 150,* 275–285. https://doi.org/10.1677/joe.0.1500275

Holick, M. F. (2003). Prostatic 25-hydroxy vitamin D-1-α-hydroxylase and its implication in prostate cancer. *Journal of Cellular Biochemistry, 88,* 315–322.

Kanis, J. A., Cundy, T., Earnshaw, M., Henderson, R. G., Heynen, G., Naik, R., & Russell, R. G. G. (1979). Treatment of renal bone disease with 1α derivatives of vitamin D3. *An International Journal of Medicine, 48,* 289–322.

Katzburg, S., Lieberherr, M., Ornoy, A., Klein, B. Y., Hendel, D., & Somjen, D. (1999). Isolation and hormonal responsiveness of primary cultures of human bonederived cells: Gender and age differences. *Bone, 25,* 667–673. https://doi.org/10.1016/S8756-3282(99)00225-2

Katzburg, S., Hendel, D., Waisman, A., Posner, G. H., Kaye, A. M., & Somjen, D. (2005). Treatment with "non-hypercalcemic" analogs of 1,25 dihydroxy vitamin D3 increases responsiveness to 17-β estradiol, dihydrotestosterone or raloxifene in primary human osteoblasts. *The Journal of Steroid Biochemistry and Molecular Biology, 88,* 213–219.

Miksicek, R. J. (1994). Interaction of urally occurring nonsteroidal estrogens with expressed recombinant human estrogen receptor. *The Journal of Steroid Biochemistry and Molecular Biology, 49,* 153–160. https://doi.org/10.1016/0960-0760(94)90005-1

Posner, G. H. (2004). Modulation of response to estrogens in cultured human female bone cells by A Non- calcemic vitamin D analog: Changes in nuclear and membranal binding. *The Journal of Steroid Biochemistry and Molecular Biology, 89–90,* 393–395.

Posner, G. H., Lee, J. K., Wang, Q., Peleg, S., Burke, M., Brom, H., … Kensler, T. W. (1998). Icemic, antiproliferative, transcriptionally active, 24- fluorinated hybrid analogues of the hormone 1α, 25- dihydroxyvitamin D3. Synthesis and preliminary biological evaluation. *Journal of Medicinal Chemistry, 41,* 3008–3014.

Pryke, A. M., Duggan, C., White, C. P., Posen, S., & Mason, R. S. (1990). Tumor necrosis factor-α induces vitamin D-1-hydroxylase activity in normal human alveolar macrophages. *Journal of Cellular Physiology, 142,* 652–656. https://doi.org/10.1002/(ISSN)1097-4652

Reichel, H., Koeffler, H. P., & Norman, A. W. (1987). Synthesis *in vitro* of 1,25-dihydroxyvitamin D3 and 24,25-dihydroxyvitamin D3 by interferon–stimulated normal human bone marrow and alveolar macrophages. *Journal of Cellular Biochemistry, 262,* 10931–10937.

Reiss, N., & Kaye, A. M. (1981). Identification of the major component of the estrogen induced protein of rat uterus as the BB isozyme of creatine kinase. *The Journal of Biological Chemistry, 256,* 5741–5749.

Setchell, K. D., & Lydeking-Olsen, E. (2003). Dietary phytoestrogens and their effect on bone: Evidence from *in vitro* and *in vivo,* human observational, and dietary intervention studies. *The American Journal of Clinical Nutrition, 78,* 593S–609S.

Sömjen, D., Harell, A., Jaccard, N., Weisman, Y., & Kaye, A. M. (1990). Reciprocal modulation by sex steroids and calciotropic hormones of skeletal cell proliferation. *The Journal of Steroid Biochemistry and Molecular Biology, 37,* 491–499. https://doi.org/10.1016/0960-0760(90)90392-X

Somjen, D., Katzburg, S., Lieberherr, M., Hendel, D., & Yoles, I. (2006). DT56a stimulates gender- specific human cultured bone cells *in vitro. The Journal of Steroid Biochemistry and Molecular Biology, 98,* 90–96. https://doi.org/10.1016/j.jsbmb.2005.08.002

Somjen, D., Katzburg, S., Stern, N., Kohen, F., Sharon, O., Limor, R., … Weisman, Y. (2007). 25 hydroxy-vitamin D3-1α hydroxylase expression and activity in cultured human osteoblasts and their modulation by parathyroid hormone, estrogenic compounds and dihydrotestosterone. *The Journal of Steroid Biochemistry and Molecular Biology, 107,* 238–244. https://doi.org/10.1016/j.jsbmb.2007.03.048

Sömjen, D., Kaye, A. M., Harell, A., & Weisman, Y. (1989). Modulation by vitamin D status of the responsiveness of rat bone to gonadal steroids. *Endocrinology, 125,* 1870–1876. https://doi.org/10.1210/endo-125-4-1870

Somjen, D., Kohen, F., Lieberherr, M., Gayer, B., Schejter, E., Katzburg, S., … Stern, N. (2005). The effects of phytoestrogens, and their carboxy derivatives on human vascular and bone cells: New insights based on studies with carboxy- biochainin A. *The Journal of Steroid Biochemistry and Molecular Biology, 93,* 293–303. https://doi.org/10.1016/j.jsbmb.2004.12.029

Somjen, D., Weisman, Y., Harell, A., Berger, E., & Kaye, A. M. (1989). Direct and sex specific stimulation by sex steroids of creatine kinase activity and DNA synthesis in rat bone. *Proceedings of the National Academy of Sciences, 86,* 3361–3365. https://doi.org/10.1073/pnas.86.9.3361

Somjen, D., Waisman, A., Lee, J.-K., Posner, G. H., & Kaye, A. M. (2001). A noncalcemic, 25 dihydroxy vitamin D3 (JKF) up regulates the induction of creatine kinase B in osteoblast-like ROS 17/2.8 cells and in rat diaphysis by estradiol-17β. *The Journal of Steroid Biochemistry and Molecular Biology, 77*, 205–212. https://doi.org/10.1016/S0960-0760(01)00065-6

Sömjen, D., Waisman, A., Weisman, J., & Kaye, A. M. (1998). Nonhypercalcemic analogs of vitamin D stimulate creatine kinase activity in osteoblast-like ROS 17/2.8 cells and upregulate their responsiveness to estrogens. *Steroids, 63*, 340–343. https://doi.org/10.1016/S0039-128X(98)00026-9

Somjen, D., Waisman, A., Weisman, Y., & Kaye, A. M. (2000). "Nonhypercalcemic" analogs of, 25 dihydroxy vitamin D augment the induction of creatine kinase B by selective estrogen receptor modulators (SERMS) in osteoblast-like cells and rat skeletal organs. *The Journal of Steroid Biochemistry and Molecular Biology, 72*, 79–88. https://doi.org/10.1016/S0960-0760(00)00028-5

Sömjen, D., Weisman, Y., & Kaye, A. M. (1995). Pretreatment with 1,25 (OH)2 vitamin D or 24,25 (OH)$_2$ vitamin D increases synergistically responsiveness to sex steroids in skeletal derived cells. *The Journal of Steroid Biochemistry and Molecular Biology, 55*, 211–217. https://doi.org/10.1016/0960-0760(95)00175-Y

Somjen, D., & Yoles, I. (2003a). DT56a (Tofupill/Femarelle) selectively stimulates creatine kinase specific activity in skeletal tissues of rats but not in the uterus. *The Journal of Steroid Biochemistry and Molecular Biology, 86*, 93–98. https://doi.org/10.1016/S0960-0760(03)00252-8

Somjen, D., & Yoles, I. (2003b). DT56a stimulates creatine kinase specific activity in vascular tissues of rats. *Journal of Endocrinological Investigation, 26*, 966–971. https://doi.org/10.1007/BF03348193

Tham, D. M., Gardner, C. D., & Haskell, W. L. (1998). Potential health benefits of dietary Phytoestrogens: A review of the clinical, epidemiological, and mechanistic evidence. *The Journal of Clinical Endocrinology & Metabolism, 83*, 2223–2235.

Walters, M. R. (1992). Newly identified actions of the vitamin D endocrine system. *Endocr Rev, 13*, 719–764.

Yoles, I., Yogev, Y., Frenkel, Y., Hirsch, M., Nahum, R., & Kaplan, B. (2004). Efficacy and safety of standard versus low-dose Femarelle (DT56a) for the treatment of menopausal symptoms. *Clinical and Experimental Obstetrics and Gynecology, 31*, 123–126.

Yoles, I., Yogev, Y., Frenkel, Y., Nahum, R., Hirsch, M., & Kaplan, B. (2003). Tofupill/femarelle (DT56a): A new phyto-selective estrogen receptor modulator-like substance for the treatment of postmenopausal bone loss. *Menopause, 10*, 522–525. https://doi.org/10.1097/01.GME.0000064864.58809.77

Zehnder, D., Bland, R., Walker, E. A., Bradwell, A. R., Howie, A. J., Hewison, M., & Stewart, P. M. (2001). Expression of 25 - hydroxyvitamin D3-1 the human kidney. *Journal of the American Society of Nephrology, 10*, 2465–2473.

Zehnder, D., Bland, R., Williams, M. C., McNinch, R. W., Howie, A. J., Stewart, P. M., & Hewison, M. (2001). Extrarenal expression of 25-hydroxyvitamin d(3)-1 α-hydroxylase. *The Journal of Clinical Endocrinology & Metabolism, 86*, 888–894.

14

An essential oil blend significantly modulates immune responses and the cell cycle in human cell cultures

Xuesheng Han[1]*, Tory L. Parker[1] and Jeff Dorsett[1]

*Corresponding author: Xuesheng Han, dōTERRA International, LLC, 389 S. 1300 W, Pleasant Grove, UT 84062, USA
E-mail: lhan@doterra.com
Reviewing editor: Maté Biro, University of New South Wales, Australia
Additional information is available at the end of the article

Abstract: In the current study, we examined the biological activities of an essential oil blend (EOB) in validated human cell cultures, which model the molecular biology of autoimmune diseases and chronic inflammation. EOB is primarily composed of essential oils from wild orange, clove, cinnamon, eucalyptus, and rosemary. These disease models allow the measurement of changes in protein biomarkers induced by EOB treatment. Four T cell autoimmune disease systems and one skin cell system were used for biomarker analysis. Biomarkers levels were measured both before and after EOB treatment for statistic analysis. EOB exhibited significant effects on the levels of protein biomarkers that are critically involved in inflammation, immune modulation, and tissue remodeling processes. The overall inhibitory effect of EOB on these protein biomarkers suggests that it has anti-inflammatory and immune modulating properties. EOB also showed significant anti-proliferative activity against these cells. We next investigated the effect of EOB on genome-wide gene expression in a skin disease model. EOB significantly modulated global gene expression in the skin disease model. Further analysis showed that EOB robustly affected signaling pathways related to inflammation, immune function, and cell cycle control. This study documents the biological activities of EOB in complex human disease models, and indicates that EOB affects various biological and physiological processes in

ABOUT THE AUTHORS

Dr Han's group primarily studies the health benefits of essential oils. We are specifically interested in the efficacy and safety of essential oils and their active components. We work closely with research institutes, hospitals, and clinics to develop quality essential oil products with therapeutic benefits. The research work discussed in this paper represents one part of a large research project, which was designed to extensively examine the effect of essential oils on human cells. This study, along with others, will further the understanding of the health benefits of essential oils for a wide research audience. Besides essential oils, we are also interested in studying the health benefits of herbal supplements and skin care products. Dr Han holds a PhD in Biological Sciences and is an elected Fellow of the American College of Nutrition. Parker holds a PhD in Nutritional Sciences. Dorsett holds a MS in Health Sciences.

PUBLIC INTEREST STATEMENT

Essential oils have become popular globally for health reasons. Our study examined the effects of an essential oil blend (EOB) on human cell systems that mimic different diseases. These effects of the EOB were determined by measuring biomarkers that are linked to inflammation, immune function, and wound healing. We found that the EOB had strong anti-proliferative, anti-inflammatory, immune modulatory, and wound healing activities. The effect of the EOB on gene expression in human skin cells was also studied. The EOB robustly affected genes and processes related to inflammation, immune function, and cell cycle control. The study findings suggest that essential oils affect various biological and physiological processes in human cells. Exploration of the health benefits of essential oils may lead to viable options for fighting many diseases. Thus, this study provides an important stepping stone for further research on essential oils and their health benefits for humans.

human cells. This study suggests that EOB possesses significant anti-inflammatory and immune modulating properties.

Subjects: Biochemistry; Pharmaceutical Science; Pharmacology; Immunology

Keywords: anti-proliferation; inflammation; genome-wide gene expression; immune response; cell cycle control; wild orange; clove; cinnamon; eucalyptus; rosemary

1. Introduction

Traditional medicine has long employed the use of botanical preparations to address human ailments. "Aromatherapy," a term coined by René Maurice Gattefossé in 1920, is one such modality. It involves the therapeutic use of essential oils, complex mixtures of volatile aromatic compounds naturally found in plants, to improve physical, emotional, and spiritual well-being (Vigan, 2010). In addition to being used as natural remedies, essential oils and their respective major constituents are commonly used in cleaners, perfumes, cosmetics, dentistry, agriculture, and food preservatives. Essential oils can be diffused aromatically, consumed internally, or applied topically to the skin to achieve a desired benefit. Emerging scientific evidence supports these traditional uses of essential oils as effective therapeutic tools. A better understanding of the biological activities of essential oils is needed, especially regarding their effects on the human immune system and their potential anti-cancer properties.

Many studies have examined the biological activities of individual essential oils or their isolated constituents. For example, clove oil has been shown *in vitro* to increase both primary and secondary humoral responses (Halder, Mehta, Mediratta, & Sharma, 2011) and exhibit antiviral activity against herpes simplex virus (Tragoolpua & Jatisatienr, 2007). Several other essential oils, as well as their major constituents, have been found to possess medicinal properties, including antibacterial, anti-fungal, anti-proliferative, anti-inflammatory, antioxidant, and anesthetic properties. However, many of these studies have only utilized single cell lines or rodent models (Chong, Alegre, Miller, & Fairchild, 2013). Cell lines alone do not model primary disease biology, and rodent models do not accurately reflect regulatory complexities of human disease (Mak, Evaniew, & Ghert, 2014). Human cell coculture systems can help address these limitations by combining healthy host cells, diseased cells (e.g. tumor cells), and disease-relevant stimuli to mimic host-disease micro-environments (Bergamini et al., 2012). In this study, we evaluated the biological activities of a commercially available essential oil blend (EOB), in well-validated human cell coculture systems. EOB is primarily a mixture of essential oils from wild orange (*Citrus sinensis*), clove (*Eugenia caryophyllata*), cinnamon (*Cinnamomum zeylanicum*), eucalyptus (*Eucalyptus globulus*), and rosemary (*Rosmarinus officinalis*).

2. Materials and methods

All experiments were conducted in the Biologically Multiplexed Activity Profiling (BioMAP) platform, a set of primary human cell systems designed to model disease biology in a robust and reproducible way. The systems consist of three components, a cell type or cell types (many systems involve cocultures), stimuli to create the disease environment, and then a set of biomarker (protein) readouts to examine how treatments affect that disease environment. See Table S1 in Supplementary Material for a glossary of BioMAP systems used in the study.

2.1. Cell cultures

Primary human (H) cells (i.e. neonatal dermal fibroblasts (HNDFs), umbilical vein endothelial cells (HUVECs), peripheral blood mononuclear cells (PBMCs), and B cells) were obtained as previously described (Bergamini et al., 2012). HNDFs were plated in low serum conditions 24 h before stimulation with cytokines. HUVECs were obtained from Cascade Biologics (Portland, OR, USA) and were cultured in endothelial cell growth medium-2 containing manufacturer-provided supplements and 2% heat-inactivated fetal bovine serum (Hyclone, Logan, UT, USA). HUVECs were subcultured with 0.05% trypsin-0.53 mM ethylenediaminetetraacetic acid (EDTA, Mediatech, Herndon, VA, USA) according to the manufacturer's instructions. PBMCs were prepared from buffy coats (Stanford Blood Bank, Stanford, CA, USA) by centrifugation over Histopaque-1077 (Sigma-Aldrich, St. Louis, MO, USA).

Stimulatory molecules for these cell coculture systems were as follows: A mixture of interleukin (IL)-1β, tumor necrosis factor (TNF)-α, interferon (IFN)-Υ, basic fibroblast growth factor (bFGF), epidermal growth factor (EGF), and platelet-derived growth factor (PDGF) for the HDF3CGF system (HNDFs); T-cell receptor (TCR) ligands (1×) for the SAg system (PBMCs + HUVECs); immunoglobulin M antigens and TCR ligands (0.001×) for the BT system (B cells + PBMCs); TCR ligands (0.001×) for the HDFSAg system (HNDFs + PBMCs); and IL-2 and TCR ligands (0.1×) for the /TH2 system (T helper cell 2 blasts + HUVECs). Cell culture and stimulation conditions for the HDF3CGF, SAg, BT, HDFSAg, and / TH2 coculture assays have been described in detail elsewhere, and were performed in 96-well format (Bergamini et al., 2012).

2.2. Protein-based readouts

An enzyme-linked immunosorbent assay (ELISA) was used to measure the biomarker levels of cell-associated and cell membrane targets. Soluble factors from supernatants were quantified using homogeneous time-resolved fluorescence detection, bead-based multiplex immunoassay, or capture ELISA. Overt adverse effects of test agents on cell proliferation and viability (cytotoxicity) were measured by sulforhodamine B (SRB) for adherent cells, and alamarBlue staining for cells in suspension. For proliferation assays, individual cell types were cultured at subconfluence and measured at time points optimized for each system (48–96 h). Detailed information has been described elsewhere (Bergamini et al., 2012). Measurements were performed in triplicate wells, and a glossary of the biomarkers used in this study is provided in Supplementary Table S2.

Quantitative biomarker data are presented as the mean $\log_{10}$ relative expression level (compared to the respective mean vehicle control value) ± standard deviation of triplicate measurements. Differences in biomarker levels between EOB- and vehicle-treated cultures were tested for significance with the unpaired Student's t test. A p-value < 0.05, outside of the significance envelope, with an effect size of at least 10% (more than 0.05 $\log_{10}$ ratio units), was considered statistically significant.

2.3. RNA isolation

Total RNA was isolated from cell lysates using the Zymo *Quick-RNA* MiniPrep kit (Zymo Research Corporation, Irvine, CA, USA) according to the manufacturer's instructions. RNA concentration was determined using a NanoDrop ND-2000 (Thermo Fisher Scientific, Waltham, MA, USA). RNA quality was assessed with a Bioanalyzer 2100 (Agilent Technologies, Santa Clara, CA, USA) and an Agilent RNA 6000 Nano Kit. All samples had an A260/A280 ratio between 1.9 and 2.1 and an RNA integrity number score greater than 8.0.

2.4. Microarray analysis for genome-wide gene expression

A 0.01% (*v/v*) concentration of EOB was tested for its effect on expression of 21,224 genes in the HDF3CGF system after 24 h treatment. Samples for microarray analysis were processed by Asuragen, Inc. (Austin, TX, USA) according to the company's standard operating procedures. Biotin-labeled cRNA was prepared from 200 ng of total RNA with an Illumina TotalPrep RNA Amplification kit (Thermo Fisher Scientific) and one round of amplification. The cRNA yields were quantified using ultraviolet spectrophotometry, and the distribution of transcript sizes was assessed using the Agilent Bioanalyzer 2100. Labeled cRNA (750 ng) was used to probe Illumina Human HT-12 v4 Expression BeadChips (Illumina, Inc., San Diego, CA). Hybridization, washing, staining with streptavidin-conjugated Cyanine-3, and scanning of the Illumina arrays were carried out according to the manufacturer's instructions. Illumina BeadScan software was used to produce the data files for each array; raw data were extracted using Illumina BeadStudio software.

Raw data were uploaded into R and analyzed for quality-control metrics using the beadarray package. Data were normalized using quantile normalization, then were re-annotated and filtered to remove probes that were non-specific or that mapped to intronic or intragenic regions (Barbosa-Morais et al., 2010). The remaining probe sets comprised the data-set for the remainder of the analysis. Fold-change expression for each was calculated as the $\log_2$ ratio of EOB to vehicle control. These fold-change values were uploaded to Ingenuity Pathway Analysis (IPA, Qiagen, Redwood City, CA, www.qiagen.com/ingenuity) to generate the networks and pathway analyses.

2.5. Reagents

The EOB (commercial name On Guard, dōTERRA International LLC, Pleasant Grove, UT, USA) was diluted in dimethyl sulfoxide (DMSO) to 8X the specified concentrations (final DMSO concentration was no more than 0.1% [v/v]); 25 μL of each 8 × solution was added to the cell culture to obtain a final volume of 200 μL. DMSO (0.1%) served as the vehicle control. Gas chromatography–mass spectrometry (GC-MS) analysis of EOB showed that its main (≥2%) chemical constituents were d-limonene, eugenol, 1,8-cineole, and cinnamaldehyde. GC-MS results of EOB will be made public via the supplier's website www.sourcetoyou.com, as part of its transparency program.

3. Results

We studied EOB's activities in disease-mimicking cell culture systems. Key activities were designated if biomarker values were significantly different ($p < 0.05$) from vehicle controls at studied concentrations, outside of the significance envelope, with an effect size of at least 10% (0.05 log units). Only biomarkers designated as key activities are discussed below.

3.1. Bioactivity profile of EOB in the autoimmune systems, SAg, BT, HDFSAg, and /TH2

To explore the biological activities of EOB in immune systems, we analyzed four different concentrations (0.01, 0.004, 0.0014, and 0.0004% [v/v]) of EOB in four different T cell autoimmune systems (i.e. SAg, BT, HDFSAg, and /TH2). EOB at 0.01% was overtly cytotoxic and was excluded from further analyses. The lower two concentrations of EOB did not have a significant effect on the biomarkers, so the 0.004% EOB was used for further analysis and is discussed below. For subsequent cell systems, the highest EOB concentration that did not demonstrate cytotoxicity was utilized.

In the SAg system, EOB reduced the level of cell differentiation 40 (CD40), which is a cell surface adhesion receptor related to immunomodulatory activity (Figure 1(A)). In the BT system, EOB significantly decreased the level of an inflammation biomarker, soluble tumor necrosis factor alpha (sTNFα). In the same system, it also significantly decreased the levels of three immunomodulatory biomarkers, namely secreted immunoglobulin G (sIgG), soluble interleukin 17A (sIL-17A), and soluble interleukin 17F (sIL-17F) (Figure 1(B)). In both systems, EOB showed significant anti-proliferation activity and cytotoxic activity against PBMCs (Figure 1(A) and (B)).

In the HDFSAg system (Figure 1(C)), EOB reduced the levels of three inflammation-related biomarkers, vascular cell adhesion molecule 1 (VCAM-1), interferon gamma-induced protein 10 (IP-10), and sTNF-α, as well as the levels of four immunomodulatory biomarkers, sIL-17A, sIL-17F, sIL-2, and sIL-6 (Figure 1(C)). EOB also slightly increased the level of vascular endothelial growth factor (sVEGF). In the /TH2 system, EOB significantly decreased the levels of Eotaxin-3 and P-Selectin, which are both important inflammation-related biomarkers (Figure 1(D)). The immunomodulatory biomarker CD40 was also significantly decreased by EOB in the /TH2 system (Figure 1(D)).

3.2. Bioactivity profile of EOB in the dermal fibroblast system, HDF3CGF

The study results from autoimmune systems showed promising anti-inflammatory and immune function-modulating effects of EOB. We then analyzed its activity in a dermal fibroblast cell system, HDF3CGF, which features the microenvironment of inflamed human skin cells. Four different concentrations (0.01, 0.0033, 0.0011, and 0.00037% [v/v]) of EOB were initially analyzed for viability assays. None of them was overtly cytotoxic to these cells, and thus, the 0.01% EOB was included for further analysis and is discussed below.

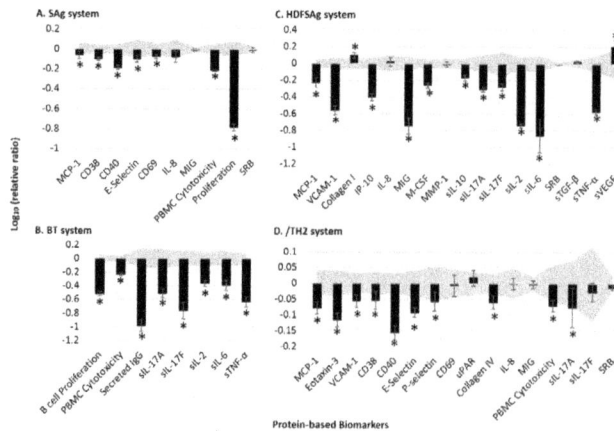

Figure 1. The bioactivity profile of the essential oil blend (EOB, 0.004% v/v) in autoimmune systems SAg (A), BT (B), HDFSAg (C), and /TH2 (D).

Notes: Each X-axis denotes protein-based biomarker readouts in the respective system. Each Y-axis denotes the log relative expression levels of these biomarkers compared to respective vehicle control values. The 95% confidence interval of the mean vehicle control values are marked by the gray shaded area. Each black bar represents the mean ± standard deviation of three measurements. *$p < 0.05$ vs. vehicle control, outside of 95% confidence interval, with an effect size of at least 10% (more than 0.05 log ratio units). MCP-1, monocyte chemoattractant protein 1; CD, cluster of differentiation; PBMC, peripheral blood mononuclear cell; SRB, sulforhodamine B; sIgG, secreted immunoglobulin G; sIL, soluble interleukin; sTNF-α, soluble tumor necrosis factor-alpha; VCAM-1, vascular cell adhesion molecule 1; IP- 10, interferon gamma-induced protein 10; MIG, monokine induced by gamma interferon; M-CSF, macrophage colony-stimulating factor; MMP-1, matrix metalloproteinase 1; sTGF-β1, soluble transforming growth factor-beta1; sVEGF, soluble vascular endothelial growth factor; uPAR, urokinase plasminogen activator receptor.

EOB showed significant anti-proliferative activity in dermal fibroblast cells. The levels of several inflammation-related biomarkers, including VCAM-1, IP-10, interferon-inducible T-cell alpha chemoattractant (I-TAC), and monokine induced by gamma interferon (MIG), decreased in response to EOB (Figure 2). Macrophage colony-stimulating factor (M-CSF) and plasminogen activator inhibitor 1 (PAI-1), also decreased in response to EOB.

3.3. Effects of EOB on genome-wide gene expression in the HDF3CGF system

To further explore the effect of EOB in human cells, we analyzed the effect of 0.01% (v/v) (i.e. the highest tested concentration that was non-cytotoxic to these cells) on the RNA expression of 21,224 genes in the HDF3CGF system. The results showed a robust and diverse effect of EOB on regulating human genes, with many genes being upregulated and many others being downregulated (Table S3). Among the 200 most-regulated genes by EOB (log$_2$ the fold change ratio of expression over vehicle control ≥1.5), the vast majority (172 out of 200 genes) were significantly downregulated.

IPA showed that the bioactivity of EOB significantly overlapped with many canonical pathways from the literature-validated database (Figure 3). Many of these pathways are critically involved in the processes of inflammation, immune response, cell cycle control, DNA damage response, and cancer biology. The significant matches of EOB bioactivity suggest that it affects various biological and physiological functions in human cells. For example, all the top four matched signaling pathways play important roles in the process of inflammation and cell cycle control (Tables S4–S7).

4. Discussion

Although essential oils have been used to prevent and treat diseases for many centuries, the scientific understanding of their biological action remains elusive. Therefore, we chose to study an EOB in the BioMAP platform to explore the biological activities of EOB in human cells. The BioMAP platform is a set of primary human cell systems designed to robustly model disease biology and ascertain the effect of therapeutic interventions on that biology.

4.1. EOB's effects on inflammation, immune function, and tissue remodeling

Inflammation is a protective response that involves immune cells, blood vessels, and molecular mediators. The purpose of inflammation is to eliminate the initial cause of cell injury, remove

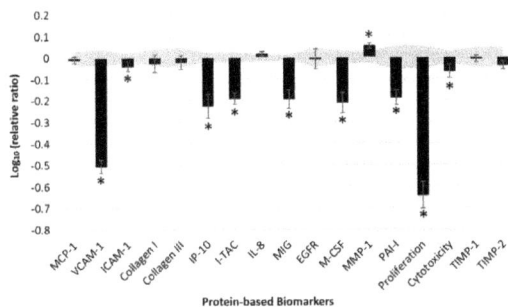

Figure 2. The bioactivity profile of the essential oil blend (EOB, 0.01% v/v) in the dermal fibroblast system, HDF3CGF.

Notes: X-axis denotes protein-based biomarkers readouts. Y-axis denotes the log relative expression levels of these biomarkers compared to respective vehicle control values. The 95% confidence interval of the mean vehicle control values are marked by the gray shaded area. Each black bar represents the mean ± standard deviation of three measurements. *$p < 0.05$ vs. vehicle control, outside of 95% confidence interval, with an effect size of at least 10% (more than 0.05 log ratio units). MCP-1, monocyte chemoattractant protein; VCAM-1, vascular cell adhesion molecule 1; ICAM- 1, intracellular cell adhesion molecule 1; IP-10, interferon gamma-induced protein 10; I-TAC, interferon-inducible T-cell alpha chemoattractant; IL-8, interleukin-8; MIG, monokine induced by gamma interferon; EGFR, epidermal growth factor receptor; M-CSF, macrophage colony-stimulating factor; MMP-1, matrix metalloproteinase 1; PAI-1, plasminogen activator inhibitor 1; TIMP, tissue inhibitor of metalloproteinase.

necrotic cells and tissues damaged by the injury, and initiate tissue repair. Chronic inflammation may lead to various diseases, ranging from hay fever to periodontitis, atherosclerosis, rheumatoid arthritis, and cancer. Wound healing is a complex process composed of several phases: blood clotting (hemostasis), inflammation, growth of new tissue (proliferation), and remodeling of tissue (maturation). The wound healing process is fragile, and is susceptible to disruption, leading to the formation of non-healing chronic wounds or scar tissue. Both inflammation and wound healing processes are closely related to the immune function of the host.

EOB demonstrated its anti-inflammatory potential by reducing many inflammatory biomarkers in already highly inflamed cell systems (Figures 1 and 2). It reduced sTNFα in the BT system; VCAM-1, IP-10, and sTNF-α in the HDFSAg system; and Eotaxin-3 and P-Selectin in the /TH2 system. EOB also significantly reduced VCAM-1, IP-10, I-TAC, and MIG in the HDF3CGF system. The anti-inflammatory property of EOB is consistent with previous studies of single oils and their active constituents. Cinnamaldehyde, the major bioactive chemical constituent of cinnamon bark essential oil, has been shown to possess anti-inflammatory properties (Chen et al., 2016; Han & Parker, 2017a; Koh et al., 1998; Mendes et al., 2016; Reddy et al., 2004). Clove oil and its major constituent, eugenol, have been shown to have anti-inflammatory activities, even more effective than NSAIDs (Han & Parker, 2017b; Nogueira de Melo et al., 2011). Rosemary oil has been shown to reduce inflammation by modulating leukocyte migration (Nogueira de Melo et al., 2011). In vitro and animal studies suggest that d-limonene, the major constituent of orange oil, also possesses anti-inflammatory properties (Nogueira de Melo et al., 2011).

EOB demonstrated the potential to modulate immune responses in the highly inflamed systems (Figures 1 and 2). EOB reduced sIgG, sIL-17A, and sIL-17F in the BT system; CD40 in both SAg and /TH2 systems; sIL-17A, sIL-17F, sIL-2, sIL-10, and sIL-6 in the HDFSAg system; and M-CSF in the HDF3CGF system. EOB's inhibitory effects on these important immunomodulatory biomarkers suggest its immune-enhancing potential in these pre-inflamed systems. Many of these activities are also likely attributed to the anti-inflammatory properties of these essential oils and their active constituents, as stated above. In addition, it had been previously shown that limonene, a major constituent of EOB, may enhance immune responses and act as an immune modulator (Raphael & Kuttan, 2003).

EOB also demonstrated beneficial effects on tissue-remodeling activities, which are important for proper wound healing (Figures 1 and 2). EOB increased the level of sVEGF in the HDFSAg system. VEGF is a widely expressed growth factor that induces vascular permeability, angiogenesis,

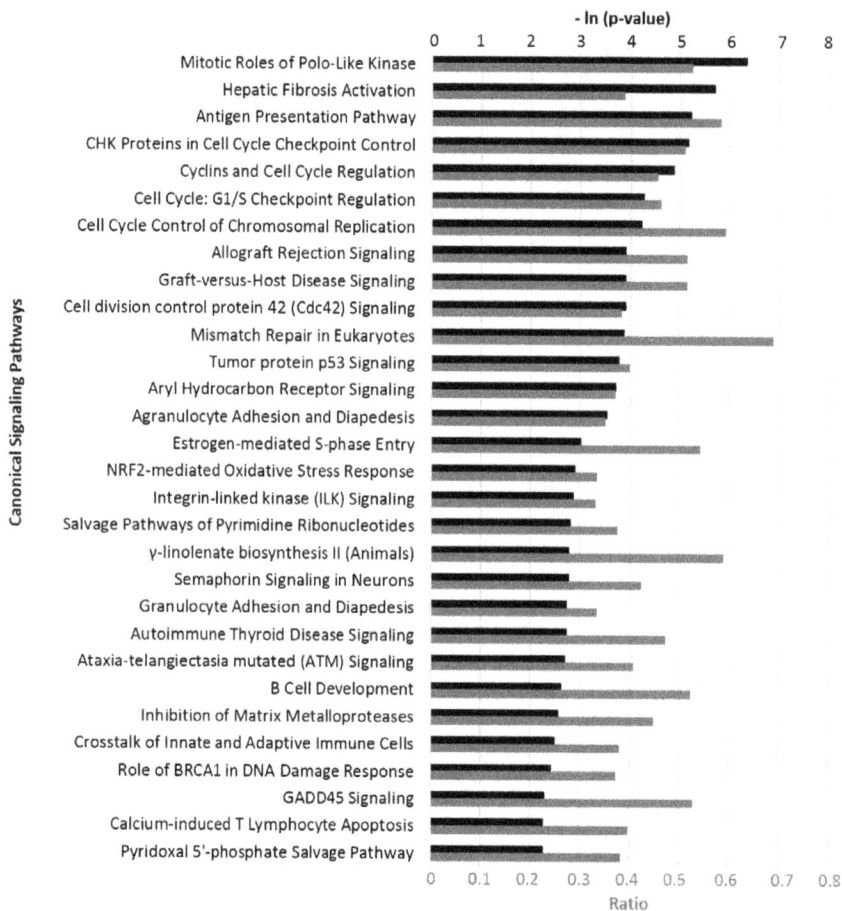

Figure 3. The 30 canonical pathways that best match the changes in gene expression caused by the essential oil blend (EOB, 0.01% v/v) in the HDF3CGF system, produced via Ingenuity Pathway Analysis (Qiagen, www.qiagen.com/ingenuity).

Notes: The p-value, which was calculated with a right-tailed Fisher's exact test, indicates how likely the observed association between a specific pathway and the data-set would be if it was only due to random chance. A larger −ln (p-value), indicated by the black bar, means that the bioactivity of the EOB more significantly matches the canonical pathway. The gray bars indicate a ratio that was calculated by taking the number of genes from the EOB data-set that participate in a canonical pathway, and dividing it by the total number of genes in that canonical pathway. The ratio indicates the percentage of genes in a pathway that were also found in the list of genes affected by EOB. GADD45, Growth Arrest and DNA Damage 45; BRCA1, breast cancer 1.

vasculogenesis, and inhibits apoptosis. Increasing VEGF may, therefore, promote better wound healing. Interestingly, EOB decreased PAI-1 in the HDF3CGF system. PAI-1 is the serine proteinase inhibitor and inhibitor of tissue plasminogen activator that is involved in tissue remodeling and fibrinolysis. Decreasing PAI-1 is also potentially beneficial to promote wound healing. A previous study showed rosemary essential oil to be effective at healing diabetic wounds by reducing inflammation and enhancing wound contraction, re-epithelialization, regeneration of granulation tissue, angiogenesis, and collagen deposition in BALB/c diabetic mice (Abu-Al-Basal, 2010).

Microarray analyses also showed evidence supporting the anti-inflammatory, wound healing, and immune modulatory effects of EOB. EOB seemed to regulate many important pathways relevant to anti-inflammation, wound healing, and immune response. For example, EOB's biological activities matched significantly with both the hepatic fibrosis/hepatic stellate cell activation canonical pathway and the antigen presentation canonical pathway. The vast majority of players (including many genes for components of the major histocompatibility complex, cytokines, and collagens) in these pathways were significantly inhibited by EOB, consistent with the potential role of EOB in reducing inflammation, modulating immune responses, and promoting wound healing.

Collectively, EOB shows various potential properties, including anti-inflammatory, immune enhancing, and wound healing promoting, in complex human cell cultures. These effects are putatively attributed to these single oils in EOB and their active constituents. Our results warrant further studies to explore the mechanisms of action in more detail.

4.2. EOB's effect on cell cycle control

In both the BT and SAg systems, 0.01% (v/v) of EOB demonstrated anti-proliferative and cytotoxic activity against PBMCs (Figure 1). This suggests a possible role of EOB in regulating the cell cycle. However, the same concentration of EOB showed little anti-proliferative or cytotoxic activity in other studied systems. Further investigation is needed to elucidate the reason for difference and the mechanisms responsible.

Microarray analyses showed significant matches between EOB's biological activities and several important cell cycle control pathways. The top two pathways were the mitotic roles of the polo-like kinase canonical pathway and the role of CHK proteins in the cell cycle checkpoint control pathway. Overall, the two pathways and the vast majority of players in these pathways were significantly inhibited by EOB. BRCA1 was one of many important genes shown to be significantly downregulated by EOB. This is consistent with the potential role of EOB to inhibit cell cycle progression in human cells. Generally, cell cycle control plays a critical role in cancer development and progression. Therefore, these results also suggest that EOB might be active in modulating cancer progression, by interacting with cell cycle control events. This property is putatively attributed to the anticancer potential of some single oils of EOB and their active constituents.

Previous studies of the individual essential oils and their major constituents also demonstrated promising anticancer properties. Cinnamon essential oil suppressed tumor growth in a Hep-2 cell xenograft model (Yang, Zheng, Ye, Li, & Chen, 2015). Cinnamaldehyde, a major constituent of cinnamon oil, has been shown to possess anti-mutagenic and anti-tumorigenic properties (de Silva & Shankel, 1987; Imai et al., 2002). In vitro studies in various cancer cell lines have also revealed cinnamaldehyde to be both anti-proliferative (Lin et al., 2013) and pro-apoptotic (Lin et al., 2013; Wu & Ng, 2007). Clove oil has been shown to possess anticancer properties against breast, colorectal, lung, and leukemia cancer cells (Kouidhi, Zmantar, & Bakhrouf, 2010; Kumar, Febriyanti, Sofyan, Luftimas, & Abdulah, 2014; Yoo et al., 2005). Rosemary oil exhibited strong cytotoxicity towards three human cancer cell lines (Wang, Li, Luo, Zu, & Efferth, 2012), prevented the spread of breast and androgen-sensitive prostate cancer cells (Hussain et al., 2010), and promoted apoptosis of liver cancer cells (Melusova, Slamenova, Kozics, Jantova, & Horvathova, 2014), presumably via cell-membrane disruption (Wei, Liu, Wang, Li, & Luo, 2008). Orange oil exhibited dose-dependent inhibition of cell proliferation and induced apoptosis in colon cancer cells (Chidambara Murthy, Jayaprakasha, & Patil, 2012). Limonene, a major component of wild orange oil, has also been reported to affect a number of cancer hallmarks (i.e. proliferation, apoptosis, and inflammation) (Miller, Thompson, Hakim, Chow, & Thomson, 2011). More recently, studies (Han & Parker, 2017a, 2017b, 2017c) of clove oil, cinnamon oil, wild orange oil, and rosemary oil in the HDF3CGF system showed that they can potentially impact the process of cancer signaling and cancer biology, suggesting promising anticancer properties.

Human cell culture systems are in vitro environments and do not reflect all of the complexities of the in vivo situation (e.g. pharmacokinetics). However, the model systems used in this study have been validated by both industrial use and regulatory acceptance. Gene expression levels were measured only after short-term exposure to EOB; how gene expression levels respond to longer-term exposure is still unknown. Finally, additional work (e.g. study in other coculture systems) is needed to better understand the complex mechanisms of action of EOB.

5. Conclusion

EOB significantly affected the levels of biomarkers related to inflammation, immune function, and tissue remodeling in various primary human cell models of diseases. EOB modulated mRNA levels of various signaling pathways (Figure 3), including inflammation, immune modulation, cell cycle regulation, and other cellular functions. To the best of our knowledge, this is the first study to explore the biological activities of this EOB in complex human cell cultures. This study provides novel and important findings of how EOB affects inflammation and immune-related biomarkers, and how it modulates genome-wide gene expression in validated human cultures.

Funding

The study was funded by dōTERRA (Pleasant Grove, UT, USA) and conducted at DiscoverX (Fremont, CA, USA).

Competing Interest

Xuesheng Han, Tory L. Parker and Jeff Dorsett are employees of dōTERRA (Pleasant Grove, UT, USA), where the studied agent, EOB, was manufactured.

Author details

Xuesheng Han[1]
E-mail: lhan@doterra.com
ORCID ID: http://orcid.org/0000-0003-2720-3011
Tory L. Parker[1]
E-mail: tparker@doterra.com
Jeff Dorsett[1]
E-mail: jdorsett@doterra.com
[1] dōTERRA International, LLC, 389 S. 1300 W, Pleasant Grove, UT 84062, USA.

References

Abu-Al-Basal, M. A. (2010). Healing potential of *Rosmarinus officinalis* L. on full-thickness excision cutaneous wounds in alloxan-induced-diabetic BALB/c mice. *Journal of Ethnopharmacology, 131,* 443–450. doi:10.1016/j.jep.2010.07.007

Barbosa-Morais, N. L., Dunning, M. J., Samarajiwa, S. A., Darot, J. F. J., Ritchie, M. E., Lynch, A. G., & Tavaré, S. (2010). A re-annotation pipeline for Illumina BeadArrays: Improving the interpretation of gene expression data. *Nucleic Acids Research, 38,* e17. doi:10.1093/nar/gkp942

Bergamini, G., Bell, K., Shimamura, S., Werner, T., Cansfield, A., Müller, K., & Neubauer, G. (2012). A selective inhibitor reveals PI3Kγ dependence of T(H)17 cell differentiation. *Nature Chemical Biology, 8,* 576–582. doi:10.1038/nchembio.957

Chen, Y.-F., Wang, Y.-W., Huang, W.-S., Lee, M.-M., Wood, W. G., Leung, Y.-M., & Tsai, H.-Y. (2016). Trans-cinnamaldehyde, an essential oil in cinnamon powder, ameliorates cerebral ischemia-induced brain injury via inhibition of neuroinflammation through attenuation of iNOS, COX-2 expression and NFκ-B signaling pathway. *Neuromolecular Medicine.* doi:10.1007/s12017-016-8395-9

Chidambara Murthy, K. N., Jayaprakasha, G. K., & Patil, B. S. (2012). D-limonene rich volatile oil from blood oranges inhibits angiogenesis, metastasis and cell death in human colon cancer cells. *Life Sciences, 91,* 429–439. doi:10.1016/j.lfs.2012.08.016

Chong, A. S., Alegre, M.-L., Miller, M. L., & Fairchild, R. L. (2013). Lessons and limits of mouse models. *Cold Spring Harbor Perspectives in Medicine, 3,* a015495. doi:10.1101/cshperspect.a015495

de Silva, H. V., & Shankel, D. M. (1987). Effects of the antimutagen cinnamaldehyde on reversion and survival of selected Salmonella tester strains. *Mutation Research/*

Genetic Toxicology, 187, 11–19. doi:10.1016/0165-1218(87)90071-1

Halder, S., Mehta, A. K., Mediratta, P. K., & Sharma, K. K. (2011). Essential oil of clove (*Eugenia caryophyllata*) augments the humoral immune response but decreases cell mediated immunity. *Phytotherapy Research, 25,* 1254–1256. doi:10.1002/ptr.3412

Han, X., & Parker, T. L. (2017a). Antiinflammatory activity of cinnamon (*Cinnamomum zeylanicum*) bark essential oil in a human skin disease model. *Phytotherapy Research, n/a-n/a..* doi:10.1002/ptr.5822

Han, X., & Parker, T. L. (2017b). *Eugenia caryophyllata*Anti-inflammatory activity of clove (*Eugenia caryophyllata*) essential oil in human dermal fibroblasts. *Pharmaceutical Biology, 55,* 1619–1622. doi:10.1080/13880209.2017.1314513

Han, X., & Parker, T. L. (2017c). Essential oils diversely modulate genome-wide gene expression in human dermal fibroblasts. *Cogent Medicine, 4,* 1307591. doi:10.1080/2331205X.2017.1307591

Hussain, A. I., Anwar, F., Chatha, S. A. S., Jabbar, A., Mahboob, S., & Nigam, P. S. (2010). *Rosmarinus officinalis* essential oil: Antiproliferative, antioxidant and antibacterial activities. *Brazilian Journal of Microbiology, 41,* 1070–1078. doi:10.1590/S1517-838220100004000027

Imai, T., Yasuhara, K., Tamura, T., Takizawa, T., Ueda, M., Hirose, M., & Mitsumori, K. (2002). Inhibitory effects of cinnamaldehyde on 4-(methylnitrosamino)-1-(3-pyridyl)-1-butanone-induced lung carcinogenesis in rasH2 mice. *Cancer Letters, 175,* 9–16. doi:10.1016/S0304-3835(01)00706-6

Koh, W. S., Yoon, S. Y., Kwon, B. M., Jeong, T. C., Nam, K. S., & Han, M. Y. (1998). Cinnamaldehyde inhibits lymphocyte proliferation and modulates T-cell differentiation. *International Journal of Immunopharmacology, 20,* 643–660. doi:10.1016/S0192-0561(98)00064-2

Kouidhi, B., Zmantar, T., & Bakhrouf, A. (2010). Anticariogenic and cytotoxic activity of clove essential oil (*Eugenia caryophyllata*) against a large number of oral pathogens. *Annals of Microbiology, 60,* 599–604. doi:10.1007/s13213-010-0092-6

Kumar, P. S., Febriyanti, R. M., Sofyan, F. F., Luftimas, D. E., & Abdulah, R. (2014). Anticancer potential of *Syzygium aromaticum* L. in MCF-7 human breast cancer cell lines. *Pharmacognosy Research, 6,* 350–354. doi:10.4103/0974-8490.138291

Lin, L.-T., Tai, C.-J., Chang, S.-P., Chen, J.-L., Wu, S.-J., & Lin, C.-C. (2013). Cinnamaldehyde-induced apoptosis in human hepatoma PLC/PRF/5 cells involves the mitochondrial death pathway and is sensitive to inhibition by cyclosporin A and z-VAD-fmk. *Anti-Cancer Agents in Medicinal Chemistry, 13,* 1565–1574. doi:10.2174/18715206113139990144

Mak, I. W., Evaniew, N., & Ghert, M. (2014). Lost in translation: animal models and clinical trials in cancer treatment. *American Journal of Translational Research, 6,* 114–118.

Melusova, M., Slamenova, D., Kozics, K., Jantova, S., & Horvathova, E. (2014). Carvacrol and rosemary essential oil manifest cytotoxic, DNA-protective and pro-apoptotic effect having no effect on DNA repair. *Neoplasma, 61,* 690–699. doi:10.4149/neo_2014_084

Mendes, S. J. F., Sousa, F. I. A. B., Pereira, D. M. S., Ferro, T. A. F., Pereira, I. C. P., Silva, B. L. R., & Fernandes, E. S. (2016). Cinnamaldehyde modulates LPS-induced systemic inflammatory response syndrome through TRPA1-dependent and independent mechanisms. *International Immunopharmacology, 34,* 60–70. doi:10.1016/j.intimp.2016.02.012

Miller, J. A., Thompson, P. A., Hakim, I. A., Chow, H.-H. S., & Thomson, C. A. (2011). d-Limonene: a bioactive food component from citrus and evidence for a potential role in breast cancer prevention and treatment. *Oncology Reviews, 5*, 31. doi:10.4081/oncol.2011.31

Nogueira de Melo, G. A., Grespan, R., Fonseca, J. P., Farinha, T. O., Silva, E. L., Romero, A. L., & Cuman, R. K. N. (2011). *Rosmarinus officinalis L.* essential oil inhibits *in vivo* and *in vitro* leukocyte migration. *Journal of Medicinal Food, 14*, 944–946. doi:10.1089/jmf.2010.0159

Raphael, T. J., & Kuttan, G. (2003). Immunomodulatory activity of naturally occurring monoterpenes carvone, limonene, and perillic acid. *Immunopharmacology and Immunotoxicology, 25*, 285–294. doi:10.1081/IPH-120020476

Reddy, A. M., Seo, J. H., Ryu, S. Y., Kim, Y. S., Kim, Y. S., Min, K. R., & Kim, Y. (2004). Cinnamaldehyde and 2-methoxycinnamaldehyde as NF-κB inhibitors from *Cinnamomum cassia. Planta Medica, 70*, 823–827. doi:10.1055/s-2004-827230

Tragoolpua, Y., & Jatisatienr, A. (2007). Anti-herpes simplex virus activities of *Eugenia caryophyllus* (Spreng.) Bullock & SG Harrison and essential oil, eugenol. *Phytotherapy Research, 21*, 1153–1158. doi:10.1002/ptr.2226

Vigan, M. (2010). Essential oils: Renewal of interest and toxicity. *European Journal of Dermatology, 20*, 685–692.

Wang, W., Li, N., Luo, M., Zu, Y., & Efferth, T. (2012). Antibacterial activity and anticancer activity of *Rosmarinus officinalis* L. essential oil compared to that of its main components. *Molecules, 17*, 2704–2713. doi:10.3390/molecules17032704

Wei, F.-X., Liu, J.-X., Wang, L., Li, H.-Z., & Luo, J.-B. (2008). Expression of bcl-2 and bax genes in the liver cancer cell line HepG2 after apoptosis induced by essential oils from *Rosmarinus officinalis. Journal of Chinese Medicinal Materials, 31*, 877–879.

Wu, S.-J., & Ng, L.-T. (2007). MAPK inhibitors and pifithrin-alpha block cinnamaldehyde-induced apoptosis in human PLC/PRF/5 cells. *Food and Chemical Toxicology, 45*, 2446–2453. doi:10.1016/j.fct.2007.05.032

Yang, X.-Q., Zheng, H., Ye, Q., Li, R.-Y., & Chen, Y. (2015). Essential oil of Cinnamon exerts anti-cancer activity against head and neck squamous cell carcinoma via attenuating epidermal growth factor receptor - tyrosine kinase. *Journal of B.U.ON, 20*, 1518–1525.

Yoo, C.-B., Han, K.-T., Cho, K.-S., Ha, J., Park, H.-J., Nam, J.-H., & Lee, K.-T. (2005). Eugenol isolated from the essential oil of *Eugenia caryophyllata* induces a reactive oxygen species-mediated apoptosis in HL-60 human promyelocytic leukemia cells. *Cancer Letters, 225*, 41–52. doi:10.1016/j.canlet.2004.11.018

15

Synthesis, cytotoxicity, and long-term single dose anti-cancer pharmacological evaluation of dimethyltin(IV) complex of N(4)-methylthiosemicarbazone (having ONS donor ligand)

Md. Shamsuddin Sultan Khan[1], Md. Abdus Salam[2], Rosenani S.M.A. Haque[2], Amin Malik Shah Abdul Majid[1]*, Aman Shah Bin Abdul Majid[3], Muhammad Asif[1], Mohamed Khadeer Ahamed Basheer[1] and Yaseer M. Tabana[1]

*Corresponding author: Amin Malik Shah Abdul Majid, EMAN Research and Testing Laboratory, School of Pharmaceutical Sciences, Universiti Sains Malaysia, Minden, Pulau Pinang, Malaysia

E-mail: aminmalikshah@gmail.com

Reviewing editor: Tsai-Ching Hsu, Chung Shan Medical University, Taiwan

Additional information is available at the end of the article

Abstract: Background and Objective: Toxicity of the chemotherapeutic compounds is widely investigated. An organotin (IV) derivative was designed to modulate the toxicity and long-term anticancer efficacy of the single dose. Materials and Methods: The reaction of dimethyltin(IV) dichloride with N(4)-methylthiosemicarbazone derived by condensation of 4-methylthiosemicarbazone with 5-bromo-2-hydroxybenz-aldehyde was prepared in 1:1 M ratio in absolute methanol. The newly synthesized complex was characterized by elemental analysis, FT-IR, electronic, and ^{1}H, ^{13}C and ^{119}Sn NMR spectroscopy. In vitro cytotoxicity (MTT, (3-(4,5-Dimethylthiazol-2-yl)-2,5-Diphenyltetrazolium Bromide)), anticancer (migration, clonogenic, 3D tumor aggregation, nucleus condensation and mitochondrial membrane potential) activity, and in silico QSAR and molecular docking studies were performed. Results: The title

ABOUT THE AUTHORS

The angiogenesis-based drug discovery project in EMAN RESEARCH is a kind of multidisciplinary and translational research not for profit. EMAN team is performing to design and characterize the efficacy of the candidate drug with highest safety and minimum toxicity. EMAN core focus is on the natural organic molecule and their effect in metabolic diseases. The metallic compounds have highest toxicity in comparison to other small molecules. EMAN team is investigating how to minimize the toxicity of the metallic compound and designing scaffold hope and pharmacophore not only using Tin, also Silver, Gold, and Platinum. This project is a continuation of screening of thiosemicarbazone-containing molecule with halogen, metal, aromatic rings, and alkyl substituents for the discovery of non-toxic molecule with cytostatic activity in cancer cells.

Caption: Md. Shamsuddin Sultan Khan is the leading researcher of EMAN Research and Testing Laboratory for the discovery of molecules with non-toxic, micronutrient, and complementary activity against cancer.

PUBLIC INTEREST STATEMENT

Organotin (IV) molecules have broader therapeutic range in the pharmaceutical industry. This complex is very flexible to design a safest drug for the cancer patient. Our designed molecule (BHBM) is one of proof of concept for anticancer therapy with wider safety margin. The chemical functional group of the BHBM compound has shown anticancer activity with little single dose for more than a week which will reduce the cost of the cancer treatment because of its high efficacy.

compound was observed to be potent and selective toxic against MCF-7, HCT-116, and A549 human cancer cell lines. Moreover, this derivative was found to be less-cytotoxic and higher cytostatic at the single dose than other organotin (IV) complexes due to modulation of chelation of ligand with Sn(IV) ion. The anticancer activities against A549 cancer cells, however, were only moderate. The reason for this could be due to inhibition of enzymatic reaction in the cells for glucose uptake, DNA and protein synthesis. Discussion and Conclusion: The resonance impact of aromatic rings, hydrogen bonding, and ROS reduction, NO generation, caspase induction showed potential impact to the cancer cell apoptosis, antimigration, and inhibition of tumor aggregation of this compound.

Subjects: Biochemistry; Bioinformatics; Pharmaceutical Science; Pharmacology

Keywords: Synthesis; Organotin(IV) complex; NO; ROS, caspase, antioxidant, anticancer; QSAR

1. Introduction

Cytotoxic compounds are widely investigated and used to stop the proliferating cells nonspecifically (i.e. cancer cells and normal cells). The side effects of these compounds were severe and tolerated because of no alternative options before developing the targeted therapy. As a result, for undesirable effects in preclinical models is high, which a cause to terminate the toxic compounds. Therefore, searching a compound with the high safety margin with not only tolerable undesirable side effects, but also a good efficacy profile is still involved in cancer drug discovery. To reduce the safety attrition and selectivity to the efficacy and toxicity is one of the second strategy to design a small molecule which could bind to the desired target exclusively. For example, the designing of organotin complexes by replacing the cisplatin from cisplatin-based metallic chemotherapeutic drugs brings much opportunity in the anticancer field to develop it further as selective compound due to their flexible structural modification for strong coordination ability, multidentate chelate mode with monomeric or dimeric structure (Yin & Chen, 2006).

Organotin complexes have anticancer activity and the potential for active metallopharmaceuticals. The antiproliferative activity of different types of Organotin complex is not new in the cancer drug discovery and development. The antitumor properties of tin complexes was first discovered in 1929 (Collier, 1929). The apoptosis activity of the Organotin(IV) complex was established by Cima and Ballarin (1999), Pellerito et al. (2005), Tabassum and Pettinari (2006). The organotin(IV) complexes have both cytotoxic and cytostatic roles in anticancer field. Most of them are highly toxic and mechanism of action remains unknown. Organotin(IV) compounds with carboxylates as ligands such as vinyltin and phenyltin complexes, for example, $[Sn(CH\ CH_2)_3\{-OOCC_6H_3-3,4-(NH_2)_2\}]n$, $[Sn(C_6H_5)_3\{OOCC_6H_3-3,4-(NH_2)_2\}]$, $[Sn(C_6H_5)_3\{OOC-2-C_6H_4\ N\ NC_6H_4\ N-4-(CH_3)_2\}]$ and $[Sn(CH\ CH_2)_3\{OOC-2-C_6H_4\ N\ NC_6H_4\ N-4-(CH_3)_2\}]$ have been found (Pruchnik et al., 2002) as strong cytostatic against human cancer cell lines (IC_{50} range 0.1–3.0 µg/ml). In vitro antiproliferative activity of the organotin (IV) complexes is compared against several human tumor cell lines with doxorubicin, cisplatin, 5-fluorouracil, methotrexate, and etoposide. From these investigations, it was established that organotin (IV) complexes showed their activity from highly toxic and static to moderate toxic and static based on the nature (alkyl/phenyl/aryl) and size of covalently attached R groups of Sn(IV) atom, partition coefficients, bulkiness of the functional groups R attached to Sn(IV), organotin moiety (R), the ligand (L), the number of tin atoms, and the number of free coordination positions against the tumor cell lines used. Organotin (IV) complex with α-amino acids and their derivatives shows higher cytotoxic and cytostatic effects because of the alkyl groups bound to tin (Tian et al., 2005) than those of the clinically widely used cisplatin. Organotin(IV) complex with the chloro-substituted benzohydroxamato and with the length of the carbon chain of the alkyl ligand also shows higher toxicity that is revealed in organotin(IV) complexes with thione/thiol, too (Collery & Perchery, 1993; Gielen, 2003; Saxena & Huber, 1989; Xanthopoulou et al., 2008).

Hydrazone ligands and their coordination Sn(IV) complexes have an important impact on the inhibition of cancer cell growth. The growth mechanism of cancer cell can be blocked by the formation of chelate to metal, i.e. Sn(IV) from the cell. The hydrazone ligand molecule is versatile as azomethine complex and depicted by the structure of $R_2C=NNR_2$. They are interlinked by nitrogen atoms and different from imines and oximes. This metal coordinated ligand forms the stable chelates with transition metal or metal present in the cell can prevent any enzymatic reaction to the growth of cancer cell. Cancer cell proliferation is inhibited for the chelation of metal complex (Richardson, 1997) due to cytotoxic oxygen radicals by preventing uptake of metal from transferrin of the neoplastic cells. For example, NF-κB is activated in most of the cancer cells and promotes proliferation, inflammation and inhibits apoptosis. The DNA synthesis in the NF-κB can be blocked by the alkylation of glutamine amino acid. NF-κB is produced by phosphorylation of κB subunit and stimulates the transcription gene (Reddy et al., 2011). Chelation of the metal tin complex can be used to prevent the activation of NF-κB through the alkylation of glutamine or other active amino acids involved in cancer cell proliferation. Due to the formation of stable chelate, the metabolic and enzymatic reactions cannot introduce with hydrazone. For example, pyridoxal isonicotinoyl hydrazone ligands and their tin(IV), iron(III), gallium(III), and copper(II) complexes are synthesized for antitumor activity. These ligands are specific to inhibit mammary tumors and leukemias in mice due to high interaction with the metal in the cell (Richardson, 1997). 2-hydroxy-1-naphthaldehyde isonicotinoyl hydrazone and its iron complexes (Lovejoy & Richardson, 2002; Richardson & Bernhardt, 1999; Yuan, Lovejoy, & Richardson, 2004) and di-2-pyridyl ketone isonicotinoyl hydrazone and its analogs (Bernhardt, Caldwell, Chaston, Chin, & Richardson, 2003; Becker, Lovejoy, Greer, Watts, & Richardson, 2003) have anticancer activity for their iron chelation efficiency. Aroyl hydrazones of pyridoxal and salicylaldehyde hydrazone, 2-Benzoxazolyl, and 2-benzimidazolyl hydrazones, cyano acetic acid hydrazones of 3- and 4-acetylpyridine, 2,6-dimethylimidazo[2,1-b]-[1,3,4]thiazole-5-carbohydrazide, organotin(IV) complexes of 2-hydroxy-1-naphthaldehyde 5-chloro-2-hydroxy benzoylhydrazone are active against in several cancer cell lines (Cocco, Congiu, Lilliu, & Onnis, 2006; Easmon et al., 2001; EL-Hawash, Abdel Wahab, & El-Demellawy, 2006; Johnson, Murphy, Rose, Goodwin, & Pickart, 1982; Min et al., 2013; Terzioglu & Gürsoy, 2003; van Reyk, Sarel, & Hunt, 2000).

Thiosemicarbazone derivative having halogen (i.e. Cl) substitutions has been studied for anticancer activity in our lab by Mouayed et al. (2015) with cytotoxicity 5 μg/ml in HCT-116 and MCF7 cell lines. The organtotin (IV) thiosemicarbazone derivatives (R = Me, Ph, o-ClPhCH2) have higher cytotoxicity 0.09 μg/ml to <10 μg/ml in A549 cells and 0.5–4.1 μg/ml in HCT-8 cells (Min et al., 2013). The long-term pharmacological evaluation has not been conducted for other related derivative of this group. Therefore, our present study is conducted to screen the further anticancer pharmacological activity of the organotin (IV) derivatives (Scheme 1) as it may be a suitable candidate for modifications in order to improve tolerable cytotoxicity, dissolution properties and long-term effect of the single dose in cancer. Further, in the present study a derivative of hydrazone-based organotin (IV) compound is redesigned as thiosemicarbazone derivative having halogen (i.e. Br) substitutions based on the cancer cell pharmacology. The derivative is designed to modulate the cytotoxicity in the cancer cells as well as to reduce the toxic effects to normal cell lines. Quantitative structure activity relationship (QSAR) and molecular docking were performed to determine the possible mechanism of action related to the structure.

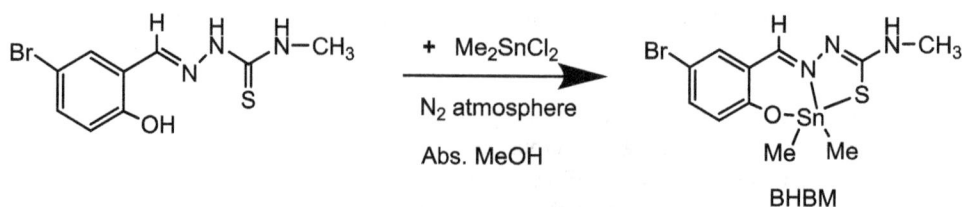

Scheme 1. Synthesis of dimethyltin(IV)-5-bromo-2-hydroxybenzaldehyde-N(4)-methylthiosemicarbazide complex.

2. Experimental

2.1. Materials and methods
All reagents were purchased from Fluka, Aldrich and JT Baker. All solvent were used without further purification. Melting point was measured by the Stuart Scientific SMP1 melting point apparatus. UV–Vis spectra were recorded in DMSO solution with a Perkin Elmer Lambda 25 UV-vis spectrophotometer. Infrared (IR) spectra were recorded by the Perkin Elmer System 2000 spectrophotometer using the KBr disk method in the range of 4,000–400 at room temperature. ^{1}H, ^{13}C, and ^{119}Sn NMR spectra were recorded on a Bruker 500 MHz-NMR spectrophotometer relative to SiMe$_4$ and Me$_4$Sn in DMSO solvent. Elemental analysis was conducted by the Perkin Elmer 2400 Series-11 CHN analyzer. Molar conductivity measurements were carried out with a Jenway 4510 conductivity meter using a DMSO solvent.

2.2. Synthesis of 5-bromo-2-hydroxybenzaldehyde-N(4)-methylthiosemicarbazone (H$_2$BHBM)
A solution of 5-bromo-2-hydroxybenzaldehyde (0.95 g, 4.75 mmol) in ethanol (20 ml) was added to a solution of 4-methyl-3-thiosemicarbazide (0.5 g, 4.75 mmol) in ethanol (20 ml). The resulting yellow solution was refluxed with stirring for 2 h and then filtered. A yellow product (formed at room temperature) was then filtered, washed with ethanol, and air dried. M.p: 185–187°C, (1.06 g, 78%). UV-vis (DMSO) $\lambda_{max/nm}$: 260, 322, 361. FT-IR (KBr, cm^{-1})ν_{max}: 3332 (s, OH), 3195 (s, NH), 1620 (m, C=N), 1556 (s, C$_{aro}$-O), 1368, 856 (w, C–S), 988 (m, N–N). ^{1}H NMR (DMSO-d$_6$, ppm): 11.42 (s, 1H, OH), 10.35 (s, 1H, N-NH), 8.57 (s, 1H, CS-NH), 8.57 (s, 1H, CH = N), 8.15, 7.83, 6.83, (s, d, d, aromatic- H), 3.01 (s, CH$_3$). ^{13}C NMR (DMSO-d$_6$,ppm): 187.48 (C=S), 155.44 (C=N), 137.01–111.04 (aromatic-C), 30.80 (CH$_3$). Anal. Calc. for C$_9$H$_{10}$BrN$_3$OS: C, 37.47; H, 3.47; N, 14.57. Found: C, 37.60; H, 3.82; N, 14.45%.

2.3. Synthesis of [Me$_2$Sn(BHBM)] complex
5-bromo-2-hydroxybenzaldehyde-N(4)-methylthiosemicarbazide (0.228 g, 1.0 mmol) was dissolved in absolute methanol (10 mL) under nitrogen atmosphere in a round bottom reaction flask. Methanol solution of dimethyltin(IV) dichloride (0.219 g, 1.0 mmol) was added dropwise and resulted into a yellow solution. The resulting reaction mixture was refluxed for 4 h (Figure 1(A)) and cooled to room temperature. The yellow microcrystals were obtained from slow evaporation of the resulting solution at room temperature. The microcrystals were filtered, and washed with a small amount of cold methanol, and dried in vacuum over silica gel. Yield: 0.35 g, 78%: Mp.: 240–242°C: Molar conductance (DMSO) Ω^{-1} cm^2 mol^{-1}: 10.36: UV-Visible (DMSO) λ_{max} (nm): 274, 338, 384, 417: FT-IR (KBr, cm^{-1}) ν_{max}: 3,252 (s, NH), 1,597 (m, C=N), 1,537 (s, C$_{aro}$-O), 1,022 (w, N–N), 1,322, 828 (m, C–S), 652 (w, Sn–C), 565 (w, Sn–O), 455 (w, Sn–N). ^{1}H NMR (DMSO-d$_6$, ppm): 9.81 (s, 1H, CS-NH), 8.43 (s, 1H, s, CH=N), 7.25–7.42 (m, 3H, aromatic-H), 3.05 (m, 3H, CH$_3$), 1.07 (s, 6H, Sn–CH$_3$). ^{13}C NMR (DMSO-d$_6$, ppm): 178.11 (C=S), 147.32 (C=N), 110.54–145.43 (aromatic-C), 31.45 (CH$_3$), 18.46 (Sn–CH$_3$). Anal. Calc. for C$_{11}$H$_{14}$N$_3$BrSOSn: C, 30.38; H, 3.24; N, 9.66%. Found: C, 30.34; H, 3.20; N, 9.61%.

2.4. Thermal and chemical stability studies
BHBM powder (50 mg) was transferred into the glass vials and applied 60°C in the oven for 48 h. After that, some portion of the compound was incubated with LiOH and the amount of BHBM was measured. The degraded quantity of the compound was calculated using mass balance. The effect of pH on the compound and its chemical stability was determined by incubation of the compound in an aqueous HCl (pH 2) and Krebs-Heneseleit bicarbonate buffer (pH 7.4) at 37°C for 2 h followed by the analytical HPLC assay for BHBM.

2.5. Stability in human serum
BHBM solution (10 mM) with water was added to preheated (37°C) human serum (Sigma) with the resulting concentration 0.5 mM. The 500 μL solution then withdrawn at appropriate intervals and added to 500 μL acetonitrile containing 0.1% trifluoroacetic acid to deproteinize the serum. The solution was vortexed and centrifuged for 10 min at 2,000 rpm and filtered by 0.45 μm PTFE filters (Biofil). The filtrate was analyzed by RP-HPLC method.

(A)

(E)-2-(5-bromo-2-hydroxybenzylidene)-
N-methylhydrazine-1-carbothioamide

5-bromo-2-hydroxy benzaldehyde-
N(4)-methyl thiosemicarbazone

(B)

(C)

(D)

(E)

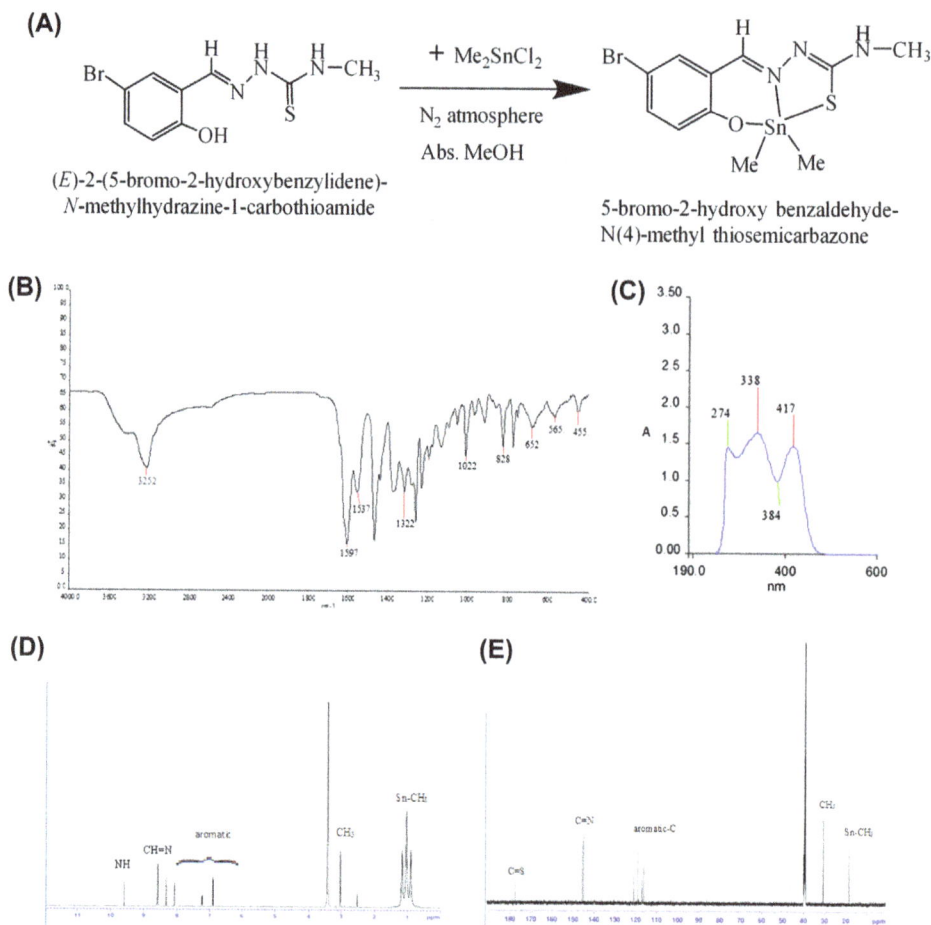

Figure 1. (A) Synthesis of dimethyl tin(IV): 5-bromo-2-hydroxy benzaldehyde-N(4)-methyl thiosemicarbazone; (B) IR spectrum; (C) UV-visible spectrum of dimethyltin(IV)-N(4)-methylthiosemicarbazone complex; (D) ¹H NMR spectrum; (E) ¹³C NMR spectrum.

The HPLC analytical method was developed on Agilent 1200 series coupled to a photodiode array detector (Agilent, CA, USA). Chromatographic separation was achieved at room temperature (25oC) with a C4 reversed-phase column (Thermo Scientific™, USA 2504.6 mm, 5-μM particle size) using a gradient solvent system that consisted of mobile phase acetonitrile and 0.05% formic acid, 50:50 (v/v), pH 5, running at an isocratic mode at flow rate 1.0 ml/min. The injected volume of the sample was 20 μl and the total run time was 15 min with BHBM eluting at 3.2 min. The absorbance was set at 200–450 nm and the UV was achieved at 220 nm which was used for quantification. Quantification was carried out using HPLC retention time and UV spectrum of the analyte. The method was specific and sensitive with a lower limit of quantification of 1 ng/mL and ChemStation LC3D software was used to ensure quantification of BHBM. The variation of intraday and interday was minimum < 10%.

2.6. Antioxidant activity

2.6.1. DPPH radical scavenging assay
DPPH radical scavenging assay was conducted by method of Khan et al. (2013). In brief, 100 μL of molecule 26 was dissolved in MeOH:H2O (1:1) and 100 μL DPPH (200 μmol L−1) prepared in methanol and incubated at room temperature for 30 min. Ascorbic acid was used as reference standards. The amount of remaining DPPH was evaluated at 517 nm. The results were expressed as EC_{50} and the results (I%) were calculated by following equation:

$$I\% = 1 - \frac{A_{sample}}{A_{blank}} \times 100$$

where, A_{blank} is the absorbance of the control reaction (containing all reagents except the test material). EC_{50} was calculated from the graph plotted by percentage inhibition of BHBM.

2.6.2. Ferric-reducing antioxidant power (FRAP) assay

FRAP was used to determine the antioxidant capacity of BHBM using the method of Benzie and Strain (1996). Single concentration of BHBM was mixed with the 150 μL FRAP working solution (300 mmol L^{-1} acetate buffer, pH 3.6, 10 mmol L^{-1} TPTZ in 40 mmol L^{-1} HCl and 20 mmol L^{-1} FeCl$_3$ in a ratio of 10:1:1). The mixed solution was incubated for 8 min and read at 600 nm by Tecan microplate reader. The results were calculated from the standard curve of Ferrous sulfate (FeSO4.7H2O) as reference standard and expressed as nmol Fe^{2+} equivalent μg^{-1} compound.

2.6.3. Intracellular ROS activity

HCT116 cells were seeded in a 48-well plate for 2 days. After 80% confluent of cells, treated the cells by BHBM for 24 h. Cells were washed by PBS and 100 μl ROS reagents were added in each well. Cells were incubate at 37°C for 30 min. Cell lysis buffer was added to the cells after the incubation period and centrifuged at 10,000 rpm for 10 min. Supernatant was collected and diluted with PBS and read in Tecan fluorescence microplate reader at 485–520 nm. Standard curve of the ROS standard was prepared and concentration was determined using the standard curve.

2.7. Ionization constants (pKa) and lipophilicity (logD 7.4) descriptors

The ionization constants of compounds were determined using UV-metric methods. pKa is the 50% protonated state of the molecule that is measured by analyzing changes in multi-wavelength UV spectra during acid-base titration of the sample. This method is more suitable for colored compound due to pH-sensitive chromophores. Potentiometric titration was conducted by at least four separate titrations for BHBM compound. BHBM solutions (20 ml with 1 mM) were prepared by adjusting the ionic strength to 0.15 M with KCl. The solutions' pH was stated with 1.8 using 0.5 N HCl. The solutions were titrated with 0.5 N KOH to the pH of 12.2. Titrations were performed in inert gas system (N2). Ionization constants were determined by spectral analysis as described by Carlos, Martínez, and Dardonville (2013). Briefly, the raw OD values of UV-spectra recorded and processed as following using Microsoft Excel program. At first, OD values of the BHBM were subtracted from blank solutions. Second, data were normalized as Abs400 nm = 0. The spectral difference plot provides the wavelengths of maximum positive and negative absorbance. Absorbance difference Vs pH was plotted from the acid spectra and the spectra at every other pH at the chosen wavelengths. Grahpad Prism program was used to determine the pKa of BHBM using these data worked out by nonlinear regression using Equation (1):

$$\text{Absorbance total} = \frac{\left(\varepsilon_{HA}-\varepsilon_{A^-}\right) \times [10^{(pH-pKa)}]}{1 + 10^{(pH-pKa)}} \times S_t \tag{1}$$

ε_{HA} is the extinction coefficients of the acid forms of the compound and the minimum absorbance difference curve,

ε_{A^-} is the extinction coefficients of the base forms of the compound and maximum absorbance difference curve, S_t is the total compound concentration

Lipophilicity (LogD7.4) of the BHBM was determined by shake-flask procedure at pH 7.4. Phosphate buffer solution was prepared with ionic strength adjusted to 0.15 M with KCl. N-octanol was added to the buffer and shaking for four hour to be saturated of the two phases. BHBM was added to the saturated solution and shaken for 20 min until the partitioning equilibrium of solutes was reached. The solution was centrifuged at 10,000 rpm, 10 min. After separation, BHBM was quantified in the aqueous phase and octanol phase by UV spectrophotometer (Shimadzu) at 230 nm and the following equation was used to calculate the Log D7.4.

$$LogD = Log\left(\frac{C_{oct}}{C_{aq}}\right)$$

2.8. Measurement of the generation of NO from BHBM and its apoptotic effect in HCT116

The amount of NO generated from BHBM was determined in the cultured cancer cells with DMEM media and DMEM without cells. BHBM was added to the media for 5 or 60 min at 37°C. The generation of NO from BHBM was measured by quantification of nitrite (NO_2) using Griess reagent. HCT116 Cell culture supernatant was harvested and mixed with Griess reagent. The mixture was incubated for 10 min at room temperature. The absorbance was recorded at 550 nm using an ELISA reader. Standard curve was constructed from the serial dilutions of sodium nitrite (0.5–100 M) and was used to determine the nitrite concentrations.

Effect of NO donors on HCT116 cancer cell survival was determined by seeding the cells in 96-well plate. Cells were treated with or without BHBM as NO donors. The plate with the BHBM was incubated for 1 h at 37°C. MTT solution (5 mg/ml) was then added to the each well and incubated for another hour at 37°C. Live cells converted into the dark-blue formazan product from the yellow tetrazolium salt of MTT. Formazan was solubilized with DMSO (100 µl per well) and the absorbance at 570 nm was measured in an ELISA reader (Multiscan RC). The absorption of formazan is directly related to the number of viable cells.

3. Anticancer study design

After synthesizing the compound, anticancer activity was screened to determine the short-term and long-term efficacy of BHBM (Figure 2). This strategy determined the dose repetition necessity and effect of single dose for long-term activity.

3.1. Antiproliferative assay

Human cancer cell lines MCF-7 (breast), A549 (lung, non-small cells), HCT-116(colon), and EAhy926 (normal) endothelial cells were collected from ATCC USA. Normal epithelial MDCK cell line was a kind gift from Michael M. Gottesman, Laboratory of Cell Biology, National Cancer Institute, USA. We cultured the cells in RPMI 1640 (GIBCO BRL, Life Technologies) and DMEM supplemented with 5% of fetal bovine serum and penicillin (final concentration of 1 mg/mL) and streptomycin (final concentration of 200 µg/mL). A confluence of 70–80% cells were used for seeding in 96-well plates (100 µL cells/well) and incubated (0.25–250 µg/mL) for 48 h at 37°C and 5% of CO2.

Before seeding, old medium was aspirated from the cell culture flask and added freshly prepared media after washing by sterile phosphate buffered saline (PBS) (pH 7.4), 2–3 times. Trypsin was added followed by discarding PBS and distributed evenly onto cell surfaces. Cells were incubated at 37°C in 5% CO2 for 2–5 min. To detach the cells properly, the flasks were gently tapped and observed under inverted microscope. About 5 mL of fresh media (10% FBS) was added to neutralize trypsin activity. Cells were counted and seeded 10,000 cells per well. The plate was incubated for 24 h to be ready for the treatment of cells. Six low to high doses were used in the treatment. After 48 h of treatment, MTT reagents were added and incubated for 4 h. 20 µl MTT lysis solution was added to each well and read the plate at 570 and 620 nm wavelengths using a high-end TecanM200 Pro multimode microplate reader. Obtained data were analyzed to determine the concentration of compound which inhibited the cell growth by 50% (IC_{50}) from the optical density. Tamoxifen and 5-FU were used as the standard drug. The results are calculated from two independent experiments, each done in triplicate.

3.2. Migration assay

We performed migration assay using three cancer cell lines. Cells were cultured and 20,000 cells were seeded per well in six-well plate. We kept cells to become 95% confluent monolayer to create wound at the middle of the well from corner to corner. The easiest and cheapest way to create wound

Experimental Design

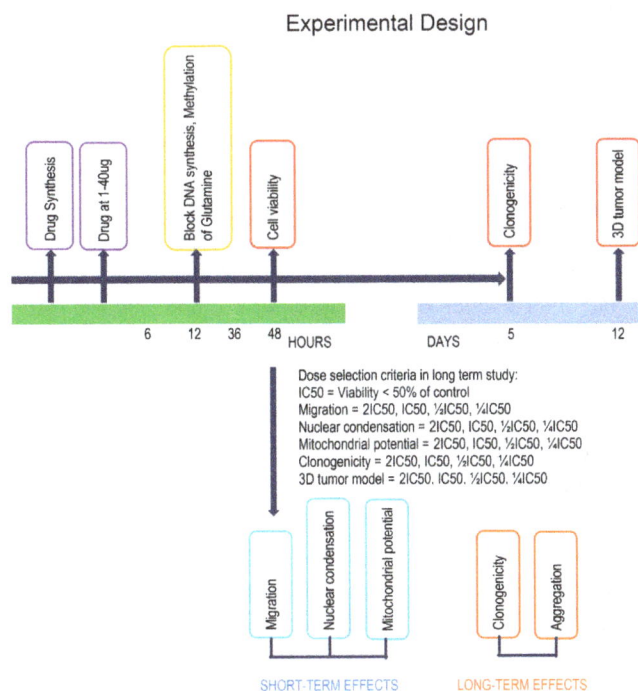

Figure 2. The compound screening procedure: first, synthesized compound was tested in three cancer cell lines, MCF7, HCT116, A549, and normal cell lines EAhy926, at multiple concentrations of 6.25 to 200 μg/ml.

Note: The percentage viability of cells was measured as short-term effects and changes in clonogenic potential and 3D tumor spheroid assay as long-term effects of the synthesized compound. Two criteria were used to select the concentration of compound for further analysis: IC_{50} of the compound that reduce cell viability to 50% of control and high–low quantity of IC_{50}. Next, the compound was used as double, half, and quarter IC_{50} concentrations to investigate short-term effects (migration, nuclear condensation, mitochondrial potential) for the ability to induce apoptosis and long-term effects (clonogenic, 3D tumor) to influence anticlonogenic, antiaggregation (3D tumor model) activity.

manually is sketch a straight-line using 200-μl micropipette tips with the aid of sterile ruler. The cells were gently washed two times by PBS. Cells in six wells were treated with four doses of compounds, standard drug, and without treatment for negative control, respectively. Five to six pictures were captured by inverted microscope for wound of each well at 0, 6, 12, 18, and 24 h. The width of the wound was measured using Image-J software version-1. Thirty readings were taken for each well. The results were calculated to determine the percentage of migration inhibition by following equation:

$$\% \text{ wound closure} = 1 - \frac{\text{width at the indicated times(h)}}{\text{width at zero time}} \times 100$$

3.3. Colony formation Assay

The colony formation assay was performed in three cancer cell lines. Cells were seeded at 1,000 cells per well in 6 wells plate. After 24 h of seeding, cells were treated with 4 doses of compound, 1 dose of standard drug and 1 well used as negative control. After 48 h of treatment, old media was aspirated and fresh media added after washing with PBS. Cells were under observation until 50 colonies were formed in the negative control. After 5–7 days, the experiment was stopped and washed the cells with PBS before adding 4% paraformaldehyde. About 300 μl paraformaldehyde was used in each well to fix the cells. After incubation for 30 min, cells were washed with PBS and 0.2% crystal violet was added to stain the cells. Plating efficiency (PE) and survival fraction were calculated by following equation:

$$PE = \frac{number\ of\ colonies\ formed}{number\ of\ cell\ seeded} \times 100$$

3.4. Apoptotic assay

3.4.1. Determination of nuclear condensation by Hoechst 33258 stain
20,000 cells were seeded per well in 48-well plate. Cells were treated after 48 h of incubation with 70–80% confluent monolayer. Lung cancer cell (A549) was treated with four concentrations of BHBM time dependently at 6 h, 12 h, and 24 h. Cells were fixed by adding 4%, 300 µl paraformaldehyde in each well and incubate for 20 min. After incubation, the cells were washed with PBS and added 100-µl hoechst 33258 stain (1 µg/ml in PBS). After 30 min of incubation, cells were observed under an inverted microscope to capture the pictures to analyze irregular-shaped cells and regular-shaped cells. Cells with bright and shiny presents the condition of condensed nucleus and shrinked cyto-plasm denoted to apoptotic morphology. These apoptotic cells were counted for four fields per well. The percentage of apoptotic cells were determined by following equation:

$$\% \text{ Apoptotic index} = \frac{\text{number of apoptotic cell}}{\text{total number of live cell}} \times 100$$

3.4.2. Determination of mitochondrial potential
Lung cancer cell line A549 was used to study the mitochondrial inhibitory potential of BHBM. Cells were treated as described in a hoecht staining assay. During the addition of hoecht stain, rhodamine 123 (5 µg/ml PBS) was added, followed by hoecht at the same quantity. Definition and calculation of mitochondrial potential condition are similar to nuclear condensation as it is bright and shiny with irregular shape.

3.5. Spheroid assay
We optimize the spheroid assay through hanging drop aggregation method. Hanging drops were prepared using lung cancer cell (A549) suspension containing glucose. Per hanging drops containing 20,000 cells were placed on the petridish lid (100 mm) and incubated it as inverted position over the dishes for 3 days. Spheroid was formed inside the hanging drops and transferred them from lid to 48-well plate after coated with 1% agarose. Diameter of spheroid was calculated to determine the inhibitory potential of BHBM by Image-J soft.

3.6. Effect of BHBM in apoptotic targets of Caspase 3/7, 8, 9
The assay was carried out according to manufacturer's instructions (Promega, USA). The cells were seeded in 96-well plates at 20 x 10⁴ cells/well in 200-µl medium and incubated for 24 h to allow the attachment of cells. The treated cells were incubated for approximately 18 h. Then, 100 µl of the caspase reagent containing cell lysis buffer and specific luminogenic substrate for caspase 3/7, 8, and 9. After incubating the plates for 30 min, the luminescence intensity was measured using micro-plate reader (Tecan, Switzerland). The activity of the BHBM in caspase was represented by fold change of the caspase activity upon treatment with BHBM. The fold changes in the caspase activity were measured by the following formula:

$$\text{Fold change} = \frac{L_s - L_b}{L_c - L_b}$$

where: L_s = Luminescence reading of wells treated with tested extract

L_b = Luminescence reading of blank

L_c = Luminescence reading of wells treated with vehicle

The results were stated as mean of fold changes in caspase activity in comparison to control ± SD ($n = 3$).

3.7. Computational QSAR

CODESSA was used to generate a set of molecular descriptors used in the QSAR analysis (da Silva et al., 2012; Katritsky, Lobanov, & Karelson, 1996; Pacheco et al., 2013). In the statistical analysis for the QSAR study, the following parameters were employed: maximum number of descriptors per model, 1; the criterion of significance of 15 parameters, 0.01; level of high correlation, 0.99; and level of intercorrelation significance, 0.8.

3.8. Computational molecular docking

The synthesized compound was designed and the structure was analyzed using Chembiodraw Ultra v.13.0 and CODESSA. The protein structure file is taken from PDB (www.rcsb.org/pdb) was cleaned from polymers, water molecules, and heteroatoms. Automated docking is used to determine the orientation of BHBM binding with NFκB. A genetic algorithm method implemented in the program Auto Dock Vina (Trott & Olson, 2010) was employed. Gasteiger charges were added, merged non-polar hydrogens, and set TORSDOF to six rotatable bonds to BHBM. For docking calculations, Gasteiger Marsili partial charges are assigned to the BHBM and nonpolar hydrogen atoms were merged (Gasteiger & Marsili, 1980). All torsions are allowed to rotate during docking. The grid map is centered at the protein active site and is adjusted such that it can accommodate the binding site of the protein at the torsional degree of freedom with 0.5 units. The genetic algorithm is applied for minimization, using default parameters. The number of docking runs, the population in the genetic algorithm, the number of energy evaluations, and the maximum number of interactions are 50, 250, 100,000, and 10,000, respectively.

4. Results and discussion

4.1. Chemistry and characterization

The reaction of Me_2SnCl_2 with 5-bromo-2-hydroxybenzaldehyde-N(4)-methylthiosemicarbazone in 1:1 mol ratio takes place according to Figure 1. This newly synthesized complex is yellow solids, air stable and soluble in DMSO, DMF, MeOH, $CHCl_3$, and CH_2Cl_2 solvents. The composition of this complex was confirmed by their analytical data and structures were suggested by spectroscopic investigations. The electronic spectrum of complex showed bands at 274, 338, and 384, nm corresponding to n-π* transition of the azomethine portion and π-π* transitions of the aromatic ring and thiolate function. The band at 417 nm was suggested ligand → metal charge transfer (LMCT) (Figure 1(C)). In IR spectrum, the sharp, strong, single peak at 3,252, 1,597, 1,537, 1,022, and 1,322, 828 cm^{-1} for complex gives evidence for the presence of $\nu(NH)$, $\nu(C=N)$, $\nu(C_{aro}-O)$, $\nu(N-N)$, and $\nu(C-S)$ bands, respectively (Figure 1(B)). The appearance of new bands of the complexes at 565 and 455 cm^{-1} assigned to $\nu(Sn-O)$ and $\nu(Sn-N)$ supports the bonding of nitrogen and oxygen to tin(IV) (Haque & Salam, 2015), which are not observed in the spectra of the ligand (H_2BHBM). The peak at 652 cm^{-1} in the IR spectrum of complex supports the presence of $\nu(Sn-C)$ band in the structure. The ^1H NMR spectra of the ligand also exhibit OH and N-NH proton signals which disappear in the complex, showing the involvement of phenolic oxygen and azomethine nitrogen in bonding with the tin(IV) atom. However, in the ^1H NMR spectrum of complex, the signals of protons of the ligand from protons of the organic groups in the tin(IV) confirms metal to ligand in the complex (Figure 1(C)). Additionally, ^{13}C NMR spectral data support the proposed structures. Shifts in the positions of carbons adjacent to azomethine nitrogen (C=N) and phenolic oxygen atom clearly indicate the bonding of the azomethine nitrogen and phenolic oxygen atoms to tin(IV) atom (Figure 1(D)). In ^{119}Sn NMR spectra of complex, $\delta(^{119}Sn)$ value is found at −171.23 ppm, indicating that the tin(IV) atom is five-coordinate in this complex (Holeček, Nádvorník, Handlíř, & Lyčka, 1986).

4.2. ADME physicochemical properties of BHBM

In this study, we determined the dissociation constants (pKa) lipophilicity of the BHBM. Dissociation constants (pKa) of the BHBM were determined by the potentiometric titration in aqueous solution using a 96-well plate in multiscan microplate reader. The value of the absorbance was collected and measured the pKa. BHBM is a soluble compound and pKa values are high. Shake-flask technique was used to determine the lipophilicity of BHBM using partition pair buffered water (pH 7.4)/noctanol. The distribution coefficients (log D7.4) are shown in Table 2. BHBM showed the lower value for lipophilicity. Stability of BHBM in serum and in high temperature also was defined, but no unstable effect was found from HPLC data.

4.3. Antiproliferative mechanism of BHBM

Antiproliferative effect of the 5-bromo-2-hydroxy benzaldehyde-N(4)-methylthiosemicarbazide (BHBM) was investigated on breast (MCF7), colon (HCT 116), lung (A549), and endothelial (EAhy926) cell lines. Antiproliferative activity was determined by the MTT method and the activity of BHBM was compared with positive control of Tamoxifen and 5-fluorouracil. The concentration of BHBM that elicited 50% inhibition of proliferation of cancer cells (IC$_{50}$) (Figure 3) was varied in different cell lines. The analyzed data obtained from BHBM noted that MCF7 and HCT 116 Cell lines were most sensitive to cancer cells as attested by BHBM that positively inhibited the cells growth at IC$_{50}$ of 6 and 4 µg/ml, respectively (Figure 3). A549 cells were the least sensitive to BHBM. BHBM showed IC$_{50}$ values in the range of 20–18 µg/mL. Indeed, it also showed low selectivity index (SI) for A549 cancer cells, presenting weak antiproliferative activity against nontumorogenic endothelial EAhy926 cells and epithelial cells MDCK (Figure 3, Table 1). BHBM was also promising against HCT 116 cells by exhibiting IC$_{50}$ around 4 µg/mL (Table 1). HCT 116 cells were highly sensitive to BHBM (Table 1) may be due to high affinity of compounds to DNA (Han & Yang, 2002), the structure of the molecule and the coordination number of the tin atoms (Collinson & Fenton, 1996; Koch, Baul, & Chatterjee, 2008; Xanthopoulou et al., 2008). BHBM compromised with the MCF7, HCT 116, and A549 cells growth by 50% when used at higher concentrations 12, 8, and 40 µg/mL, respectively, while HCT16 cells were most sensitive to BHBM. Cytostatic activity was observed at the higher concentration of BHBM may be due to stopping the cell growth at synthetic (S) phase, during which time the doubling of DNA occurs. For HCT 116 cells BHBM was the most potent with IC$_{50}$ values of lower than 8 µg/mL, which may be due to the type of alkyl groups (Br) attached to the Sn moiety. Also, BHBM exhibited the largest spectrum of action. BHBM showed considerable antiproliferative activity (IC$_{50}$ between 5 and

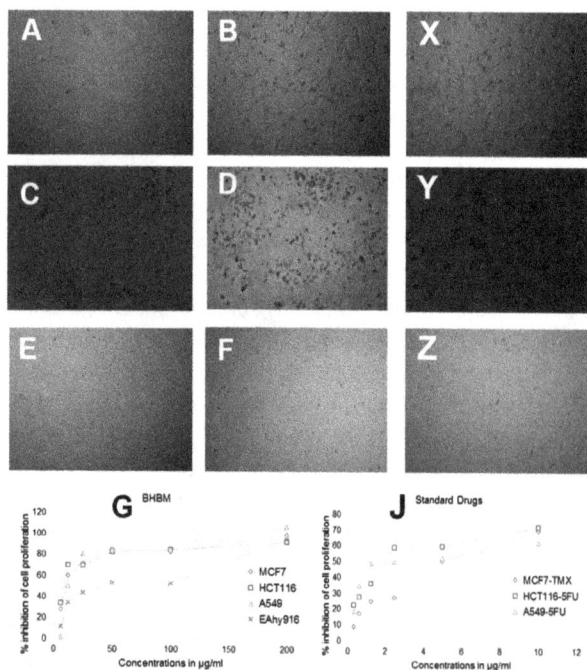

Figure 3. Photomicrographic images of different cancer cell lines and normal endothelial cells treated with BHBM tamoxifen, and 5-fluorouracil (5FU).

Note: Images were obtained from EVOS inverted microscope after 48 h of treatment. MCF7 (A: untreated, B: treated), HCT116 (C: untreated, D: treated), A549 (E: untreated, F: treated) cells were incubated with BHBM for 48 h at different doses. Percent inhibition of cell proliferation (G) was obtained and values were expressed as µg/ml. MCF7 (X), HCT116 (Y), A549 (Z) cells were incubated with tamoxifen and 5-FU for 48 h at different doses. Percent inhibition of cell proliferation (J) was obtained and values were expressed as µg/ml. Results are presented as mean ± SD of three separate experiments (n = 6).

Table 1. IC$_{50}$ and selectivity index (SI) of the various melanoma (cancer) cells

Samples	IC$_{50}$ (µg/ml) on MCF7	IC$_{50}$ (µg/ml) on HCT 116	IC$_{50}$ (µg/ml) on A549	IC$_{50}$ (µg/ml) on EAhy926	IC$_{50}$ (µg/ml) on MDCK	Selectivity Index (SI)			NO generation effect in HCT116 (IC$_{50}$ µg/ml)
						MCF7	HCT 116	A549	
BHBM	6.0±0.57	4.0 ± 0.65	20 ± 0.54	40 ± 1.2	55 ± 1.7	7 ± 2.5	10 ± 2.8	2 ± 0.8	4.5 ± 1.7
Tamoxifen	7.0±0.98	–	–	20 ± 2.5	15 ± 2.87	3 ± 1.5		–	ND
5 fluoro-uracil	–	2.0 ± 0.75	3.0 ± 0.78	20±2.9	10±2.5	–	10 ± 1.7	6 ± 0.1.9	ND

Note: SI is the ration of IC$_{50}$ for normal cell line and cancerous cell line after 48 h of treatment. IC$_{50}$ was measured after 48 h treatment with compounds using MTT staining and it was expressed as % of vehicle (0.05% DMSO)-treated control. For comparison normal cells endothelial cell line EAhy926 and normal epithelial cell line MDCK and effects of standard chemo drugs Tamoxifen and 5-fluorouracil were used. Results are presented as mean ± SD of three separate experiments ($n = 6$).

20 µg/ml) against all the cancer cell lines (Table 1). These preliminary data may provide the cytostatic effects than cytotoxic effects of the BHBM to the specific candidate target other than healthy cells (IC$_{50}$, 40 µg/ml in EAhy926 and IC$_{50}$, 55 µg/ml in MDCK) through the additive or synergistic effects of the substituents of BHBM.

Moreover, BHBM might form a stable form of the chelates with the transition metals found in the cells. This formation inhibits enzymatic reaction in the cells for glucose uptake, DNA and protein synthesis. The coordination further increases the activity of hydrazones by the additional C=N bonds. Metal ion losses the polarity due to sharing of its positive charge within chelating ring with donor atom during coordination. It results the lipophilicity of the central metal atom which provides more efficient permeation through the lipid layers of the cancer cells. The chelation form might block the metabolic pathways of the DNA synthesis of neoplastic cells and inhibit the proliferation.

Antiproliferative activity of BHBM may be achieved by two mechanisms. Chelation of hydrazone to Sn(IV) from the cells may have higher efficiency to inhibit the DNA synthesis of the cancer cell lines (Richardson & Bernhardt, 1999). The chelation may produce cytotoxic oxygen radicals that prevent the uptake of essential nutrients for cancer cells (Richardson & Bernhardt, 1999). BHBM complex is a kind of monomeric structure in which the Sn(IV) center coordinated with the enolic tridentate ligand (L) in the ONMe chelate mode and exhibits five-coordinated trigonal bipyramidal geometry. The synthesized compound revealed the anticancer activity to the human cancer cell line of HCT 116 (colon), A549 (lung), and MCF7 (breast). This result demonstrated that the alkyl group bound with Sn (IV) center and sulfur (S) present in the complex have strong activity on their *in vitro* anticancer property. Min Hong et al. have studied previously the organotin(IV) complexes with 2-hydroxy-1-naphthaldehyde 5-chloro-2-hydroxybenzoylhydrazone with IC$_{50}$ values of 0.09 µg/ml to <10 µg/ml in A549 cells and 0.5–4.1 µg/ml in HCT-8 cells (Min et al., 2013). In this study, the authors described the anticancer activity as dose dependent which was increased by high concentration because of long carbon chain butyl ligand and organic ligand R. Also, coordination geometry to tin atom may have possible role to provide the anticancer effect (Min et al., 2013). The present study of the BHBM showed long-term anticancer effect with the single dose without any repetitive dose. While the other organotin complexes with O-Sn-S coordination showed very toxic effects from low to high doses, BHBM with the methyl complex has a significant effect at lower doses only.

The cytotoxic effects of the similar group of compound named 5-bromo-2-hydroxybenzaldehyde-4, 4-dimethylthiosemicarbazone was observed with the IC$_{50}$ of 3.23–5.46 µM from the study of Haque and Salam (2015). The detailed long-term cytotoxic/cytostatic effect and long-term dose-dependent and mechanistic activity were not studied yet for this group of compound. In this study, we performed the single-dose long-term effect of the BHBM in *in vitro* 3D tumor model and other *in vitro* studies. The novel nonenolic bidentate O-Sn-S-coordination of Schiff base ligand in synthesized methyl complex was documented for the first time.

NO generation effect of BHBM in the HCT 116 cell viability was determined and found 25 µM nitrite after 5 min and 188 µM nitrite after 1 h from NO donors of BHBM in cultured RPMI, respectively. At higher concentration of nitrite BHBM showed IC_{50} of 4.5 µg/ml. This result reveals the extracellular apoptotic activity of BHBM and suggests the interaction of NO and oxygen metabolites to kill the cells.

4.4. BHBM showed antimigratory properties on different cancer cells at optimum doses

BHBM were further investigated for antimigratory potential in comparison with Tamoxifen and 5-fluorouracil, using wound-healing assay that is important hallmarks of metastasis. In a previous report, Tamoxifen and 5-fluorouracil showed inhibition of migration of cancer cells (Borley et al., 2008; Hayot et al., 2002) and thus were used as a positive control in this study. BHBM was evaluated for antimigratory activity using the IC_{50} value of the respective cell lines. Human breast, colon, and lung cancer cells of MCF7, HCT 116, and A549, respectively, were allowed to migrate in the presence of BHBM with different concentrations for 0 to 36 h and antimigratory effects were calculated. The results shown in Figure 4 indicated that BHBM had significant antimigratory properties in comparison with the negative control group and competed its activity with the inhibitory potential of Tamoxifen and 5-flurouracil. BHBM showed the percentage inhibition of migration of the cancer cells in a time-dependent manner. The lower doses of BHBM exhibited the percentage inhibition dose dependently, but higher doses unable to produce dose-dependent inhibition in lung cancer A549 cells (Figure 4). During the 12 h, BHBM showed its lower activity than at 36 h. BHBM started its inhibitory activity after 12 h and its activity was increased in course of cumulative time. In all three cancer cells, same mechanism of action was observed from BHBM that was substantial increase of its activity during the period of 12–36 h. This result reveals that BHBM has less toxic effect that reproduces the safety of BHBM although IC_{50} between the range of 20–5 µg/ml obtained from MTT assay (Figure 3). Earlier published literature reported that dimethyltin was less toxic than trimethyltin and toxicity depended on the nature of the R group attached with Sn (Blunden, Cusak, & Hill, 1985). Also, If BHBM produces strong cytotoxicity by inhibition of mitochondrial oxidative phosphorylation, it could not reveal its antimigratory activity (Aldridge, 1976). The degree of toxic action may be controlled by the interaction of amino acid of cancer cells at certain active sites (Hall & Zuckerman, 1977) of imidazole N-H of histidine residue and an S-H group (Aldridge, 1976). Both of these groups present in the BHBM. Recently, published study synthesized the isonicotinoyl hydrazone derivatives as 2-hydroxy-1-naphthaldehyde isonicotinoyl hydrazone and its iron complexes (Lovejoy & Richardson, 2002; Yuan et al., 2004), and di-2-pyridyl ketone isonicotinoyl hydrazone (Bernhardt et al., 2003; Becker et al., 2003), Aroyl hydrazones of pyridoxal and salicylaldehyde hydrazone (Johnson et al., 1982; van Reyk et al., 2000), 2-Benzoxazolyl and 2-benzimidazolyl hydrazones (Easmon et al., 2001), Cyano acetic acid hydrazones of 3- and 4-acetylpyridine (EL-Hawash et al., 2006), 2,6-dimethylimidazo[2,1-b]-(Cima & Ballarin, 1999; Pellerito et al., 2005; Yin & Chen, 2006) thiazole-5-carbohydrazide and the derivatives of hydrazine pyrimidines (Cocco et al., 2006; Terzioglu & Gürsoy, 2003), Organotin(IV) complexes of 2-hydroxy-1-naphthaldehyde 5-chloro-2-hydroxy benzoylhydrazone, and isothiazolehydrazones for anticancer activity. These compounds have lack of S-H group, but this group is presented in the BHBM to produce an optimum effect to kill the cancer cell and its growth with slow and less toxic mechanism. In a comparison with past synthesized organotin (IV) complex, most of the organotin complex was highly toxic than BHBM (Table 1 and supplementary Table 1). Oppositely, BHBM could kill the cells at the optimum dose and the higher doses are not toxic. Thus, the results are suggesting the linear correlation between cytotoxicity and antimigratory activity. In case of lung cancer cells, the migration may be through caveolin-1 and protein kinase B (Akt) signaling pathway (Brazil & Hemmings, 2001). BHBM may block the activation of this signaling molecule and produce antimigratory activity. For breast cancer cells, PGE2 plays a vital role in the migration. It is assumed that BHBM may intervene with the migrational capacity of this signaling molecule to inhibit the migration at the dose of 12 µg/ml till 36 h. During the first 12 h, BHBM was unable to control the migration, but it stopped the migration after 12 h. The reason may be due to delay of weakening the signaling molecules specifically. The opposite effect was found from Tamoxifen and lower doses of BHBM at 6 µg/ml and 1.5 µg/ml which were highly active at first 12 h, but not active for a longer period of time. BHBM significantly showed its prolonged activity and it

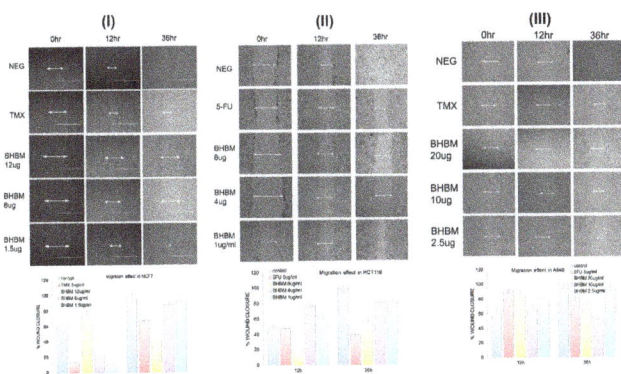

Figure 4. Antimetastatic functional assay: migration.

Note: Migration Effect of BHBM on proliferation and migration of MCF7 (I). Effect of BHBM on MCF7 proliferation. BHBM inhibited MCF7 proliferation in dose-dependent manner with IC_{50} 6 µg/ml ($n = 6$, values are in mean ± SD). Due to the successful migration of MCF7 in untreated group, the wound is almost closed after 18 h, whereas in BHBM-treated group, the wound remained open even aft4er 24 h incubation. BHBM (3 µg/ml) caused significant inhibition of MCF7 migration. At a concentration of 6 µg/ml, BHBM caused dislodgement of monolayer of MCF7 (indicated by the arrows) with almost complete inhibition of migration. BHBM also successfully arrested the colon cancer HCT116 cells (II) and lung cancer cell line A549 (III) proliferation and migration for more than 36 h with an administration of a single dose. Antimigratory effects of BHBM were found as dose dependently. This effect showed a linear correlation between cytotoxicity and percentage wound closure. Values were expressed as means ± SD.

seems it may continue its inhibition without further dose top-up. The reason of this effect is, perhaps, BHBM stops the abnormal performance of signaling molecule and tune its performance in cancer cell as well as normal cells. This result was not reproduced with the colon cancer HCT 116 cells treated by BHBM at the dose of 8, 4, and 1 µg/ml (Figure 4-II). At the duration of 12 h, the migration was inhibited but it was fast at 36 h due to inactivity of BHBM and 5-flurouracil.

The results clearly indicate that BHBM could successfully inhibit the migration of MCF7 cells with the administration of a single dose by reducing the COX-2 derived PGE2 synthesis and delay the migration of HCT 116 and A549 (Figure 4-III). BHBM might have a good interfere with metastatic process by the inhibition of migration as the first step of invasion of cancer cells. However, this claim will be rationalized through the investigation of murine cancer model.

4.5. Inhibitory effect of BHBM on clonogenicity

Clonogenic survival is an important hallmark of metastasis. This assay was used to measure the long-term cytotoxicity of BHBM (Yalkinoglu, Schlehofer, & zur Hausen, 1990). BHBM showed strong inhibition in colonization of cancer cells dose-dependently in the breast cancer MCF7 cells and colon cancer HCT 116. Percentage survival fraction and plating efficiency (PE) of BHBM of the clonogenic assay in MCF7 and HCT 116 at different concentrations are shown in Figure 5-I and 5-II. Tested samples successfully affected the self-renewing capacity of the melanoma cells at the dose of 3–13 µg/ml for MCF7 and 2–8 µg/ml for HCT 116. This result showed the linear correlation between cytotoxicity and cellular damage for aggregation. Tamoxifen and 5-flurouracil reproduced this activity. In both cell lines, negative control treated with DMSO (vehicle) showed cytostatic activity. BHBM may form the stable chelate to stop the DNA synthesis of the cancer cells and loose its proliferation and aggregation ability.

4.6. BHBM inhibits tumor aggregation property in hanging drop assay

Cancer cell suspension started to aggregate in a three-dimensional (3D) shape in the hanging drop within 48 h. The lung cancer cell line was used to develop the 3D *in vitro* tumor model to screen the effect of BHBM at different doses. Optimal seeding densities (Del Duca, Werbowetski, & Del Maestro, 2004), time and well shape promotes the good aggregation of cancer cells with the size of 300—500 µM at day 4. This size is appropriate to study the antiproliferative potential and cell cycle for BHBM. The cell culture plate coated with agar is maintained for 12 days by a routine change of culture medium to observe the prolonged period of BHBM activity. The layer of the cells is folded due to non-permissive agar. BHBM showed percentage inhibition of the density of tumor aggregation based on the optimal dose at 20 µg/ml till 12 days (Figure 6). Rest of the doses of 40, 10, and 5 µg/ml produced inhibition of aggregation

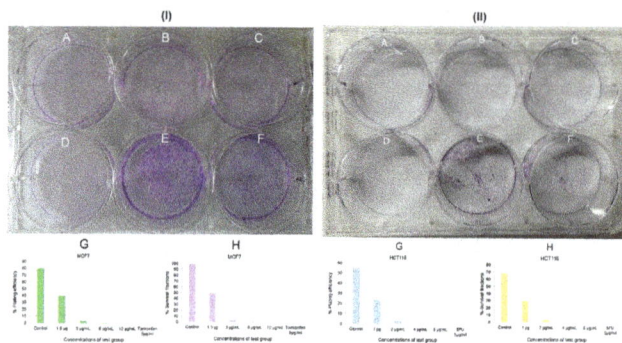

Figure 5. BHBM at 1–12 µg/ml negatively influenced the clonogenic growth of melanoma cells.

Note: MCF7 cells were treated with BHBM 12 µg/ml (I-A), 6 µg/ml (I-B), 3 µg/ml (I-C), and 1.5 µg/ml (I-F), DMEM media with DMSO as a Negative Control (I-E), and 5 µg/mL of tamoxifen as a Positive Control (I-D). Cell colonies were stained with crystal violet and counted. Anticlonogenic activity of BHBM was expressed as percentage of control treated with vehicle (0.05% DMSO). Experiments were performed in triplicate. HCT116 cells were treated with BHBM 8 µg/ml (II-A), 4 µg/ml (II-B), 2 µg/ml (II-C), and 1 µg/ml (II-F), DMEM media with DMSO as a negative control (II-E), and 5 µg/mL of 5-FU as a positive control (II-D). Cell colonies were stained with crystal violet and counted. Anticlonogenic activity of BHBM was expressed as percentage of control treated with vehicle (0.05% DMSO). Experiments were performed in triplicate.

similar to 5-flurouracil at 5 µg/ml. BHBM with 20 µg/ml showed significantly higher activity than 5-fluro-uracil. Oppositely, untreated spheroids in negative control showed dense aggregation containing 5,000 cells/drop at the beginning and it was developed to a highly dense colony of compact cells at 12 days. Due to the presence of S-H and N-H group of BHBM, it may provide negative effect in cancer cell aggregation and therefore spheroids were brittle and light dense. This result replicated the antimigratory and clonogenic inhibition activity by following the described mechanism above.

4.7. Effect of BHBM on morphological changes and nuclear condensation of MCF7 cells

MCF7 cells were stained with hoechst 33258 to evaluate the percent apoptotic index of BHBM at the dose of 12–1.5 µg/ml at 6 h and 12 h. BHBM showed chromatin condensation of the nucleus time dependently and at optimal dose of 6 µg/ml (Figure 7-I). Cells in negative control showed uniform staining of nucleus. The apoptotic indices of BHBM were lower at 12–1.5 µg/ml at 6 h and 12 h treatment that was lower than Tamoxifen 5 µg/ml. At 24 h of treatment with BHBM, 98% cells were dead which showed higher lethality than Tamoxifen (data not shown). The reason of this effect of BHBM may be for the morphological, chemical, and biological transformation of cellular molecules in the cells with the interaction of chemical groups. N-H, S-H, and C-N bond of the BHBM may shrink the nuclei and collapse the chromatin into high density shape. As a result, cells formed the membrane blebbing, nuclear condensation, fragmentation, and chromatin dissolution (Hayot et al., 2002).

4.8. BHBM reduces mitochondrial membrane potential

The effect of BHBM on mitochondrial membrane potential of the breast cancer MCF7 cell was evaluated using Rhodamine 123 fluorescent dye through the uptake of intensity of the dye by the cells (Rahn, Bombeck, & Doolittle, 1991). BHBM-treated cells showed higher intensity than untreated cells as negative control (Figure 7-II). BHBM showed the higher apoptotic indices on the cells at 6 µg/ml after 12 h than Tamoxifen at 5 µg/ml (Figure 7-II). The results were dose dependent at 12 h period for BHBM with all doses which was not found during the treatment of 6 h. BHBM revealed the lower activity at its highest dose of 12 µg/ml, but at 24 h, 98% cells were dead (data not shown). It clearly indicates that the mechanism of action of BHBM may be started after 12 h. Therefore, low signal of fluorescence is found from BHBM-treated cells at the higher dose of 12 µg/ml, which implies the loss of membrane potential to cause the apoptotic cell death. Therefore, the fluorescent intensity was dependent on the time, optimal dose, and the duration of mechanism of action of BHBM.

4.9. Strong binding activity of BHBM with anticancer target NF-κB

Molecular docking was performed to define the possible clash between the active site of the NF-κB and BHBM. This will provide correct confirmation and orientation of the collision of the compound and stereochemical organization of the target. The binding capacity of the BHBM with NF-κB was determined by the hydration energy descriptor and hydrogen bonding.

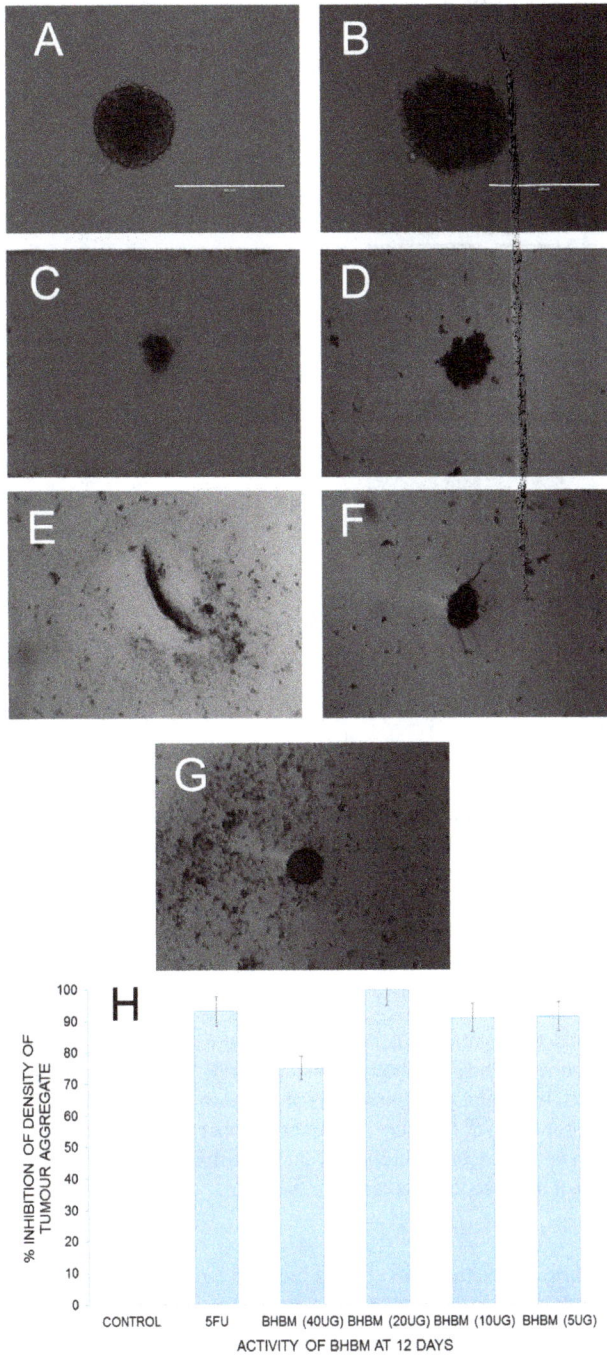

Figure 6. Three-dimensional (3D) tumor spheroid-based functional assay: invasion. 48-well agar-coated flat-bottomed plates were used to generate A549 spheroids (a single spheroid per well).

Note: Images were captured at four days intervals (data not shown) using an EVOS microscope. Analysis was carried out using Image-J software and percent inhibition tumor aggregation was obtained in 12 days. The comparative effects of BHBM on in vitro A549 (lung) tumor in hanging drop assay. Values were measured as means ± SD. Results are presented as mean ± SD of three separate experiments ($n = 6$).

The anticancer activity of BHBM was determined in docking analysis through the hydrogen bonding with the amino acids of the NF-κB transcription factor subunit p65 (Figure 8-I-A, I-B). The hydration energy of this molecule is small. From the binding energy of BHBM (Figure 8-I) with this big

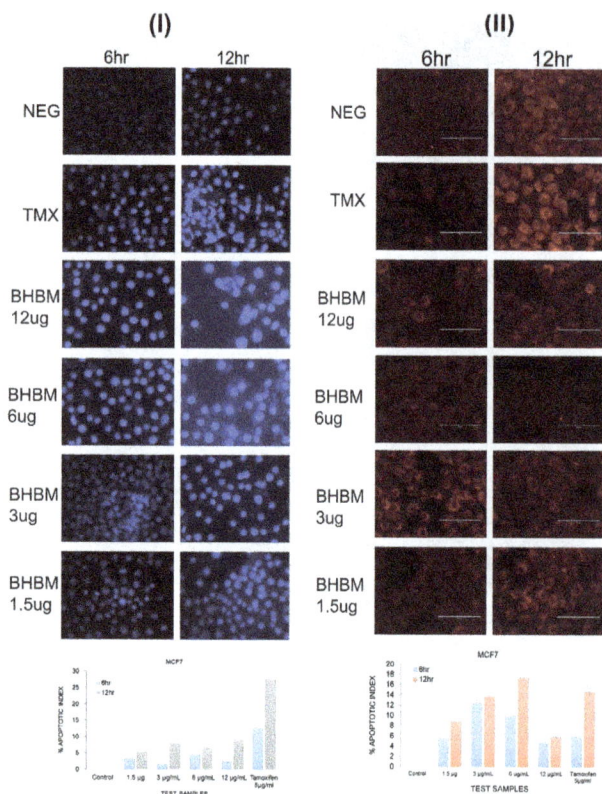

Figure 7. Photomicrographic images of the MCF7 in nuclear condensation assay (I). Staining of cultures by the use of Hoechst revealed the damaged DNA as an indicator of apoptosis. Higher cell death was found at 24 h (data not shown) than the tamoxifen. BHBM showed almost similar apoptotic index with mid and high doses at 12 h. The medium dose 6 μg/ml was more effective than the higher dose. BHBM at very high dose was not pronounced as anticancer activity. Values were determined as means ± SD. Photomicrographic images of the MCF7 in mitochondrial membrane potential (II).

Note: The rhodamine staining of nuclei reflects the intracellular generation of the cancer cells after treated with BHBM. Photomicrographic images of the cells showed the higher efficiency of the BHBM (higher intensity of rhodamine uptake cells) to disrupt the cancer cells at the dose of 6 μg/ml during 6 h treatment than 12 h. Percent apoptotic index was measured from the number of apoptotic cells (shiny cells) to the total number of cells in each selected microfield. Values are measured as means ± SD. Results are presented as mean ± SD of three separate experiments (n = 6).

molecule, it was obvious that the signaling capacity of NF-κB was blocked by the dimethyl tin (IV) complex BHBM either on transcriptional level or translational level. BHBM produced five hydrogen bonds with the NF-κB by Sn(IV), nitrogen, oxygen, and methyl group (Figure 8-I-C, I-D). Hydrogen bonds were found between the GLU222, GLU49, and GLN241. Strong capacity of binding of BHBM with the glutamine can block the alkylation and inhibit the DNA synthesis which leads antiproliferative activity of cancer cells. In addition, BHBM may also have lower hydration energy that provides good hydrogen bonding. Further, Tamoxifen and 5-flurouracil showed hydrogen bonding with the glutamine and strong anticancer activity (Figure 8-I-E, I-F, I-G, I-H). Of course, the binding energy levels were not similar among the tested standard reference drugs and BHBM because of the different substitute and functional groups. The activity conducted in the in vitro studies and the molecular docking binding of the BHBM with the NF-κB is quite similar to standard reference drugs. The atom level binding interactions results (Figure 8-I-I) between BHBM and NF-κB may provide rational prediction of mechanism for the single dose long-term effect and tolerable cytotoxicity as well as cytostatic that is different than the toxicity of the 5-flurouracil and Tamoxifen. In this study, we have conducted ROS, DPPH, FRAP, and caspase assay to investigate the influence of BHBM in in vitro. BHBM showed significant antioxidant capacity in DPPH and FRAP assay. Similarly, we found decreased ROS

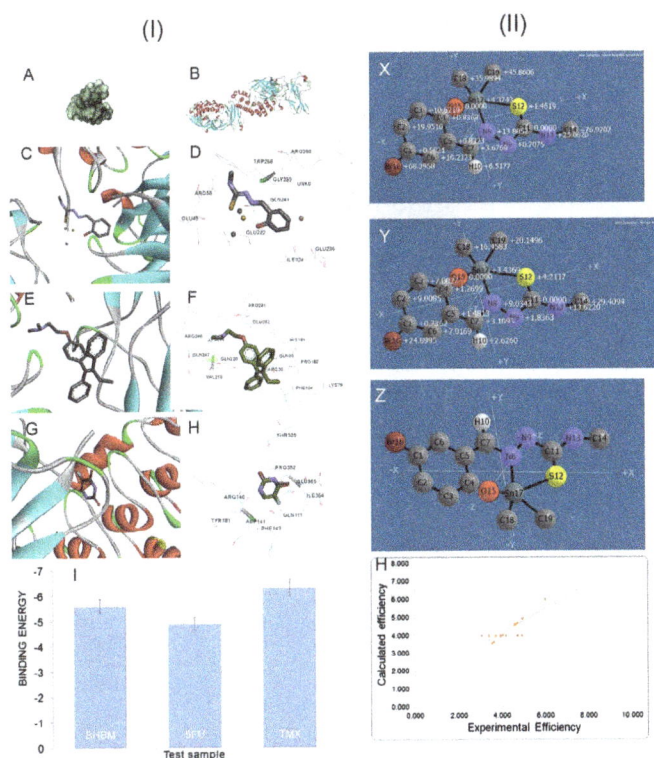

Figure 8. Visualization of Ligand and protein interaction profile: surface visualization (I-A) and cartoon display (I-B) of NF-κB and active site residue interaction of protein NF-κB (I-D, I-F, I-H).

Note: Molecular docking was performed in Autodock Vina. The binding energy obtained from docking analysis reflected the anticancer activity that correlates QSAR model. Descriptor Solvent accessible surface area (SASA), surface area (SA) of BHBM was generated by CODESSA software (X, Y, Z). SASA is very important to correlate the molecule structure with the anticancer activity. QSAR model was developed in CODESSA using this SASA descriptor by the multiple linear regression (MLR) method. Calculated efficiency versus the experimental efficiency of the training set molecules were plotted by MLR method (H). Results are presented as mean ± SD of three separate docking experiments.

level in the BHBM-treated HCT 116 cells at the dose of 4 and 8 µg/ml. HCT 116 cells become under stress due to this reduced level of ROS (Table 2). The reason of the antioxidant activity of the BHBM may be due to form of chelation of Sn(IV), resonance impact of aromatic rings, hydrogen bonds, and the accumulation of Sn in the cancer cells than normal cells (Figure 9). Concurrently, due to the ROS stress in HCT 116 cells by BHBM, caspases and NF-κB have been modulated to kill the cells. The effects of BHBM on caspases 3/7, 8, and 9 were significantly increased in the levels of activated caspases observed in the hoecht and rhodamin staining at the nucleus morphology by caspase 3/7, while initiation of cell death by caspase 8 and 9 through ligation of cell death receptors and disruption of mitochondrial membrane integrity, respectively. The results showed that there was significant increase in caspase activities due to the influence of BHBM at 4 and 8 µg/ml in compared to control (Table 2).

4.10. QSAR of BHBM for anticancer activity

QSAR studies were performed to find the relationship between anticancer activity and molecular structure. This correlation was determined by the multiple linear regression model between molecular descriptors and the efficiency of the BHBM. Anticancer activity was dependent on the nature of the substitution of different parts of the structure. The essential chemical features of the test compound for anticancer activity were identified by the QSAR of Hansch approach. The structural analysis of BHBM was carried out in CODESSA software. This software was used to determine the set of molecular descriptors because of its ability to generate more than 400 molecular descriptors such as molecular constitution, geometry, and topology and with electrostatic, thermodynamic, and

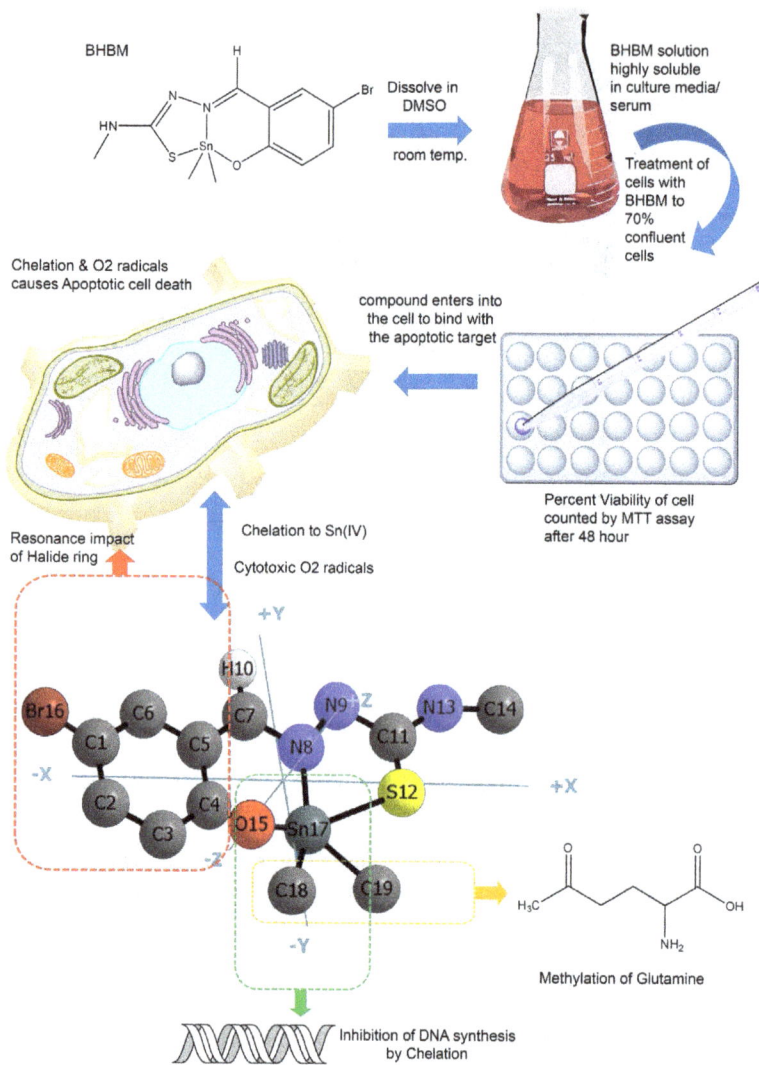

Figure 9. Anticancer activity of BHBM through the induction of apoptotic cell death.

Notes: In addition, the illustrated mechanism of action depicts that Dimethyl tin(IV) complex BHBM could produce multiple site of action to stop the nutrient supply to the cancer cells and weaken the cancer cells with the specific interaction to the active amino acids. BHBM is not highly toxic to the cells, but the cell death occurs due to its multiple hit of action. Chelation, cytotoxic O2 radicals, methylation of glutamine, preventing signals to synthesize the DNA, resonance impact of halide group, hydrogen bonding capacity and so on.

quantum chemical parameters. Biological efficiency and the molecular descriptors of BHBM were correlated using the heuristic method (Xia, Ma, Zheng, Zhang, & Fan, 2008). Biological activity of BHBM in the cancer cell lines and the calculated efficiency based on molecular descriptors HDCA-1 was used to determine the correlation. The hydrogen bond donor atom is described from the HDCA-1 descriptor. The value of the HDCA-1 descriptor is the substitution of the accessible surface area of H bond donor atoms by following equation (Stanton & Jurs, 1990; Stanton, Egolf, Jurs, & Hicks, 1992),

$$HDCA1 = \sum_{D} S_D D \in H_{H-donor}$$

where, S_D– solvent-accessible surface area of H-bonding donor H atoms.

This model was developed for the most sensitive cell line HCT 116 and showed in Figure 8 plotted as calculated efficiency versus experimental efficiency. From the correlation it is assumed that good

Compounds	pKa	LogD	Antioxidant			Caspase		
			DPPH EC50 (µg/mL)	FRAP (nmol Fe+2 eq./ mg compound)	ROS	Caspase 3	Caspase 8	Caspase 9
BHBM	7.38 ± 0.12	−0.92 ± 1.5	320.50 ± 12.18	54.83 ± 0.48	66.23% decrease	0.735 ± 0.87	0.71 ± 0.58	0.661 ± 0.97
Ascorbic acid	–	–	7.30 ± 0.01	–	–	–	–	–
control	–	–	–	–	–	0.588 ± 0.43	0.651 ± 1.2	0.551 ± 0.35

Table 2. pKa and LogD physicochemical properties of BHBM

Note: Effect of BHBM on levels of caspases 3/7, 8, and 9. A significant increase in the apoptotic markers, caspases 3/7, 8 and 9 was observed after 48 h to HCT116 cells. BHBM at 4 and 8 µg/mL significantly increased the levels of caspases 8 and 9. Higher concentration of BHBM had an effect on caspase 8 and 9 induction that was comparable to the control. The results are mean values ± SEM (n=3). Free radical scavenging activity of the BHBM measured by DPPH assay. Results are expressed as mean ± SEM (n=3). Effect of the BHBM on ROS level. The relative ROS fluorescence intensity in cultures of HCT116 cells treated with 4 and 8 µg/mL of BHBM over 48 h was measured using the ROS assay reagent. HCT116 cells incubated with two different concentrations of BHBM showed a dose-dependent decrease ROS levels compared to the control. Results are means ± SEM of three experiments (n = 6).

solvent accessible surface area leads to the biological activity. The reason may be due to the presence of hydroxyl groups in the BHBM's substitute of the meta- and para-positions.

The used compounds in developing the QSAR model consisted with a series of test set molecules, training set molecules, and validation set molecules were labeled in CODESSA during multiple linear regression analysis. The quality of this analysis was justified by the parameters of correlation coefficient (R), standard error of the estimate (SEE), variance ratio (F) at specified degrees of freedom (df) (Table 3), (Figure 8-II).

Additionally, anticancer activity of the BHBM was led by the substitution of N-H, S-H, and aromatic ring with the Sn(IV) molecule. It was observed that BHBM inhibited the activity of cancer cells with 100% in comparing with the standard drugs of Tamoxifen and 5-flurouracil at the optimal doses for different cancer cell lines. In most of the anticancer experiments, it was found that BHBM showed its 100% capacity in killing the cancer cells after a certain period of time, i.e. after few hours. This may be due to the presence of electron donating groups in the Sn(IV) ring coordinated with the oxygen, nitrogen, and sulfur atom to form the stable chelate with the active amino acid of the receptor site which has electron deficiency. In this reaction, the resonance of the molecule is important too. It is clearly observed from the molecular docking analysis that positive steric parameter at O15, Sn17, C18, and C19 position explains why the small groups of the BHBM has better fit with the receptor cavity (Figure 8-I). The possible hydrophobic interaction due to presence of the aromatic ring, and aliphatic group plays a vital role for proper entry of BHBM into the hydrophobic pocket of the receptor (Figure 8-I).

Hydration energy is a useful physicochemical parameters to define the discharged energy in contact with the water molecules for a molecule. The higher value of the hydration energy of the molecule indicates better solubility in water. Hydration is also depends on the size of the molecule and small molecules performs the high solubility. The solvent accessible surface is of BHBM and the developed model provides an insight of hydration energy. The good correlation with training set molecules indicates the presence of hydration energy is high in BHBM. It provides the mechanism of action of BHBM as anticancer activity through the hydration process and solubility of the molecule. The hydrogen bonding acceptors atoms in cancer cell lines MCF7, HCT 116, and A549 impose a greater effect in increasing the hydration energy which is highly negative. As a result, the IC_{50} of the BHBM was increased and showed 100% anticancer activity as well as other standard drugs of Tamoxifen and 5-flurouracil. In a comparison with the BHBM, it had a lower anticancer activity in A549 cells than MCF7 and HCT 116. The reason may be due to low hydration energy from the cells. The

Table 3. QSAR model generated by CODESSA for predicted pIC$_{50}$ and calculated IC$_{50}$

Method	Property	No. of Descriptor	R^2	CVR^2	F	Structure Set	Descriptor Set	s^2	SEE	p (F)	CVSE
MLR	Efficiency	20	0.711	−3932.84	0.11	22 training set	Solvent accessible surface area	2.75	0.35	0.99	40.45

Note: Statistical data shown the correlation coefficient R^2 value 0.711 indicates structure activity relationship with the training set molecules were strongly agree to quantify the anticancer activity of the BHBM. Standard error of the estimate (SEE), variance ratio (F), p(F), s^2, CVR^2, CVSE also were shown in the table.

molecular docking analysis also revealed that the greater impact of hydration energy due to either role of BHBM or NF-κB (subunit P65 and P50) for the hydrogen bond acceptor atoms during interaction.

5. Conclusion

We provide a comprehensive suite of simple, and reproducible method for synthesis of [Me$_2$Sn(BHBM)] that recapitulates in vitro hallmarks of anticancer activity and that at the same time QSAR with molecular docking analysis provides a quantitative mechanism of action with high-throughput preclinical studies. We provide evidence that by manipulating the substituents of Sn(IV) complex with S-H, N-H, C-N, Me-Sn-Me, aromatic ring with halogen atoms have the potential to enhance the targeted anticancer activity that may be selective to the specific target active site through chelation and alkylation of glutamine and narrow down the nutrient supply to the cancer cells with prolonged duration of mechanism in a harmless manner of chemotherapeutic treatment. The strategy of single-dose screening by designing a organotin(IV) compound with good water and human fluid solubility by adding the polar substituents, and halogen atoms might be useful for candidate anticancer drug prior to in vivo studies.

Abbreviations:

DMEM   Dulbecco's Modified Eagle Medium

MTT   dimethyl thiazolyl diphenyl tetrazolium

Acknowledgments
The Authors are grateful to the USM doctoral fellowship and TWAS postdoctoral fellowship.

Funding
We are thankful to the EMAN Lab management who helped during the analysis and experimentation. The entire study is supported by Universiti Sains Malaysia (USM) under the Research University (RU) [grant number 1001/PKIMIA/811217] and [grant number RUT 1001/PFARMASI/851001].

Competing interests
The authors declare no competing interest.

Author details
Md. Shamsuddin Sultan Khan[1]
E-mails: rafatherouf@gmail.com, jupitex@gmail.com
Md. Abdus Salam[2]
E-mail: salambpx@yahoo.com
Rosenani S.M.A. Haque[2]
E-mail: rosenani@usm.my
Amin Malik Shah Abdul Majid[1]
E-mail: aminmalikshah@gmail.com
Aman Shah Bin Abdul Majid[3]
E-mail: amanshah75@googlemail.com
ORCID ID: http://orcid.org/0000-0001-9679-122X
Muhammad Asif[1]
E-mail: asif_pharmacist45@yahoo.com
Mohamed Khadeer Ahamed Basheer[1]
E-mail: dr.khadeer2014@gmail.com
Yaseer M Tabana[1]
E-mail: yasser.tabana@hotmail.com
[1] EMAN Research and Testing Laboratory, School of Pharmaceutical Sciences, Universiti Sains Malaysia, Minden, Pulau Pinang, Malaysia.
[2] The School of Chemical Sciences, Universiti Sains Malaysia (USM), Penang 11800, Malaysia.
[3] Advanced Medical and Dental Institute (IPPT), Universiti Sains Malaysia, Bertam, Kepala Batas, Malaysia.

Citation information
Cite this article as: Synthesis, cytotoxicity, and long-term single dose anti-cancer pharmacological evaluation of dimethyltin(IV) complex of N(4)-methylthiosemicarbazone (having ONS donor ligand), Md. Shamsuddin Sultan Khan, Md. Abdus Salam, Rosenani S.M.A. Haque, Amin Malik Shah Abdul Majid, Aman Shah Bin Abdul Majid, Muhammad Asif, Mohamed Khadeer Ahamed Basheer & Yaseer M Tabana, Cogent Biology (2016), 2: 1154282.

References

Aldridge, W. N. (1976). The influence of organotin compounds on mitochondrial functions. In J. J. Zuckerman (Ed.), *Organotin compounds, new chemistry and applications* (Vol. 157, Chapter 13, pp. 186–196). New York, NY: American Chemical Society.

Becker, E. M., Lovejoy, D. B., Greer, J. M., Watts, R., & Richardson, D. S. (2003). Identification of the di-pyridyl ketone isonicotinoyl hydrazone (PKIH) analogues as potent iron chelators and anti-tumour agents. *British Journal of Pharmacology, 138*, 819–830.

Benzie, I. F. F., & Strain, J. J. (1996). The ferric reducing ability of plasma (frap) as a measure of "antioxidant power": The FRAP assay. *Analytical Biochemistry, 239*, 70–76. http://dx.doi.org/10.1006/abio.1996.0292

Bernhardt, P. V., Caldwell, L. M., Chaston, T. B., Chin, P., & Richardson, D. R. (2003). Cytotoxic iron chelators: Characterization of the structure, solution chemistry and redox activity of ligands and iron complexes of the di-2-pyridyl ketone isonicotinoyl hydrazone (HPKIH) analogues. *Journal of Biological Inorganic Chemistry, 8*, 866–880. http://dx.doi.org/10.1007/s00775-003-0486-z

Blunden, S. J., Cusak, P. A., & Hill, R. (1985). *The industrial uses of tin chemicals*. London: Royal Society of Chemistry.

Borley, A. C., Hiscox, S., Gee, J., Smith, C., Shaw, V., Barrett-Lee, P., & Nicholson, R. I. (2008). Anti-oestrogens but not oestrogen deprivation promote cellular invasion in intercellular adhesion-deficient breast cancer cells. *Breast Cancer Research, 10*, R103. http://dx.doi.org/10.1186/bcr2206

Brazil D. P., & Hemmings, B. A. (2001). Ten years of protein kinase B signalling: A hard Akt to follow. *Trends in Biochemical Sciences., 26*, 657–664. http://dx.doi.org/10.1016/S0968-0004(01)01958-2

Carlos, H., Martínez, R., & Dardonville, C. (2013). Rapid Determination of Ionization Constants (pKa) by UV Spectroscopy Using 96-Well Microtiter Plates. *ACS Medicinal Chemistry Letters, 4*, 142–145.

Cima, F., & Ballarin, L. (1999). TBT-induced apoptosis in tunicate haemocytes. *Applied Organometallic Chemistry, 13*, 697. http://dx.doi.org/10.1002/(ISSN)1099-0739

Cocco, M. T., Congiu, C., Lilliu, V., & Onnis, V. (2006). Synthesis and *in vitro* antitumoral activity of new hydrazinopyrimidine- 5-carbonitrile derivatives. *Bioorganic & Medicinal Chemistry, 14*, 366–372. http://dx.doi.org/10.1016/j.bmc.2005.08.012

Collery, P., & Perchery, C. (1993). Clinical experience with *tumor*-inhibiting gallium complexes. In B. K. Keppler (Ed.), *Metal complexes in cancer chemotherapy* (pp. 249–258). Weinheim: VCH.

Collier, W. A. (1929). Zur experimentellen Therapie der Tumoren. *Zeitschrift für Hygiene und Infektionskrankheiten, 110*, 169–174. http://dx.doi.org/10.1007/BF02175963

Collinson, S. R., & Fenton, D. E. (1996). *Coordination Chemistry Reviews, 148*, 19–40. http://dx.doi.org/10.1016/0010-8545(95)01156-0

da Silva, D. L., Reis, F. S., Muniz, D. R., Ruiz, A. L. T. G., de Carvalho, J. E., Sabino, A. A., & Modolo, L. V. (2012). Free radical scavenging and antiproliferative properties of Biginelli adducts. *Bioorganic & Medicinal Chemistry, 20*, 2645–2650. http://dx.doi.org/10.1016/j.bmc.2012.02.036

Del Duca, D., Werbowetski, T., & Del Maestro, R. F. (2004). Spheroid preparation from hanging drops: Characterization of a model of brain tumor invasion. *Journal of Neuro-Oncology, 67*, 295–303. http://dx.doi.org/10.1023/B:NEON.0000024220.07063.70

Easmon, J. G., Puerstinger, G., Roth, T., Fiebig, H. H., Jenny, M., Jaeger, W., … Hofmann, J. (2001). 2-benzoxazolyl and 2-benzimidazolyl hydrazones derived from 2-acetylpyridine: A novel class of antitumor agents. *International Journal of Cancer, 94*, 89–96. http://dx.doi.org/10.1002/(ISSN)1097-0215

EL-Hawash, S. A. M., Abdel Wahab, A. E., & El-Demellawy, M. A. (2006). Cyanoacetic acid hydrazones of 3-(and 4-)acetylpyridine and some derived ring systems as potential antitumor and anti-hcv agents. *Archiv der Pharmazie, 339*, 14–23. http://dx.doi.org/10.1002/(ISSN)1521-4184

Gasteiger, J., & Marsili, M. (1980). Iterative partial equalization of orbital electronegativity - a rapid access to atomic charges. *Tetrahedron, 36*, 3219–3228. http://dx.doi.org/10.1016/0040-4020(80)80168-2

Gielen, M. (2003). An overview of forty years organotin chemistry developed at the Free Universities of Brussels ULB and VUB. *Journal of the Brazilian Chemical Society, 14*, 870–877.

Hall, W. T., & Zuckerman, J. J. (1977). Tin(II) and dimethyltin(IV) derivatives of amino acids. *Inorganic Chemistry, 16*, 1239–1241. http://dx.doi.org/10.1021/ic50171a053

Han, G., & Yang, P. (2002). Synthesis and characterization of water-insoluble and water-soluble dibutyltin(IV) porphinate complexes based on the tris(pyridinyl)porphyrin moiety, their anti-tumor activity *in vitro* and interaction with DNA. *Journal of Inorganic Biochemistry, 91*, 230. http://dx.doi.org/10.1016/S0162-0134(02)00369-0

Haque, R. A., & Salam, M. A. (2015). Synthesis, spectroscopic properties and biological activity of new mono organotin(IV) complexes with 5-bromo-2-hydroxybenzaldehyde-4, 4-dimethylthiosemicarbazone. *Cogent Chemistry, 1*, 1045212. http://dx.doi.org/10.1080/23312009.2015.1045212

Hayot, C., Farinelle, S., De Decker, R., Decaestecker, C., Darro, F., Kiss, R., & Van Damme, M. (2002). In vitro pharmacological characterizations of the anti-angiogenic and anti-tumor cell migration properties mediated by microtubule-affecting drugs, with special emphasis on the organization of the actin cytoskeleton. *International Journal of Oncology, 21*, 417–425.

Holeček, J., Nádvorník, M., Handlíř, K., & Lyčka, A. (1986). 13C and 119Sn NMR spectra of Di-n-butyltin(IV) compounds. *Journal of Organometallic Chemistry, 315*, 299. http://dx.doi.org/10.1016/0022-328X(86)80450-8

Johnson, D. K., Murphy, T. B., Rose, N. J., Goodwin, W. H., & Pickart, L. (1982). The experimental therapy for the tumors. *Journal of Hygiene and Infectious Diseases, 67*, 159–165. http://dx.doi.org/10.1016/S0020-1693(00)85058-6

Katritsky, A. R., Lobanov, V. S., & Karelson, M. (1996). *CODESSA: Reference manual, University of Florida* (Version 2). Gainesville, FL: Semichem, Terrace.

Khan, M. S. S., Syeed, S. H., Uddin, M. H., Akter, L., Ullah, M. A., Jahan, S., & Rashid, M. H. (2013). Screening and evaluation of antioxidant, antimicrobial, cytotoxic, thrombolytic and membrane stabilizing properties of the methanolic extract and solvent-solvent partitioning effect of Vitex negundo Bark. *Asian Pacific Journal of Tropical Disease, 3*, 393–400. http://dx.doi.org/10.1016/S2222-1808(13)60090-0

Koch, B., Baul, T. S. B., & Chatterjee, A. (2008). Cell proliferation inhibition and antitumor activity of novel alkyl series of diorganotin(IV) compounds. *Journal of Applied Toxicology, 28*, 430–438. http://dx.doi.org/10.1002/(ISSN)1099-1263

Lovejoy, D. B., & Richardson, D. R. (2002). Novel "hybrid" iron chelators derived from aroylhydrazones and thiosemicarbazones demonstrate selective antiproliferative activity against tumor cells. *Blood, 100*,

666–676.
http://dx.doi.org/10.1182/blood.V100.2.666

Min, H., Yin, H., Zhang, X., Chuan, L., Caihong, Y., & Shuang, C. (2013). Di- and tri-organotin(IV) complexes with 2-hydroxy-1-naphthaldehyde 5-chloro-2-hydroxybenzoylhydrazone: Synthesis, characterization and in vitro antitumor activities. *Journal of Organometallic Chemistry, 724*, 23–31.

Mouayed, A. H., Muhammad, A. I., Muhammad, A., Rosenani, A. H., Mohammed, B. K. A., Amin, M. S. A. M., & Teoh, S. G. (2015). A synthesis, crystal structures and in vitro anticancer studies of new thiosemicarbazone derivatives. *Phosphorus Sulfur and Silicon and the Related Elements, 190*, 1498–1508.

Pacheco, S. R., Braga, T. C., da Silva, D. L., Horta, L. P., Reis, F. S., Ruiz, A. L., ... de Fatima, Â (2013). Biological activities of eco-friendly synthesized hantzsch adducts. *Medicinal Chemistry, 9*, 889–896.
http://dx.doi.org/10.2174/1573406411309060014

Pellerito, C., D'Agati, P., Fiore, T., Mansueto, C., Mansueto, V., Stocco, G., ... Pellerito, L. (2005). Synthesis, structural investigations on organotin(IV) chlorin-e6 complexes, their effect on sea urchin embryonic development and induced apoptosis. *Journal of Inorganic Biochemistry, 99*, 1294–1305.
http://dx.doi.org/10.1016/j.jinorgbio.2005.03.002

Pruchnik, F. P., Bańbuła, M., Ciunik, Z., Chojnacki, H., Latocha, M., Skop, B., ... Nasulewicz, A. (2002). Structure, properties and cytostatic activity of triorganotin (aminoaryl)carboxylates. *European Journal of Inorganic Chemistry*, 3214–3221.
http://dx.doi.org/10.1002/1099-0682(200212)2002:12<3214::AID-EJIC3214>3.0.CO;2-K

Rahn, C. A., Bombeck, D. W., & Doolittle, D. J. (1991). Assessment of mitochondrial membrane potential as an indicator of cytotoxicity. *Fundamental and Applied Toxicology, 16*, 435–448.
http://dx.doi.org/10.1016/0272-0590(91)90084-H

Reddy, D. M., Qazi, N. A., Sawant, S. D., Bandey, A. H., Srinivas, J., Shankar, M., ... Sampath Kumar, H. M. (2011). Design and synthesis of spiro derivatives of parthenin as novel anti-cancer agents. *European Journal of Medical Chemistry, 46*, 3210–3217

Richardson, D. R. (1997). Cytotoxic analogs of the iron(III) chelator pyridoxal isonicotinoyl hydrazone: Effects of complexation with copper(II), gallium(III), and iron(III) on their antiproliferative activities. *Antimicrobial Agents & Chemotherapy, 41*, 2061–2063.

Richardson, D. R., & Bernhardt, P. V. (1999). Crystal and molecular structure of 2-hydroxy-1-naphthaldehyde isonicotinoyl hydrazone (NIH) and its iron(III) complex: An iron chelator with anti-tumour activity. *Journal of Biological Inorganic Chemistry, 4*, 266–273.
http://dx.doi.org/10.1007/s007750050312

Saxena, A. K., & Huber, F. (1989). Organotin compounds and cancer chemotherapy. *Coordination Chemistry Reviews, 95*, 109–123.
http://dx.doi.org/10.1016/0010-8545(89)80003-7

Stanton D. T., & Jurs, P. C. (1990). Development and use of charged partial surface area structural descriptors in computer-assisted quantitative structure-property

relationship studies. *Analytical Chemistry, 62*, 2323–2329.
http://dx.doi.org/10.1021/ac00220a013

Stanton D. T., Egolf, L. M., Jurs, P. C., & Hicks, M. G. (1992). Computer-assisted prediction of normal boiling points of pyrans and pyrroles. *Journal of Chemical Information and Modeling, 32*, 306–316.
http://dx.doi.org/10.1021/ci00008a009

Tabassum, S., & Pettinari, C. (2006). Chemical and biotechnological developments in organotin cancer chemotherapy. *Journal of Organometallic Chemistry, 691*, 1761–1766.
http://dx.doi.org/10.1016/j.jorganchem.2005.12.033

Terzioglu, N., & Gürsoy, A. (2003). Synthesis and anticancer evaluation of some new hydrazone derivatives of 2,6-dimethylimidazo[2,1-b][1,3,4]thiadiazole-5-carbohydrazide. *European Journal of Medicinal Chemistry, 38*, 781–786.
http://dx.doi.org/10.1016/S0223-5234(03)00138-7

Tian, L., Qian, B., Sun, Y., Zheng, X., Yang, M., Li, H., & Xueli, L. (2005). Synthesis, structural characterization and cytotoxic activity of diorganotin(IV) complexes ofN-(5-halosalicylidene)-α-amino acid. *Applied Organometallic Chemistry, 19*, 980–987.
http://dx.doi.org/10.1002/(ISSN)1099-0739

Trott, O., & Olson, A. J. (2010). AutoDock Vina: Improving the speed and accuracy of docking with a new scoring function, efficient optimization and multithreading. *Journal of Computational Chemistry, 31*, 455–461.

van Reyk, D. V., Sarel, S., & Hunt, N. (2000). Inhibition of *in vitro* lymphoproliferation by three novel iron chelators of the pyridoxal and salicyl aldehyde hydrazone classes. *Biochemical Pharmacology, 60*, 581–587.
http://dx.doi.org/10.1016/S0006-2952(00)00347-6

Xanthopoulou, M. N., Hadjikakou, S. K., Hadjiliadis, N., Milaeva, E. R., Gracheva, J. A., Tyurin, V. Y., ... Charalabopoulos, K. (2008). Biological studies of new organotin(IV) complexes of thioamide ligands. *European Journal of Medicinal Chemistry, 43*, 327–335.
http://dx.doi.org/10.1016/j.ejmech.2007.03.028

Xia, B., Ma, W., Zheng, B., Zhang, X., & Fan, B. (2008). Quantitative structure-activity relationship studies of a series of non-benzodiazepine structural ligands binding to benzodiazepine receptor. *European Journal of Medicinal Chemistry, 43*, 1489–1498.
http://dx.doi.org/10.1016/j.ejmech.2007.09.004

Yalkinoglu, A. O., Schlehofer, J. R., & zur Hausen, H. (1990). Inhibition of N-methyl-N''-nitro-n-nitrosoguanidine-induced methotrexate and adriamycin resistance in CHO cells by adeno-associated virus type 2. *International Journal of Cancer, 45*, 1195–1203.
http://dx.doi.org/10.1002/(ISSN)1097-0215

Yin, H. D., & Chen, S. W. (2006). Synthesis and characterization of organotin(IV) compounds with Schiff base of o-vanillin-2-thiophenoylhydrazone. *Journal of Organometallic Chemistry, 691*, 3103–3108.
http://dx.doi.org/10.1016/j.jorganchem.2006.03.003

Yuan, J., Lovejoy, D. B., & Richardson, D. R. (2004). Novel di-2-pyridyl-derived iron chelators with marked and selective antitumor activity: In vitro and in vivo assessment. *Blood, 104*, 1450–1458.
http://dx.doi.org/10.1182/blood-2004-03-0868

16

RaFoSA: Random forests secondary structure assignment for coarse-grained and all-atom protein systems

Emmanuel Oluwatobi Salawu[1,2,3]*

*Corresponding author: Emmanuel Oluwatobi Salawu, TIGP Bioinformatics Program, Institute of Information Science, Academia Sinica, Taipei, Taiwan; Institute of Bioinformatics and Structural Biology, National Tsing Hua University, Hsinchu, Taiwan; School of Computer Science, University of Hertfordshire, Hertfordshire, UK

E-mails: tools@bioinformatics.center, emmanuel.s@iis.sinica.edu.tw

Reviewing editor: Yasser Gaber, Beni-Suef University, Egypt

Additional information is available at the end of the article

Abstract: Secondary structures (SS) of proteins are of great importance to structural, molecular, and computational biology and chemistry. Accurate and reliable method for automatic SS assignment when only coarse-grained (CG) information is available is needed. RaFoSA, a novel, accurate, and reliable method for automatic SS assignment based on coordinates of alpha carbon (CAC) atoms alone is presented here. Results from RaFoSa have been rigorously compared to those from Dictionary of Protein SS (DSSP, the acclaimed gold-standard for automatic SS assignment) and STRIDE. Requiring only CAC, RaFoSA achieves an agreement of 96% (and 94%) with DSSP (and STRIDE) that require all-atom and hydrogen-bonding information. No known automatic SS assignment method based on CG system has ever achieved such agreement with DSSP and STRIDE. Furthermore, RaFoSA has been applied to a real-life problem and its possible use for ranking proteins in their order of SS-based stability is shown in this paper. Overall, RaFoSA's abilities to accurately and reliably assign SS to CG or all-atom protein systems make this work important. Furthermore, it must be emphasized that SS assignment by RaFoSA is different from (and is more rigorous than) SS prediction from amino acids sequence. Indeed, SS assignment by RaFoSA can differentiate between frames from molecular dynamics simulations

ABOUT THE AUTHOR

Emmanuel Oluwatobi Salawu received bachelor of technology (with Honors) in Physiology at Ladoke Akintola University of Technology, and later proceeded to study Computer Science at the University of Hertfordshire where he earned master of science (with Distinction). He is currently a Bioinformatics and Structural Biology PhD candidate at National Tsing Hua University. He has interests in molecular dynamics simulations, protein structures, machine learning, image analysis, epidemiology, and malaria research. The work presented here is part of his research activities involving secondary structures of proteins and machine learning.

PUBLIC INTEREST STATEMENT

Secondary structures (SS) of proteins are of great importance to structural, molecular, and computational biologists and chemists. Therefore, there is a need to develop accurate and reliable method for SS assignment when only alpha carbon information is available. This is why RaFoSA, a novel and reliable method for automatic SS assignment based on the coordinates of alpha carbon (CAC) atoms alone has been developed and presented here. RaFoSA accurately and reliably assigns SS to amino acids of proteins and achieves an agreement of 96% (and 94%) with DSSP (and STRIDE) that require all-atom and hydrogen-bonding information. RaFoSA may also help in ranking proteins based on the stability of their secondary structures. Source codes of RaFoSA are available at http://bioinformatics.center/tools/rafosa. A webserver that implements RaFoSA is available at http://bioinformatics.center/servers/rafosa.

trajectories, while existing methods for SS prediction from amino acid sequence cannot. Source codes and a webserver implementation of RaFoSA are available at http://bioinformatics.center/RaFoSA.

Subjects: Biochemistry; Bioinformatics; Mathematics for Biology & Medicine; Molecular Biology; Statistics for the Biological Sciences; Structural Biology

Keywords: bioinformatics; secondary structure webserver; amino acids; protein structural stability; machine learning

1. Introduction

There are many important purposes (Cabaleiro-Lago, Szczepankiewicz, & Linse, 2012; Cino, Choy, & Karttunen, 2012; Emr & Silhavy, 1983; Ji & Li, 2010; Konvalinová et al., 2015; Myers & Oas, 2001) for which structural, molecular, and computational biologists (SMCB) need SS of proteins even when only coarse-grained (CG) information of the protein is available. Proper visualization of the structure and dynamics of proteins following CG molecular dynamics simulations (which continue to gain popularity) is one of the most notable of such purposes (Humphrey, Dalke, & Schulten, 1996). Without SS information, the molecular systems (MS) of interest are more difficult to study and less appealing to sight (Figure 1). Availability of SS information removes the ordeal and allows proper visualization and smoother study of the static and dynamic properties of the MS of interest. In addition, SS changes of MS are frequently used to support evidences for, or against, structural stabilities of MS (Camilloni, De Simone, Vranken, & Vendruscolo, 2012; Pires, Ascher, & Blundell, 2014; Provencher & Gloeckner, 1981).

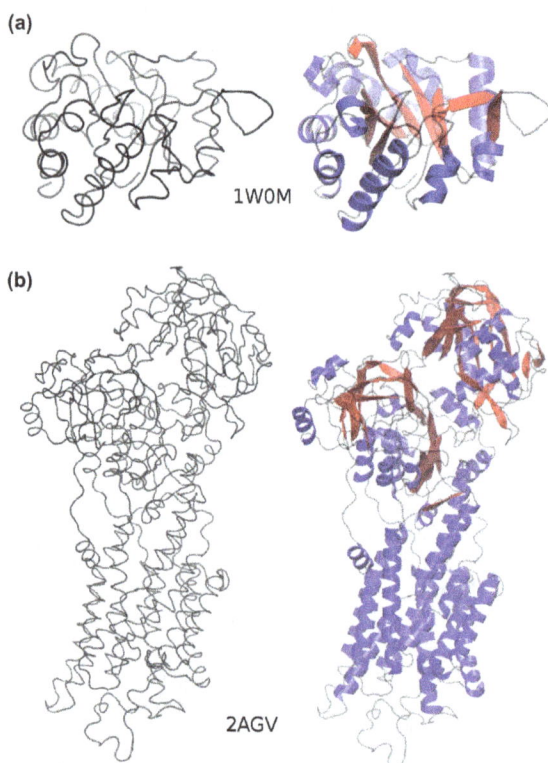

(a)

(b)

Figure 1. RaFoSA improves visualization of molecules when only coarse-grained information is available. (a) 1WOM, and (b) 2AGV are protein molecules arranged in an apparent order of increasing SS complexities. The visualizations on the left side (when all-atom and SS information are lacking) are not helpful and do *not allow one to clearly differentiate (more structurally stable) sheets from (less structurally stable) coils.* In contrast, the visualizations on the right side when RaFoSA provides SS information are more useful/helpful and are visually appealing. With the SS assigned, one can see the residues that form more structurally stable sheets (red) and helixes (blue), or less structurally stable coils (black).

Table 1. Agreements between various SS assignment methods and DSSP (Kabsch & Sander, 1983) and STRIDE (Heinig & Frishman, 2004)		
	Agreement with DSSP (%)	**Agreement with STRIDE (%)**
RaFoSA (the new method presented here)	96.2	93.9
P-SEA (Labesse et al., 1997)	83.4	84.1
VoTAP (Dupuis et al., 2004)	83.2	84.4
SEGNO (Cubellis et al., 2005)	82.4	84.1
KAKSI (Martin et al., 2005)	82.1	83.5
P-Curve (Sklenar, 1989)	79.2	78.9
DEFINE (Richards & Kundrot, 1988)	74.6	74.5

Using random forests classification (Breiman, 2001), I have developed a computer program (RaFoSA) and a webserver for automatic SS assignment. RaFoSA (which is freely available at http://bioinformatics.center/RaFoSA) requires only the *coordinates of alpha carbon atoms* (CAC). Therefore, it works for both CG and all-atom protein systems (i.e. regardless of whether all-atom information is available or not). Rather than manually looking for "human rules" (i.e. manual deterministic approach) that could guide SS assignment, the computer was allowed to learn the SS classification of proteins based on random forests algorithm (Breiman, 2001).

Direct and automatic SS assignment by RaFoSA is straightforward, and better than using a pipeline that uses a prediction approach to first reconstruct the all-atom details of the protein (Krivov, Shapovalov, & Dunbrack, 2009) and then uses the predicted all-atom structure to predict SS. Unlike RaFoSA, such pipeline would be slow and vulnerable to errors propagation and compounding errors.

Although it is shown in this paper that RaFoSA requires only CAC and achieves 96% (and 94%) agreements with DSSP (and STRIDE)—the acclaimed gold-standard for automatic SS assignment—it is important to mention that other SS assignment methods based on CAC exist in the literature. Nonetheless, RaFoSA has better accuracy (and agreement with the gold-standard) than all other methods (Table 1) such as P-SEA (Labesse, Colloc'h, Pothier, & Mornon, 1997), VoTAP (Dupuis, Sadoc, & Mornon, 2004), SEGNO (Cubellis, Cailliez, & Lovell, 2005), KAKSI (Martin et al., 2005), P-Curve (Sklenar, 1989), and DEFINE (Richards & Kundrot, 1988). A review of the SS assignment algorithms for these methods has been published elsewhere (Andersen & Rost, 2009).

2. System and methods

2.1 Data-set and machine learning

3D structures of non-homologous protein molecules (with less than 30% sequence similarity) were obtained from Brookhaven Protein Data Bank (Berman et al., 2000), and energy-minimized using AMBER (Salomon-Ferrer, Case, & Walker, 2013) force-fields (ff12SB). "Coordinates of alpha carbon (CA) atoms" (CAC) were extracted from the energy-minimized structures. For each residue in a given protein, 30 CAC-based features (Figure 2, Supplementary Table S1) were extracted and used for training random forests (RFC) (Breiman, 2001) machine-learning classifiers. Target labels were DSSP-assigned (Andersen, Palmer, Brunak, & Rost, 2002; Kabsch & Sander, 1983) SS classes. Eight RFC, one RFC for each of the seven SS classes and one for unknown class, were trained. With multidimensional grid search in Scikit-learn (Pedregosa et al., 2011), parameter spaces were explored using 10-fold cross-validation. The optimal model parameters found are summarised in the README file in RaFoSA's source code downloadable from RaFoSA's webpage.

The features (i.e. variables, Figure 2) used for the machine learning models are able to describe the local conformation and geometry of every five consecutive amino acids –i–2, i–1, i, i+1, and i+2

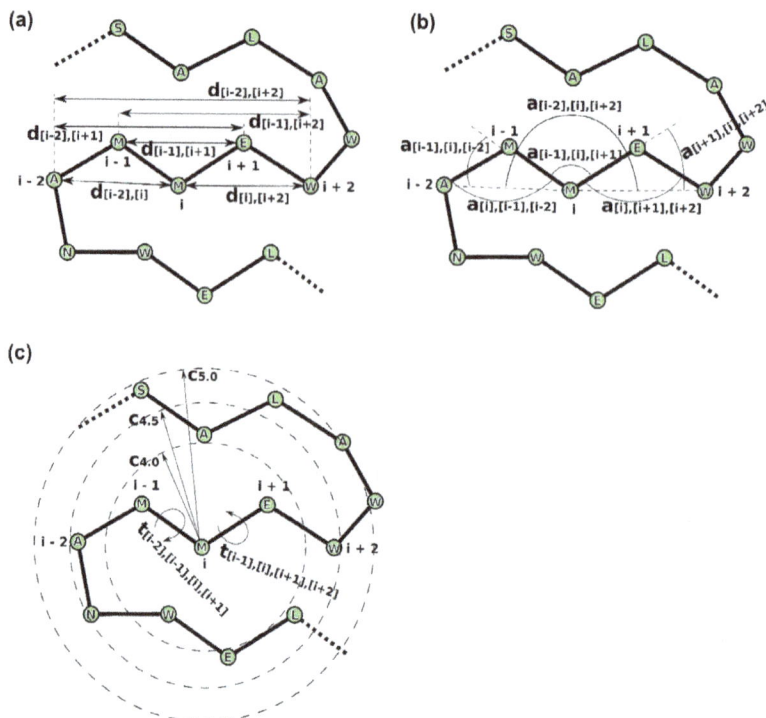

Figure 2. Features used in RaFoSA. One of the features is the residue type (which is any of the 20 standard amino acids or "X" for any non-standard amino acid). Other features are related to alpha carbon (CA) atoms. Six of the features are CA-CA distances (a), such as $d_{i-1,i+1}$. Other six of the features are CA-CA-CA angles (b), such as $a_{i-2,i,i+2}$. Four of the features are sign or angle of CA-CA-CA-CA torsional angles (c), such as $t_{i-1,i,i+1,i+2}$. While the remaining features are based on residue–residue contacts (c), such as $C_{i,4.0}$. "i" is the index of the current residue. "$i - 1$" (or "$i + 1$") is the index of the residue immediately before (or immediately after) the current residue.

(Figure 2). Such conformation and geometry are able to capture the secondary structures (SS) of proteins.

At this point, it is important to give the basis of random forests classification. A random forest classifier is an ensemble learning method that uses various sub-samples of the data-set to fit a pre-defined number of classifiers for the classification task of interest based on decision trees. Together, the decision trees make up the forest called the random forest classifier. When a trained random forest classifier is used for classification, each of the decision trees that make up the forest is used to predict the class of the current sample. The mode of the classes predicted by all decision trees is reported as the class for the current sample in the classification problem. It has been shown that random forest classification (as well as the analogous random forest regression) has many advantages (Amaratunga, Cabrera, & Lee, 2008; Biau, 2012; Breiman, 2001). Notably, with random forest, over-fitting is not an issue (Amaratunga et al., 2008; Biau, 2012; Breiman, 2001). It is very fast and its accuracy is overall better than those of other current machine learning algorithms (Amaratunga et al., 2008; Biau, 2012; Breiman, 2001).

2.2 Secondary structure assignment

Each trained RFC predicts the assignment of the current residue to the model's SS class. For example, RFC for alpha helix (H) predicts whether the current residue should be assigned H or not; and so on. In a situation where each of two or more models reports a "true" for the current residue, "HBEGITS" order of preference is followed such that S is the least preferred. This order generally corresponds to decreasing false positivity rate and is consistent with DSSP's (Carter, Andersen & Rost, 2003; Kabsch & Sander, 1983) approach. After SS had been assigned to all residues, all the assignments are scanned for consistency. It is ensured that whenever any of H, G, or I is assigned, it is assigned to three or more consecutive residues, because a reasonable helical fragment/structure should be made up by at least three residues. This is done by either extending or removing the H, G, or I segments based on the probabilities reported by the random forest classifiers for each of the amino acid positions. In a similar way, whenever B or E is assigned, it is assigned to two or more consecutive residues. This is similar to the approach DSSP (Carter, Andersen, & Rost, 2003; Kabsch & Sander, 1983) and other SS assignment methods (Andersen & Rost, 2003; Martin et al., 2005) use.

2.3 Coarse-grained MD simulations

Using Gō (Taketomi, Ueda, & Gō, 1975) model implemented in Cafemol (Kenzaki et al., 2011), coarse-grained molecular dynamics (MD) simulations of wild type *Bacillus subtilis* LipA (PDB code: 2QXU), and its variant X mutant (PDB code: 3QZU) were carried out at three different temperatures (300 K, 350 K, and 400 K) and two replicates each. Langevin dynamics was used. Each simulation was at 40 femtoseconds time-step and 2 microseconds long.

2.4 Assessment of stability of proteins

An empirical measure for quantifying the stability of a protein molecule (SS-stability score, SSS) based on its secondary structure constituents was developed. Sheets are regarded as the most stable (and have a weight of 1.5), followed by helixes (with a weight of 1.0), while coils are the least stable (with a weight of 0.0). The overall SSS of a protein is a weighted sum of the number of residues assigned to sheets or helixes divided by fifth-sixth of its number of residues $\left(\left[\frac{1}{3}N \times 1.5\right] + \left[\frac{1}{3}N \times 1.0\right] + \left[\frac{1}{3}N \times 0.0\right] = \frac{5}{6}N\right)$. From the foregoing, a hypothetical protein that is made up only by coils is unstable (and has an overall SSS of 0.0). And the more the number of amino acids in sheets (as well as in helixes) a protein has, the higher its SSS. For purpose of illustrating the concept, I used the developed RaFoSA and SSS (derived from RaFoSA-assigned SS) to assess the SSS of wild type *B. subtilis* LipA, and its variant X mutant (Augustyniak et al., 2012). The findings are presented in the results section.

3. Results and discussion

3.1 Accuracy and reliability of RaFoSa

RaFoSA was rigorously tested on 1,000 (randomly selected, independent) protein molecules from the Protein Data Bank (Berman et al., 2000). See Supplementary Table S2 for the list of the protein molecules used. Despite working with only CAC, RaFoSA's SS assignments agree with those of DSSP (Andersen et al., 2002; Carter et al., 2003; Kabsch & Sander, 1983), STRIDE (Heinig & Frishman, 2004), and PSEA (Labesse et al., 1997). RaFoSA achieves an agreement of 96.2% (and 93.9%) with DSSP (and STRIDE) that require all-atom and hydrogen-bonding information, which are not always available. Comparison of PSEA and STRIDE (which requires all-atom details) to DSSP shows only 82% and 95.9% agreements, respectively. Such relatively low agreements between PSEA and DSSP, and between STRIDE and DSSP are already known (Klose, Wallace, & Janes, 2010; Labesse et al., 1997; Martin et al., 2005; Zhang, Dunker, & Zhou, 2008). Indeed, these make PSEA and STRIDE less desirable alternatives to RaFoSA (that is presented here), even if all-atom information is available. Furthermore, RaFoSA is able to assign SS to each of the amino acids in any given protein whatsoever, using either seven-class system or three-class system (see the next paragraph) even for CG protein models that have only CA atoms. In addition, comparison of RaFoSA to VoTAP (Dupuis et al., 2004), SEGNO (Cubellis et al., 2005), KAKSI (Martin et al., 2005), P-Curve (Sklenar, 1989), and DEFINE (Richards & Kundrot, 1988) show RaFoSA to be superior (Table 1).

The proportion of 238,216 residues (from the 1,000 proteins used for RaFoSA's evaluation) assigned to alpha helix (H), beta sheet (B), strand (E), 3-10 helix (G), pi-helix (I), turn (T), and coil/bend (S) are presented in Figure 3(a), and are compared therein to those from DSSP and STRIDE. Based on this seven-class SS information, RaFoSA and DSSP agree by 93.0%, while RaFoSA and STRIDE agree by approximately 80.0%. It must be noted that PSEA does not use seven-class system, therefore it is not included in the comparison shown in Figure 3(a).

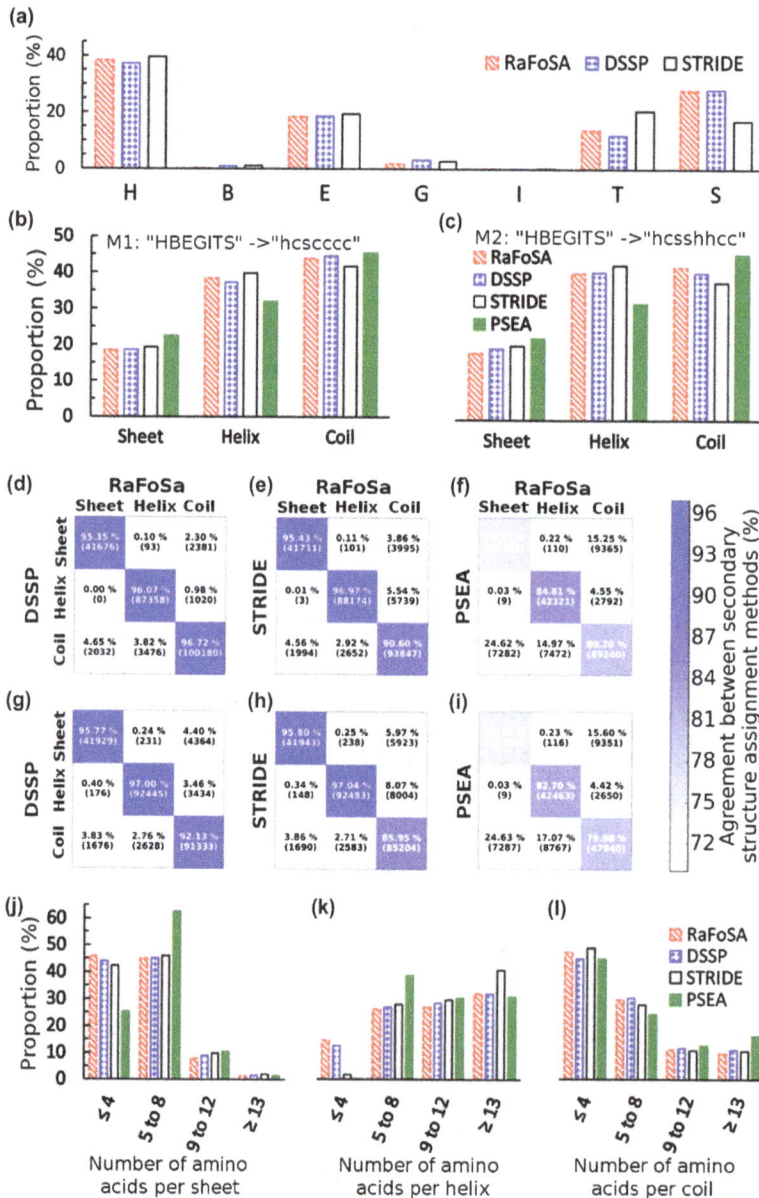

Figure. 3. Agreement between RaFoSA's SS assignment and SS assignment by some existing methods.
(a) Proportion of the assigned SS that falls in each of the seven SS classes–alpha helix (H), beta sheet
(B), strand (E), 3–10 helix (G), pi-helix (I), turn (T), and coil/bend (S). (b) Proportion of the assigned SS
that falls within each of the three SS classes–sheet, s; helix, h; coil, c–based on mapping 1, M1
("HBEGITS" → "hcscccc"). (c) Same as b, but based on mapping 2, M2 ("HBEGITS" → "hsshhcc").
Panels d to f (based on M1) and panels g to h (based on M2) show the degree of agreement between
RaFoSA and each of the other methods. Using M1, agreements between SS assignments by RaFoSA
and DSSP are shown in d, between RaFoSA and STRIDE are shown in e, and between RaFoSA and PSEA
are shown in f. Similar information are shown in g, h, and i, but based on M2. The columns in each of
the matrixes are for RaFoSA-assigned Sheets, Helixes, and Coils, respectively. The rows are for the
other SS methods RaFoSA is compared to. The intensity of the blue color in the leading diagonal of
each of the matrixes show the degree of agreement between RaFoSA and the other SS assignment
method. Number of amino acids per SS are shown in j (sheet), k (helix), and l (coil).

The seven-class SS information is often converted to three-class SS information (sheet, s; helix, h;
and coil, c) for simplicity and when being used by molecular visualization software systems (Humphrey
et al., 1996) or whenever it is used for assessing stability of MS (Camilloni et al., 2012). The seven-class
SS was therefore converted into its three-class form using the two most commonly used (Andersen &
Rost, 2003; Andersen et al., 2002; Carter et al., 2003; Heinig & Frishman, 2004; Kabsch & Sander, 1983)

mappings—Mapping 1 (M1): "HBEGITS" → "hcscccc"; Mapping 2 (M2): "HBEGITS" → "hsshhcc". The proportions of the SS assigned to sheet (s), helix (h), and coil (c) by RaFoSA, DSSP, STRIDE, and PSEA are shown in Figure 3(b) (based on M1) and Figure 3(c) (based on M2). RaFoSA agrees well with DSSP and STRIDE.

The agreements were further assessed on residue-by-residue basis using both M1 and M2. A hypothetical situation in which two methods agree perfectly will have the leading diagonal of the heatmaps (shown in Figure 3(d) to Figure 3(i)) to be deep blue (100% agreement) and the off-diagonal to be plain white. Based on M1, RaFoSA and DSSP (Figure 3(d)) agree in 96.2% of all their SS assignments (95.4% for sheets, 96.1% for helix, and 96.7% for coils). RaFoSA and STRIDE (Figure 3(e)) agree in 93.9% of all their SS assignments, while RaFoSA and PSEA (Figure 3(f)) agree in 80.8% of all their SS assignments. Based on M2, RaFoSA and DSSP (Figure 3(g)) agree in 94.8% of all their SS assignments, RaFoSA and STRIDE (Figure 3(h)) agree in 92.2% of all their SS assignments, while RaFoSA and PSEA (Figure 3(i)) agree in 80.0% of all their SS assignments.

For further comparison, the distributions of the number of residues in each sheet (Figure 3(j)), helix (Figure 3(k)), and coil (Figure 3(l)) were calculated. In all, RaFoSA shows great agreement with existing SS assignment methods, and (more importantly) works when existing gold-standard method, DSSP, cannot (namely when all-atom information is not available).

3.2 RaFoSA identifies stable variant of proteins

Since RaFoSA is accurate at, and reliable for, SS assignment regardless of whether all-atom information is available or not, RaFoSA was used to assign SS to each of the 12,000 frames (1,000 frames × 2 molecules × 2 replicates × 3 temperatures) from 24 microseconds coarse-grained (CG) molecular dynamics (MD) simulations of wild type *B. subtilis* LipA (WTLA), and its variant X mutant (XMLA). The secondary stability score (SSS, see methods section) computed from RaFoSA-assigned SS clearly identifies XMLA as being more stable than WTLA (Figure 4). This agrees with previous experiments (Augustyniak et al., 2012), and provides additional evidence that supports the accuracy and reliability of RaFoSA as well as suggests additional potential application of RaFoSA. It must be noted that analysis of stability of protein fold and identification of SS associated with protein function can be done in many other ways, most of which are more rigorous than the quick SS-based presented used in this example.

3.3 Source codes and webserver of RaFoSA

Source codes of RaFoSA are available at http://bioinformatics.center/tools/rafosa for free download. Once the user had downloaded and extracted RaFoSA into the desired directory, he/she can execute "python full/path/to/RaFoSA.py full/path/to/pdbFile.pdb" to assign SS to the amino acids of the

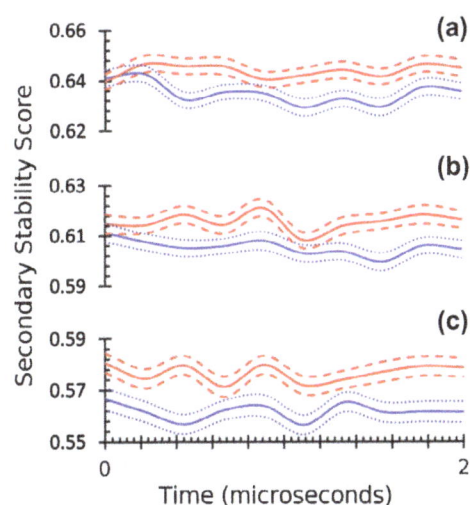

Figure. 4. Structural stability score based on RaFoSA-assigned SS identifies protein's structurally stable variant. Structural stability of wild type *B. subtilis* LipA (blue, 2QXU), and its variant X mutant (red, 3QZU) over the simulation time at three different temperatures—(a) 300 K, (b) 350 K, and (c) 400 K- are shown. The solid lines are the means, while the dotted/dashed lines are one standard deviation above or below the respective mean values.

protein contained in the specified pdb file. This is supported on all major operating systems. A working Python programing language environment (with NumPy, SciKitLearn, and BioPython) is required. Additional information on how to set up RaFoSA, the format of RaFoSA's output, and how to use RaFoSA with VMD are available on RaFoSA's webpage.

Furthermore, a webserver (summarized in Figure 5) that implements RaFoSA is made available at http://bioinformatics.center/servers/rafosa so as to further increase the user-friendliness of RaFoSA. The webserver supports all modern web browsers. User may provide input (Figure 5(a)) to the webserver in any of four ways: (1) by uploading PDB file, (2) by entering PDB content/format as text, (3) by specifying a valid PDB ID, or (4) by uploading trajectory as PSF and DCD files. Regardless of how the user submits the input, SS assignment is done for protein residues (i.e. for amino acids) alone, SS

Figure. 5. RaFoSA webserver. The webserver accepts input (a) in any of four ways: (1) PDB file, (2) PDB content/format as text, (3) PDB ID, or (4) trajectory as PSF and DCD files. SS are assigned for all frames (b) in the submitted data, and SS visualization (c) is generated for the protein for each frame. Summary statistics (line graph, (b), and doughnut chart, (c)) are also provided for the assigned SS.

are assigned for all frames (Figure 5(b)) in the submitted data, and SS visualization (Figure 5(c)) is generated for the protein for each frame. Summary statistics for the assigned SS are also provided (see the line graph and doughnut chart in Figure 5(b) and Figure 5(c), respectively).

4. Conclusions

The need to be able to study biomolecules *in silico* at biologically meaningful timescale with limited available computation power has made coarse-grained (CG) molecular dynamics (MD) simulations to increasingly gain popularity and acceptance. However, "accurate and reliable method" for SS assignment to residues of protein molecules following their CG-MD simulations (or whenever all-atom information is lacking) does not exist. RaFoSA was developed to address this problem and to cater for this need. Therefore, RaFoSA is a new method (implemented as a software and as a webserver) for protein SS assignment when only CG information is available. This paper has shown evidences and confirmed that RaFoSA produces accurate and reliable results and works well even in situations where existing methods cannot work well. For example, DSSP (Andersen et al., 2002; Kabsch & Sander, 1983) and STRIDE (Heinig & Frishman, 2004) require all-atom information and hydrogen-bonding patterns. These make them to have very limited applications, to the extent that they are completely unusable when one is directly dealing with trajectories or snapshots from CG-MD simulations. On the other hand, RaFoSA works in all cases and produces accurate and reliable results that no other SS method (including CA-only methods such as PSEA) can match.

It must, at least, be noted that the method presented in this paper is not the same as SS prediction from sequence information that is already well-documented (Cuff & Barton, 2000; Kelley & Sternberg, 2009; Rost, 2001). More importantly, it must be noted that existing SS prediction methods that use sequence information fail, and cannot serve the purposes RaFoSA serves. However, this does not necessarily imply that SS prediction from sequence information is bad or unimportant. Nonetheless, it is important to emphasize that such SS prediction method cannot serve the extended purposes for which RaFoSA has been developed. For example, sequence-based SS prediction (SBSP) methods cannot appropriately assign SS to each residue in each frame of trajectories from CG-MD simulations the way RaFoSA does. Such SBSP methods would rather assign the same (and, therefore, incorrect) static SS (that would not change over the simulation trajectory) to each of the residues in each of the frames in the trajectories. Thus, unlike RaFoSA, such SBSP methods are unable to correctly capture the structural dynamical properties of proteins at all. On the other hand, RaFoSA can identify structurally stable variant of proteins based on SSS (such as in Figure 4). SBSP methods (that are largely based on sequence homology) fail, because native SS compositions of two or more studied molecules (and thus their SSS) can be quite comparable and not differentiable in their native states, and at the beginning of the simulation (such as Figure 4(a) and Figure 4(b)). Therefore, only a method (such as RaFoSA) that can correctly assign SS to each frame in the trajectories could work for this purpose and similar purposes.

Anyone can obtain and use the source codes of non-commercial version of RaFoSA from http://bioinformatics.center/tools/rafosa. A webserver that implements RaFoSA for easy use is available at http://bioinformatics.center/servers/rafosa. User guides are also provided through the webpages.

Funding
This work was financially supported by stipends from Academia Sinica, Taipei, Taiwan, and from National Tsing Hua University, Hsinchu, Taiwan.

Acknowledgements
I thank Dr Claus Andersen (for making a copy of DSSPCont available), Dr Gilles Labesse (for making a copy of PSEA available), and Dr David Case (for providing a free academic license of AMBER14). The extensive feedback from people (Dr Jiří Koubek, Chris YC Lo, ChiHong ChangChien, etc.) who have being using RaFoSA and/or offered criticisms that helped in improving RaFoSA and this paper are acknowledged.

Competing Interests
The author declare no competing interest.

Author details
Emmanuel Oluwatobi Salawu[1,2,3]
E-mails: tools@bioinformatics.center, emmanuel.s@iis.sinica.edu.t

[1] TIGP Bioinformatics Program, Institute of Information Science, Academia Sinica, Taipei, Taiwan.
[2] Institute of Bioinformatics and Structural Biology, National Tsing Hua University, Hsinchu, Taiwan.
[3] School of Computer Science, University of Hertfordshire, Hertfordshire, UK.

Cover image
Source: Author.

References

Amaratunga, D., Cabrera, J., & Lee, Y.-S. (2008). Enriched random forests. *Bioinformatics, 24,* 2010–2014. http://dx.doi.org/10.1093/bioinformatics/btn356

Andersen, C. A. F., & Rost, B. (2003). Secondary structure assignment. *Methods of Biochemical Analysis, 44,* 341–363. Retrieved from http://doi.org/10.1002/0471721204.ch17

Andersen, C. A. F., & Rost, B. (2009). Secondary structure assignment. Structural. *Bioinformatics, 44,* 459–484.

Andersen, C. A. F., Palmer, A. G., Brunak, S., & Rost, B. (2002). Continuum secondary structure captures protein flexibility. *Structure, 10,* 175–184. Retrieved from http://doi.org/10.1016/S0969-2126(02)00700-1

Augustyniak, W., Brzezinska, A. A., Pijning, T., Wienk, H., Boelens, R., Dijkstra, B. W., … Reetz, M. T. (2012). Biophysical characterization of mutants of *Bacillus subtilis* lipase evolved for thermostability: Factors contributing to increased activity retention. *Protein Science, 21,* 487–497. Retrieved from http://doi.org/10.1002/pro.2031

Berman, H. M., Westbrook, J., Feng, Z., Gilliland, G., Bhat, T. N., Weissig, H., … Bourne, P. E. (2000). The protein data bank. *Nucleic Acids Research, 28,* 235–242. Retrieved from http://doi.org/10.1093/nar/28.1.235

Biau, G. (2012, April). Analysis of a random forests model. *Journal of Machine Learning Research, 13,* 1063–1095.

Breiman, L. (2001). Random forests. *Machine Learning,* 5–32. Retrived from http://doi.org/10.1023/A:1010933404324

Cabaleiro-Lago, C., Szczepankiewicz, O., & Linse, S. (2012). The effect of nanoparticles on amyloid aggregation depends on the protein stability and intrinsic aggregation rate. *Langmuir, 28,* 1852–1857. Retrieved from http://doi.org/10.1021/la203078w

Camilloni, C., De Simone, A., Vranken, W. F., & Vendruscolo, M. (2012). Determination of secondary structure populations in disordered states of proteins using nuclear magnetic resonance chemical shifts. *Biochemistry, 51,* 2224–2231. Retrieved from http://doi.org/10.1021/bi3001825

Carter, P., Andersen, C. A., & Rost, B. (2003). DSSPcont: Continuum secondary structure assignments for proteins. *Nucleic Acids Research, 31,* 3293–3295. http://dx.doi.org/10.1093/nar/gkg626

Cino, E. A., Choy, W. Y, & Karttunen, M. (2012). Comparison of secondary structure formation using 10 different force fields in microsecond molecular dynamics simulations. *Journal of Chemical Theory and Computation, 8,* 2725–2740. Retrieved from http://doi.org/10.1021/ct300323g

Cubellis, M. V., Cailliez, F., & Lovell, S. C. (2005). Secondary structure assignment that accurately reflects physical and evolutionary characteristics. *BMC Bioinformatics, 6*(4), 1.

Cuff, J. A., & Barton, G. J. (2000). Application of multiple sequence alignment profiles to improve protein secondary structure prediction. *Proteins: Structure, Function, and Genetics, 40,* 502–511. Retrieved from http://doi.org/10.1002/1097-0134(20000815)40:3<502::AID-PROT170>3.0.CO;2-Q, http://dx.doi.org/10.1002/(ISSN)1097-0134

Dupuis, F., Sadoc, J.-F., & Mornon, J.-P. (2004). Protein secondary structure assignment through Voronoï tessellation. *Proteins: Structure, Function, and Bioinformatics, 55,* 519–528. http://dx.doi.org/10.1002/prot.10566

Emr, S. D., & Silhavy, T. J. (1983). Importance of secondary structure in the signal sequence for protein secretion. *Proceedings of the National Academy of Sciences, 80,* 4599–4603. Retrieved from http://doi.org/10.1073/pnas.80.15.4599

Heinig, M., & Frishman, D. (2004). STRIDE: A web server for secondary structure assignment from known atomic coordinates of proteins. *Nucleic Acids Research, 32*(WEB SERVER ISS.), W500–W502. http://dx.doi.org/10.1093/nar/gkh429

Humphrey, W., Dalke, A., & Schulten, K. (1996). VMD: Visual molecular dynamics. *Journal of Molecular Graphics, 14,* 33–38. Retrieved from http://doi.org/10.1016/0263-7855(96)00018-5xz, http://dx.doi.org/10.1016/0263-7855(96)00018-5

Ji, Y.-Y., & Li, Y.-Q. (2010). The role of secondary structure in protein structure selection. *The European Physical Journal E, 32,* 103–107.Retrieved from http://doi.org/10.1140/epje/i2010-10591-5

Kabsch, W., & Sander, C. (1983). Dictionary of protein secondary structure: Pattern recognition of hydrogen-bonded and geometrical features. *Biopolymers, 22,* 2577–2637.Retrieved from http://doi.org/10.1002/bip.360221211, http://dx.doi.org/10.1002/(ISSN)1097-0282

Kelley, L. A., & Sternberg, M. J. E. (2009). Protein structure prediction on the web: A case study using the Phyre server. *Nature Protocols, 4,* 363–371. http://dx.doi.org/10.1038/nprot.2009.2

Kenzaki, H., Koga, N., Hori, N., Kanada, R., Li, W., Okazaki, K. I., … Takada, S. (2011). Cafemol: A coarse-grained biomolecular simulator for simulating proteins at work. *Journal of Chemical Theory and Computation, 7,* 1979–1989. Retrieved from http://doi.org/10.1021/ct2001045

Klose, D. P., Wallace, B. A., & Janes, R. W. (2010). 2Struc: The secondary structure server. *Bioinformatics, 26,* 2624–2625. http://dx.doi.org/10.1093/bioinformatics/btq480

Konvalinová, H., Dvořáková, Z., Renčiuk, D., Bednářová, K., Kejnovská, I., Trantírek, L., … Sagi, J. (2015). Diverse effects of naturally occurring base lesions on the structure and stability of the human telomere DNA quadruplex. *Biochimie, 118,* 15–25. Retrieved from http://doi.org/10.1016/j.biochi.2015.07.013

Krivov, G. G., Shapovalov, M. V., & Dunbrack, R. L. (2009). Improved prediction of protein side-chain conformations with SCWRL4. *Proteins: Structure, Function, and Bioinformatics, 77,* 778–795. http://dx.doi.org/10.1002/prot.v77:4

Labesse, G., Colloc'h, N., Pothier, J., & Mornon, J. P. (1997). P-SEA: A new efficient assignment of secondary structure from C alpha trace of proteins. *Computer Applications in the Biosciences: CABIOS, 13,* 291–295. Retrieved from http://doi.org/10.1093/bioinformatics/13.3.291

Martin, J., Letellier, G., Marin, A., Taly, J.-F., de Brevern, A. G., & Gibrat, J.-F. (2005). Protein secondary structure assignment revisited: A detailed analysis of different assignment methods. *BMC Structural Biology, 5,* 17. http://dx.doi.org/10.1186/1472-6807-5-17

Myers, J. K., & Oas, T. G. (2001). Preorganized secondary structure as an important determinant of fast protein folding. *Nature Structural Biology, 8*, 552–558. Retrieved from http://doi.org/10.1038/88626

Pedregosa, F., Varoquaux, G., Gramfort, A., Michel, V., Thirion, B., Grisel, O., ... Duchesnay, É. (2011, Oct). Scikit-learn : Machine learning in Python. *Journal of Machine Learning Research, 12*, 2825–2830.

Pires, D. E. V., Ascher, D. B., & Blundell, T. L. (2014). DUET: A server for predicting effects of mutations on protein stability using an integrated computational approach. *Nucleic Acids Research, 42*, 1–6. Retrieved from http://doi.org/10.1093/nar/gku411

Provencher, S. W., & Gloeckner, J. (1981). Estimation of globular protein secondary structure from circular dichroism. *Biochemistry, 20*, 33–37. Retrieved from http://doi.org/10.1021/bi00504a006

Richards, F. M., & Kundrot, C. E. (1988). Identification of structural motifs from protein coordinate data: Secondary structure and first-level supersecondary structure.

Proteins: Structure, Function, and Genetics, 3, 71–84. http://dx.doi.org/10.1002/(ISSN)1097-0134

Rost, B. (2001). Review: Protein secondary structure prediction continues to rise. *Journal of Structural Biology, 134*, 204–218. Retrieved from http://doi.org/10.1006/jsbi.2000.4336

Salomon-Ferrer, R., Case, D. A., & Walker, R. C. (2013). An overview of the Amber biomolecular simulation package. *Wiley Interdisciplinary Reviews: Computational Molecular Science, 3*, 198–210. Retrieved from http://doi.org/10.1002/wcms.1121

Sklenar, H. (1989). *Describing protein structure, 60*, 46–60.

Taketomi, H., Ueda, Y., & Gō, N. (1975). Studies on protein folding, unfolding and fluctuations by computer simulation. *International Journal of Peptide and Protein Research, 7*, 445–459.

Zhang, W., Dunker, A. K., & Zhou, Y. (2008). Assessing secondary structure assignment of protein structures by using pairwise sequence-alignment benchmarks. *Proteins: Structure, Function, and Bioinformatics, 71*, 61–67. http://dx.doi.org/10.1002/(ISSN)1097-0134

PERMISSIONS

The contributors of this book come from diverse backgrounds, making this book a truly international effort. This book will bring forth new frontiers with its revolutionizing research information and detailed analysis of the nascent developments around the world.

We would like to thank all the contributing authors for lending their expertise to make the book truly unique. They have played a crucial role in the development of this book. Without their invaluable contributions this book wouldn't have been possible. They have made vital efforts to compile up to date information on the varied aspects of this subject to make this book a valuable addition to the collection of many professionals and students.

This book was conceptualized with the vision of imparting up-to-date information and advanced data in this field. To ensure the same, a matchless editorial board was set up. Every individual on the board went through rigorous rounds of assessment to prove their worth. After which they invested a large part of their time researching and compiling the most relevant data for our readers.

The editorial board has been involved in producing this book since its inception. They have spent rigorous hours researching and exploring the diverse topics which have resulted in the successful publishing of this book. They have passed on their knowledge of decades through this book. To expedite this challenging task, the publisher supported the team at every step. A small team of assistant editors was also appointed to further simplify the editing procedure and attain best results for the readers.

Apart from the editorial board, the designing team has also invested a significant amount of their time in understanding the subject and creating the most relevant covers. They scrutinized every image to scout for the most suitable representation of the subject and create an appropriate cover for the book.

The publishing team has been an ardent support to the editorial, designing and production team. Their endless efforts to recruit the best for this project, has resulted in the accomplishment of this book. They are a veteran in the field of academics and their pool of knowledge is as vast as their experience in printing. Their expertise and guidance has proved useful at every step. Their uncompromising quality standards have made this book an exceptional effort. Their encouragement from time to time has been an inspiration for everyone.

The publisher and the editorial board hope that this book will prove to be a valuable piece of knowledge for researchers, students, practitioners and scholars across the globe.

LIST OF CONTRIBUTORS

I-Shiang Tzeng
Institute of Epidemiology and Preventive Medicine, College of Public Health, National Taiwan University, Taipei, Taiwan

Li-Shya Chen
Department of Statistics, National Chengchi University, Taipei, Taiwan

Shy-Shin Chang
Department of Family Medicine, Chang Gung Memorial Hospital, Taipei, Taiwan.

Yung-ling Leo Lee
Graduate Institute of Epidemiology and Preventive Health, College of Public Health, National Taiwan University, Taipei, Taiwan

Md. Mohabbulla Mohib, S.M. Fazla Rabby, Tasfiq Zaman Paran, Iqbal Ahmed, Nahid Hasan and Md. Abu Taher Sagor
Department of Pharmaceutical Sciences, North South University, Dhaka 1229, Bangladesh

Md. Mehedee Hasan
Department of Pharmacy, State University of Bangladesh, Dhaka 1205, Bangladesh

Sarif Mohiuddin
Department of Anatomy, Pioneer Dental College and Hospital, Dhaka 1229, Bangladesh

Lyudmila Neykova-Vasileva
Department of Emergency Toxicology, Military Medical Academy, Georgi Sofiiski 3, Sofia 1606, Bulgaria

Tony Donchev and Krasimir Kostadinov
Psychiatry Clinic, Military Medical Academy, Georgi Sofiiski 3, Sofia 1606, Bulgaria

Marie K. Holt and Stefan Trapp
Centre for Cardiovascular and Metabolic Neuroscience, Department of Neuroscience, Physiology & Pharmacology, University College London, WC1E 6BT London, UK

Muhammad Moin Uddin Mazumdar, Md. Ariful Islam, Mohammad Tanvir Hosen, Mohammad Shahin Alam, Mohammad Nazmul Alam, Md. Faruk, Md. Mominur Rahman, Mohammed Abu Sayeed and Md. Masudur Rahman
Department of Pharmacy, International Islamic University Chittagong, Chittagong, Bangladesh

Shaikh Bokhtear Uddin
Department of Botany, University of Chittagong, Chittagong, Bangladesh

Sarah M. Meeuwsen, An N. Hodac, Lauren M. Adams, Ryan D. McMunn, Maxwell S. Anschutz,Kari J. Carothers, Rachel E. Egdorf, Peter M. Hanneman, Jonathan P. Kitzrow, Cynthia K. Keonigsberg, Oscar Lopez-Martinez, Paul A. Matthew, Ethan H. Richter, Jonathan E. Schenk, Heidi L. Schmit, Matthew A. Scott, Eva M. Volenec and Sanchita Hati
Chemistry Department, University of Wisconsin – Eau Claire, Eau Claire, WI 54702, USA

Gabriela Fernandes, Andrew W. Barone and Rosemary Dziak
Department of Oral Biology, School of Dental Medicine, University at Buffalo, State University of New York, 3435 Main Street, Buffalo, NY 14201, USA

C.P. Bhunu
Department of Mathematics, University of Zimbabwe, Box MP 167, Mount Pleasant, Harare, Zimbabwe

M. Masocha
Department of Geography and Environmental Science, University of Zimbabwe, Box MP 167, Mount Pleasant, Harare, Zimbabwe.

C.W. Mahera
African Institute for Mathematical Sciences (AIMS), Box 176, Bagamoyo, Tanzania

Xuesheng Han and Tory L. Parker
dōTERRA International, LLC, 389 S. 1300 W., Pleasant Grove, UT 84062, USA

Xuesheng Han and Tory L. Parker
dōTERRA International, LLC, 389 S. 1300 W.,
Pleasant Grove, UT 84062, USA

Masahiko Ayaki, Kazuo Tsubota and Kazuno Negishi
Department of Ophthalmology, Keio University School of Medicine, 35 Shinanomachi, Shinjuku, 1608582 Tokyo, Japan

Atsuhiko Hattori and Yusuke Maruyama
Department of Biology, Tokyo Medical and Dental University, 2-8-30 Kokufudai, Ichikawa, 2720827 Chiba, Japan

P.U. Anushree, R.M. Naik, R.D. Satbhai, A.P. Gaikwad and C.A. Nimbalkar
Department of Statistic, MPKV, Rahuri 413722, Ahmednagar, India

Rajwin Raja Kanagarajah, David Charles Weerasingam Lee and Daniel Zhi Fung Lee
Perdana University – Royal College of Surgeons in Ireland, Perdana University Hall D Level 1, MAEPS Building, MARDI Complex, Jalan MAEPS Perdana, Universiti Putra Malaysia, Serdang 43400, Selangor, Malaysia

Khatijah Yusoff
Faculty of Biotechnology and Biomolecular Sciences, Department of Microbiology, Universiti Putra Malaysia, Serdang 43400, Selangor, Malaysia

Sharmini Julita Paramasivam and Kamalan Jeevaratnam
Faculty of Health and Medical Sciences, VSM Building, University of Surrey, Guildford, Surrey GU2 7AL, UK

Wai Yee Low
Perdana University – Centre of Bioinformatics, Perdana University Hall D Level 1, MAEPS Building, MARDI Complex, Jalan MAEPS Perdana, Universiti Putra Malaysia, Serdang 43400, Selangor, Malaysia
The Davies Research Centre, School of Animal and Veterinary Sciences, University of Adelaide, Roseworthy, SA 5371, Australia

Swee Hua Erin Lim
Perdana University – Royal College of Surgeons in Ireland, Perdana University Hall D Level 1, MAEPS Building, MARDI Complex, Jalan MAEPS Perdana, Universiti Putra Malaysia, Serdang 43400, Selangor, Malaysia

Perdana University – Centre of Bioinformatics, Perdana University Hall D Level 1, MAEPS Building, MARDI Complex, Jalan MAEPS Perdana, Universiti Putra Malaysia, Serdang 43400, Selangor, Malaysia
Abu Dhabi Women's College, Higher Colleges of Technology, Abu Dhabi 41012, United Arab Emirates

D. Somjen, S. Katzburg, O. Sharon, E. Knoll and D. Hendel
The Sackler Faculty of Medicine, Tel-Aviv Sourasky Medical Center, Institute of Endocrinology, Metabolism and Hypertension and Bone Diseases Unit, Tel-Aviv University, 6 Weizman Street, Tel Aviv 64239, Israel

G.H. Posner
Department of Orthopedic Surgery, Sharei-Zedek Medical Center, Jerusalem, Israel

Xuesheng Han, Tory L. Parker and Jeff Dorsett
dōTERRA International, LLC, 389 S. 1300 W, Pleasant Grove, UT 84062, USA

Md. Shamsuddin Sultan Khan Amin Malik Shah Abdul Majid Muhammad Asif, Mohamed Khadeer Ahamed Basheer and Yaseer M. Tabana
EMAN Research and Testing Laboratory, School of Pharmaceutical Sciences, Universiti Sains Malaysia, Minden, Pulau Pinang, Malaysia

Md. Abdus Salam and Rosenani S.M.A. Haque
The School of Chemical Sciences, Universiti Sains Malaysia (USM), Penang 11800, Malaysia

Aman Shah Bin Abdul Majid
Advanced Medical and Dental Institute (IPPT), Universiti Sains Malaysia, Bertam, Kepala Batas, Malaysia

Emmanuel Oluwatobi Salawu
TIGP Bioinformatics Program, Institute of Information Science, Academia Sinica, Taipei, Taiwan
Institute of Bioinformatics and Structural Biology, National Tsing Hua University, Hsinchu, Taiwan
School of Computer Science, University of Hertfordshire, Hertfordshire, UK

Index

www.ingramcontent.com/pod-product-compliance
Lightning Source LLC
Chambersburg PA
CBHW082029190326
41458CB00010B/3311